Intelligent Communication and Automation Systems

Intelligent Communication and Automation Systems

Edited by

Kamal Kumar Sharma, Akhil Gupta,
Bandana Sharma and Suman Lata Tripathi

CRC Press
Taylor & Francis Group
Boca Raton London New York

CRC Press is an imprint of the
Taylor & Francis Group, an **informa** business

First edition published 2021

by CRC Press

6000 Broken Sound Parkway NW, Suite 300, Boca Raton, FL 33487-2742

and by CRC Press

2 Park Square, Milton Park, Abingdon, Oxon, OX14 4RN

© 2021 selection and editorial matter, Kamal Kumar Sharma, Akhil Gupta, Bandana Sharma and Suman Lata Tripathi; individual chapters, the contributors.

CRC Press is an imprint of Taylor & Francis Group, LLC

The right of Kamal Kumar Sharma, Akhil Gupta, Bandana Sharma and Suman Lata Tripathi to be identified as the authors of the editorial material, and of the authors for their individual chapters, has been asserted in accordance with Sections 77 and 78 of the Copyright, Designs and Patents Act 1988.

ISBN: 978-0-367-60928-3 (hbk)
ISBN: 978-0-367-61201-6 (pbk)
ISBN: 978-1-003-10459-9 (ebk)

Typeset in Times LT Std font
by KnowledgeWorks Global Ltd.

Contents

Preface...ix
Editors xi
Contributors xiii
Introduction...xvii

Chapter 1   Role of Machine Learning in Communication Networks.................... 1

   Veerpal Kaur, Aman Singh

Chapter 2   Evolutionary Optimization Approach for Network Lifetime in
   Wireless Sensor Networks... 11

   Pooja Chaturvedi, Ajai Kumar Daniel

Chapter 3   Characterization of Reverberation Chamber...................................... 29

   Abhishek Kadri, Devendra Chandra Pande, Abhilasha Mishra

Chapter 4   Intelligent Intrusion Detection System... 47

   Anurag Singh Tomar, Aditya Bakshi

Chapter 5   A Morphological Filtering-Based Image-Enhancement Method
   for Citrus Plant Diseases 65

   Bobbinpreet Kaur, Tripti Sharma, Bhawna Goyal, Ayush Dogra

Chapter 6   MOSFET-Based Low-Power Hardware Design for Autonomous
   Applications.. 85

   Suchismita Sengupta, Ananya Dastidar

Chapter 7   Performance Analysis on Low-Power, Low-Offset, High-Speed
   Comparator for High-Speed ADC: A Review 99

   Krishan Mehra, Tripti Sharma, Simran Somal

Chapter 8   Historical Review of DRA to MIMO-DRA: Designs and
   Advances 113

   Madhusmita Mishra, Gaurav Varshney

Chapter 9   Frequency Synthesizers and Their Applications in Signal
   Processing 131

   Govind Singh Patel

Chapter 10 Design of Ultra-Low Power OTA Based on Subthreshold
Operation with High Gain, Large Transconductance
and Small Area... 139

Simran Somal, Tripti Sharma, Krishan Mehra

Chapter 11 Artificial Intelligence for Precision Medicine................................. 149

R.A. Rayan, C. Tsagkaris

Chapter 12 Review on Pupil Segmentation Using CNN-Region of Interest 157

A. Swathi, Aarti, Sandeep Kumar

Chapter 13 An Ensemble Classification-Based Model for Automatic Lung
Cancer Detection Using CT Images.. 169

Shivam Modgil, Bobbinpreet Kaur, Nitin Sharma

Chapter 14 A Revisit on the Progress of Intelligent Robotic Systems (IRS)
over the Past Three Decades 185

Saumyadip Hazra, Abhimanyu Kumar, Souvik Ganguli

Chapter 15 Intelligent Robotic Systems.. 199

Shivang Tyagi, Nthatisi Magaret Hlapisi

Chapter 16 Model Order Reduction of Some Critical Systems Using an
Intelligent Computing Technique.. 221

*Souvik Ganguli, Parag Nijhawan, Manish Kumar Singla,
Jyoti Gupta, Abhimanyu Kumar*

Chapter 17 Advanced Agricultural-Based IoT Technology............................... 239

Vikalp Joshi, Manoj Singh Adhikari

Chapter 18 Machine Learning for Solving a Plethora of Internet of Things
Problems.. 255

Sparsh Sharma, Abrar Ahmed, Mohd Naseem, Surbhi Sharma

Chapter 19 The Fusion of Blockchain and IoT Technologies with
Industry 4.0... 275

Arun Kumar Rana, Sharad Sharma

Chapter 20 Decentralized Blockchain Technology for Security
Enhancement of Internet of Things.. 291

Pranav Ratta, Amanpreet Kaur, Sparsh Sharma

Chapter 21 Identification of Background Factors Associated with
Prevalence of Common Mental Disorders among Adolescent
Students 305

Vivek Sharma, Neelam Rup Prakash, Parveen Kalra

Chapter 22 A Cloud-Based Secured IoT Framework for Log Management....... 315

Nilima Dongre, Mohammad Atique

Chapter 23 Empirical Sentiment Analysis of Social Pages 331

Aarti, Raju Pal

Index.. 343

Preface

The objective of this book is to provide a foundation of information for novices as well as research-seeking scholars working in the core and application areas of intelligent systems. This book will help individuals to comprehend the rudimentary concepts of machine learning and its applications for various aspects of engineering and medical sciences. Fundamental concepts of intelligent communication and various automation systems are illustrated in a clear and detailed manner, with explanatory diagrams where necessary. All the chapters are organized and described in a simple manner to facilitate readability.

CHAPTER ORGANIZATION

This book is organized into 23 chapters.

Chapter 1 addresses key challenges of communication networks. They are discussed thoroughly and the solutions with machine-learning algorithms are surveyed based upon the literature review.

Chapter 2 discusses enhancement of average residual energy by using the particle swarm optimization (PSO) technique in the energy efficient coverage protocol (EECP).

Chapter 3 gives details about electromagnetic compatibility (EMC), which is equipment avoiding electromagnetic interference in the EMI environment.

Chapter 4 describes various methods used in traditional intrusion detection systems (IDSs) along with details of IDSs using various machine-learning algorithms.

Chapter 5 explains spatial filter and morphological filters. The spatial filter is applied in red green blue (RGB) decomposed image planes followed by a morphological filter.

Chapter 6 highlights the design and implementation of a low-power memory bank and its application in high-speed integrated circuit design.

Chapter 7 explains a comparative review of different architecture of static and dynamic comparators for better performance in terms of delay, power dissipation and offset voltage measurements.

Chapter 8 presents an intensive review of the designs implemented on dielectric resonator antennas (DRAs) along with the evolution towards multiple-input multiple-output (MIMO) dielectric resonator antennas (DRAs).

Chapter 9 addresses the analysis and design of frequency synthesizers and their applications in the fields of communication, medical science, wireless communication, digital encoding techniques, etc.

Chapter 10 presents the design of an ultra-low power operational transconductance amplifier.

Chapter 11 explores significant discoveries in developing supporting techniques to achieve the goal of precision medicine (PM).

Chapter 12 focuses on pupil segmentation of the eye with traditional techniques and convolutional neural network (CNN) based techniques to show accuracy.

Chapter 13 describes the hybrid classification method, with a combination of k-nearest neighbour (KNN), support vector machines (SVMs) and decision trees for localizing and characterizing cancer in CT scan images.

Chapter 14 focuses mainly on the developments taking place in intelligent robotic systems for the past three decades.

Chapter 15 describes intelligent robotic systems for security purposes. The intelligent snake robot is described as an active contour model that uses energy to copy the locomotion of a biological snake.

Chapter 16 aims to simplify some critical higher-order systems with the help of a recently developed metaheuristic algorithm, called the Harris Hawks Optimization (HHO) algorithm, to analyze, simulate and subsequently design the controller.

Chapter 17 describes the ability of advanced IoT based technology. Advanced agriculture-based IoT technology provides periodical updates through historical trends and graphs.

Chapter 18 aims to develop a machine-learning model or software that can learn from experiences in a manner similar to humans for solving a plethora of IoT problems.

Chapter 19 focuses on analyzing the benefits and challenges in the fusion of blockchain and IoT technologies in the fourth revolution.

Chapter 20 analyzes the possibility of using blockchain technology with IoT-based networks for enhancement in the security of the overall network.

Chapter 21 aims to identify the prevalence (i.e., frequency) of common mental disorders (depression, anxiety and stress) among adolescent students and their association with various background factors.

Chapter 22 emphasizes the challenges in developing a working secure log management framework in IoT.

Chapter 23 discusses an efficient way to understand written sentiments based on understanding the polarity of words.

Editors

Kamal Kumar Sharma is a professor and head of the department in the Wireless Communication Department, Lovely Professional University, Punjab, India. He has 18 years of teaching and research experience and completed his Ph.D. in wireless communication and sensor networks in 2012 from IKG Punjab Technical University, Punjab. He published more than 150 research papers in reputed journals and conference proceedings. He organized numerous workshops, faculty development programmes and national and international conferences. He chaired many technical sessions in various reputed conferences. He reviewed many research papers and book chapters of reputed publishers. He presented keynote addresses in many conferences in India and abroad. In 2019, he was awarded a young leadership award from IEEE UK and Ireland chapter in London, UK. He authored the textbook *Elements of Electrical and Electronics Engineering.* His edited book *LoRA and IoT Networks for Applications in Industry 4.0* is in the publication phase with Nova Science Publishers. Dr. Sharma has obtained six patents and copyrights, and he is fellow of IETE.

Akhil Gupta is an assistant professor, School of Electronics and Electrical Engineering, Lovely Professional University, Punjab, India. He published more than 50 research papers in *IEEE Transactions on Communications*, IEEE journals and international conference papers. He is currently involved in research on massive MIMO and millimeter-wave communication. He is also working on security issues of next-generation networks of wireless communication. His research interests include the emerging technologies of the 5G wireless communication network. He is also an author of the most popular paper in *IEEE Xplore* and an active reviewer of many IEEE, Springer, Elsevier and Wiley journals. Dr. Gupta received the teaching assistantship at the Ministry of Human Resource Development from 2011 to 2013. He is a member of the IEEE, International Association of Engineers and the Universal Association of Computer and Electronics Engineers, and has more than 1400 citations to his credit.

Bandana Sharma is an assistant professor, Haryana Engineering College, India. She completed her Ph.D. in computer vision and artificial intelligence from Uttarakhand Technical University, Dehradun in 2015. She has published research papers and book chapters in Springer/UGC listed National/International Journals/Books/conference proceedings.

Suman Lata Tripathi completed her Ph.D. in microelectronics and VLSI from MNNIT, Allahabad. She earned her M. Tech. in electronics engineering from UP Technical University, Lucknow, India and her B. Tech. in electrical engineering from Purvanchal University, Jaunpur, India. She is associated with Lovely Professional University as a professor with more than 17 years of experience in academics. She has published more than 45 research papers in refereed journals and conferences. She organized workshops, summer internships and expert lectures for students. She worked as a session chair, conference steering committee member, editorial board member and reviewer in international/national IEEE journals and conferences. She received the Research Excellence Award in 2019 at Lovely Professional University. She received the best paper award at IEEE ICICS-2018. She has published an edited book, *Recent Advancement in Electronic Device, Circuit and Materials,* by Nova Science Publishers. Her edited books, *Advanced VLSI Design and Testability Issues* and *Electronic Devices and Circuit Design Challenges for IoT Application* are in production with CRC Press/Taylor & Francis Group and Apple Academic Press. She is also a book series editor in the area of green energy technology with Scrivener and Wiley. Her areas of expertise include microelectronics device modelling and characterization, low-power VLSI circuit design, VLSI design of testing and advance FET design for IoT, embedded system design and biomedical applications.

Contributors

Chapter 1

Veerpal Kaur
Department of Computer Science and
  Engineering
Lovely Professional University
Punjab, India

Aman Singh
Department of Computer Science and
  Engineering
Lovely Professional University
Punjab, India

Chapter 2

Pooja Chaturvedi
School of Management Sciences
Varanasi, India

Ajai Kumar Daniel
M.M.M. University of Technology
Gorakhpur, India

Chapter 3

Abhishek Kadri
Electromagnetic Solution Consultant
Dassault Systèmes
Pune, India

Devendra Chandra Pande
Electronics and Radar Development
Establishment (LRDE), DRDO
Bangalore, India

Abhilasha Mishra
Electronics and Telecommunication
Maharashtra Institute of Technology
Aurangabad, India

Chapter 4

Anurag Singh Tomar
School of Computer Science
University of Petroleum and Energy
  Studies
Dehradun, India

Aditya Bakshi
School of Computer Science
Shri Mata Vaishno Devi University
Katra, India

Chapter 5

Bobbinpreet Kaur
ECE Department
Chandigarh University
Gharuan, India

Tripti Sharma
ECE Department
Chandigarh University
Gharuan, India

Bhawna Goyal
ECE Department
Chandigarh University
Gharuan, India

Ayush Dogra
Ronin Institute
Montclair, New Jersey, USA

Chapter 6

Suchismita Sengupta
CMR Institute of Technology
Bangalore, Karnataka, India

Ananya Dastidar
College of Engineering and
  Technology
Bhubaneswar, Odisha, India

Chapter 7

Krishan Mehra
Chandigarh University
Mohali, Punjab, India

Tripti Sharma
Chandigarh University
Mohali, Punjab, India

Simran Somal
Chandigarh University
Mohali, Punjab, India

Chapter 8
Madhusmita Mishra
ECE Department
NIT Rourkela
Odisha, India

Gaurav Varshney
ECE Department
NIT Patna
Bihar, India

Chapter 9
Govind Singh Patel
IIMT College of Engineering Greater
  Noida
Uttar Pradesh, India

Chapter 10
Simran Somal
Chandigarh University
Mohali, Punjab, India

Tripti Sharma
Chandigarh University
Mohali, Punjab, India

Krishan Mehra
Chandigarh University
Mohali, Punjab, India

Chapter 11
R.A. Rayan
Department of Epidemiology
High Institute of Public Health
University of Alexandria
Egypt

C. Tsagkaris
Faculty of Medicine
University of Crete
Greece

Chapter 12
A. Swathi
Lovely Professional University
Punjab, India

Aarti
Lovely Professional University
Punjab, India

Sandeep Kumar
Sreyas Institute of Engineering and
  Technology
Telangana, India

Chapter 13
Shivam Modgil
Electronics and Communication
  Engineering
Chandigarh University
Gharuan, Punjab, India

Bobbinpreet Kaur
Electronics and Communication
  Engineering
Chandigarh University
Gharuan, Punjab, India

Nitin Sharma
Electronics and Communication
  Engineering
Chandigarh University
Gharuan, Punjab, India

Chapter 14
Saumyadip Hazra
Department of Electrical and
  Instrumentation Engineering
Thapar Institute of Engineering and
  Technology
Punjab, India

Abhimanyu Kumar
Department of Electrical and
  Instrumentation Engineering
Thapar Institute of Engineering and
  Technology
Punjab, India

Souvik Ganguli
Department of Electrical and
  Instrumentation Engineering
Thapar Institute of Engineering and
  Technology
Punjab, India

Chapter 15
Shivang Tyagi
School of Electronics and Electrical
  Engineering
Lovely Professional University
Phagwara, India

Nthatisi Magaret Hlapisi
School of Electronics and Electrical
  Engineering
Lovely Professional University
Phagwara, India

Chapter 16
Souvik Ganguli
Department of Electrical and
  Instrumentation Engineering
Thapar Institute of Engineering and
  Technology
Patiala, Punjab, India

Parag Nijhawan
Department of Electrical and
  Instrumentation Engineering
Thapar Institute of Engineering and
  Technology
Patiala, Punjab, India

Manish Kumar Singla
Department of Electrical and
  Instrumentation Engineering
Thapar Institute of Engineering and
  Technology
Patiala, Punjab, India

Jyoti Gupta
Department of Electrical and
  Instrumentation Engineering
Thapar Institute of Engineering and
  Technology
Patiala, Punjab, India

Abhimanyu Kumar
Department of Electrical and
  Instrumentation Engineering
Thapar Institute of Engineering and
  Technology
Patiala, Punjab, India

Chapter 17
Vikalp Joshi
Deptartment of ASS
Sara Sae PVT Limited
Dehradun, Uttarakhand, India

Manoj Singh Adhikari
School of Electronics and Electrical
  Engineering
Lovely Professional University
Phagwara, Punjab, India

Chapter 18
Sparsh Sharma
Computer Science and Engineering
Baba Ghulam Shah Badshah University
Rajouri, India

Abrar Ahmed
Computer Science and Engineering
Baba Ghulam Shah Badshah University
Rajouri, India

Surbhi Sharma
Computer Science and Engineering
Shri Mata Vaishno Devi University
Katra, India

Mohd Naseem
Computer Science and Engineering
Baba Ghulam Shah Badshah University
Rajouri, India

Chapter 19
Arun Kumar Rana
Panipat Institute of Engineering and
  Technology
Samalkha, India

Sharad Sharma
Maharishi Markandeshwar
    (Deemed to be University)
Mullana, India

Chapter 20
Pranav Ratta
University Institute of Computing
Chandigarh University
Mohali, India

Amanpreet Kaur
University Institute of Computing
Chandigarh University
Mohali, India

Sparsh Sharma
Computer Science and Engineering
Baba Ghulam Shah Badshah
    University
Rajouri, India

Chapter 21
Vivek Sharma
Department of Product and Industrial
    Design
Lovely Professional University
Phagwara, India

Neelam Rup Prakash
Department of Electronics and
    Communication
Punjab Engineering College
Chandigarh, India

Parveen Kalra
Department of Electronics and
    Communication
Punjab Engineering College
Chandigarh, India

Chapter 22
Nilima Dongre
Department of Information Technology
RAIT
Mumbai, India

Mohammad Atique
Department of Computer Science and
    Engineering
SGBAU
Amravati, India

Chapter 23
Aarti
CSE Department
Lovely Professional University
Punjab, India

Raju Pal
CSE Department
Jaypee Institute of Information
    Technology
Noida, India

Introduction

Machine intelligence (MI) and communication is a powerful buzzword and trendy technology which has directly or indirectly greatly impacted the lifestyle of every human being. In fact, MI is fast becoming an intrinsic part of our daily life and is not confined to university research labs, even if extraordinary growth has been made in this domain. The advantages of this miraculous technology are widely recognized in diversified areas, ranging from medicine to security to consumer applications and business, and resulting in improvements in the quality of life of humankind. Every new disruptive technology has its own pros and cons, and artificial intelligence (AI) is no exception to this rule. Privacy, data protection and the rights of individuals pose social and ethical challenges. This title covers interdisciplinary points including computerization, human-made consciousness and AI, the human-machine interface, control hypothesis and control frameworks, smart and responsive materials and smart sensing frameworks

The title *Intelligent Communication and Automation Systems* gives an open access home to great logical and building research on different frameworks that perceive, process, and react to improvements/directions and gain knowledge with the help of artificial intelligence and automation techniques.

SALIENT FEATURES

The scope of *Intelligent Communication and Automation Systems* stretches to the utilization of intelligent systems in industry, medical, and daily life:

- Smart sensing systems covering various applications (for example optical, temperature, gas, strain, and so forth), haptic frameworks, biosensors and picture sensors).
- Artificial intelligence, including AI hypothesis, various ways to deal with AI, discernment and rationales, language processing, AI, neural systems, multi-operator frameworks, memory and observation, neuromorphic processing, the neuroscientific parts of human-made consciousness, the human-machine interface (HMI) and brain-computer interface.
- Programmed/directed self-assembly and stimuli-responsive systems.
- Ethical/philosophical/political/economic aspects.

1 Role of Machine Learning in Communication Networks

Veerpal Kaur and Aman Singh
Lovely Professional University

CONTENTS

1.1   Introduction 1
1.2   Literature Review 2
    1.2.1   Machine Learning for Traffic Control...2
    1.2.2   Machine Learning for Routing 4
    1.2.3   Machine Learning for Intrusion Detection......................................4
1.3   Discussions 5
1.4   Conclusion 6
References...7

1.1   INTRODUCTION

Machine learning is a technology that has drawn its shape from the artificial intelligence (AI) domain upon which the present era is evolving rapidly. The need of AI-based systems is growing substantially due to which related domains are also gaining much interest of researchers. Machine learning is one popular domain of artificial intelligence which works primarily on extracting patterns from the huge database of information. The system is trained based upon the given dataset and then tested on raw data to ensure that the artificial system works well in the real world. Among many other applications, machine learning has evolved particularly in image processing and natural language processing (NLP). Various models are making use of algorithms based on probability and other concepts of mathematics, such as algebra, and are gaining support of the users when they are efficiently managing the large datasets and predicting the results as efficiently as the actual ones. Evolution of big data, image recognition and deep learning is the reason that machine learning and all the AI domains have been emerging so rapidly and are drawing the attention of economists as well as the other business leaders.

Besides the immense use of unprecedented techniques in many fields of technology, communication networks are equally touched by the growing influence of machine learning. Networks are great source of data on which machine learning is

1

dependent. As more and more users are adding to the complex chain of networks, machine learning algorithms (MLAs) can be deployed at various network layers, depending upon the need. This will help in managing the vast network.

MLAs are at an early stage of their usage in the communication networks [1] as there is a lag in the proficient personnel at service providers, and even at network vendors, who can deploy machine learning into the network as efficiently as they deploy protocols and other crucial security and information exchange paradigms. As machine learning is an active research field, the trend of its deployment in almost every field due to its autonomous nature is fascinating. This autonomous parameter is very useful for the software development industry because the ease of infusing MLAs into the software provides reusability and far better performance.

This chapter focuses on machine learning techniques that can be deployed in the communication networks at different stages of their usage, including traffic control, routing of information, intrusion detection and many more. The chapter is exploring the recent studies where machine learning has demonstrated a new trend.

1.2 LITERATURE REVIEW

In this section, the period from 2010 to 2020 has been picked to review the studies showing the usage of MLAs and techniques in communication networks for managing the resource optimally.

1.2.1 MACHINE LEARNING FOR TRAFFIC CONTROL

Traffic control and management is the most crucial part of the communication networks as the era is facing an overwhelming use of technology. This chapter will focus on the traffic flow of the network and various approaches and techniques proposed so far for addressing the issue of smooth network traffic flow.

In a paper presented in 2012 [2], a unique technique was proposed, making use of C5.0 MLA to differentiate the enormous traffic flow into various categories for seven different applications. The approach used here falls under the category of classification in machine learning domain.

A survey presented in 2016 [3] highlights various clustering approaches for traffic classification. The survey used data from various applications and conducted a performance metrics based on accuracy which may help the user/application to pick the correct algorithm for classifying its network traffic. Based upon the results shown, supervised Naïve Bayes for hypertext transfer protocol (HTTP), domain name system (DNS) and simple mail transfer protocol (SMTP) network traffic; self-learning classifier for HTTP, Gmail, real time streaming protocol (RTSP), Post Office Protocol 3 (POP3), teletype network (TELNET) and Bit Torrent network traffic; and C4.5 decision tree for SMTP, DNS network traffic have shown the highest accuracy rates.

In Reference[4], the authors studied supervised support vector machines (SVMs) and unsupervised K-means clustering for classifying the network traffic. The results depicted in the paper claimed better accuracy rate for SVM approach compared to K-means.

In a paper presented in 2019 [5], the application of five MLAs in live network for detection of distributed denial-of-service (DDoS) attack was conducted. The algorithms were judged on performance parameters such as detection rate, accuracy and precision. It was observed that the algorithms work poorly in classification of live network traffic compared to offline.

In the series of "AI for Network Traffic Control", the first edition [6] proposed a system architecture that bridges the gap of cloud computing and communication networks. The proposed traffic control algorithm helps in reducing the data to be transmitted and making optimum use of edge cloud. The second edition from the series [7] worked upon vehicular ad-hoc networks (VANETs) to address the real-time jamming detection problem by making use of AI-based techniques. The third edition focused on the classification of network traffic with an improved stacked auto-encoder [8] that displays the prominent feature learning aspect of the model. The fourth article of the series demonstrate a deep learning based technique[9] to deplete the congestion in the ultra-dense network. The proposed proactive technique worked really well in managing the traffic flow. Since then, the machine learning techniques penetrated well into the communication networks and showed impressive results. A new edition [10] in the series worked on software defines networking for managing the multimedia traffic using hybrid of deep learning and reinforcement learning. The model is efficient in learning from experience and quick decision making. The next article from the series [11] focused on the internet of things (IoT) for traffic prediction in a virtual environment. The proposed machine learning based proactive technique reduced the delay in scaling out the resources from the network. A further article from the series presented mobile edge caching [12] to manage as well as reduce the network traffic with improved quality of service. The technique has been combined with deep reinforcement learning. Various new features and challenges have been discussed for future prospects. An additional article from the series depicts the role of artificial intelligence in wireless networks [13] to significantly reduce the network traffic. The traffic control in the 5G network can be managed efficiently by supervised and unsupervised MLAs. Another article from the series worked on software-defined networking [14] combined with deep learning for traffic management in vehicular cyber-physical systems by finding optimal routes. The next article from the series focused on heterogeneous wireless network [15] and proposed a multi-dimensional traffic management technique to get better QoS. Nevertheless, the next article from the series talks about future networking [16] which will have an explosion of users and folding of information into the network. The technology should be future-ready by using artificial intelligence paradigms. The next one from the series [17] proposed a multimedia traffic classification model which works on pattern recognition of the network traffic. The model claimed good results for neural network in comparison to SVM and other methods. Moreover, some of the work has been conducted on mobile cellular networks [18] for predicting the traffic flow. The technique used in the proposal is based on deep learning.

Optical networks have been forecast as a future technology and they demand crucial management of network traffic. A paper by Mai, Nsve and Migge [19] presented machine learning classification and clustering algorithms with focus on optical networks' signal failure detection, unlabelled data detection and visualizing the monitored data.

1.2.2 MACHINE LEARNING FOR ROUTING

Another important paradigm of communication networks is the selection of routes for routing the network traffic. The routing protocols and various correlated technologies, or systems, are much in demand to coordinate the work of effective route selection demolishing the challenges faced by routing protocols. Following are the recent studies conducted in this field which require much of the focus for future prospects.

Effort-worthy work has been conducted [20], keeping in view the route selection for the traffic flow in software-defined networks. K-means and cosine similarity methods have been used to select best possible path to route the network traffic.

In the further studies on routing the information through the network, machine learning came up with resolution to all the major issues. In Reference [21], the authors proposed a machine learning based approach for routing the data packets across a wireless communication medium resulting in energy efficient use of the network.

A novel approach has been presented in Reference [22] which is dependent on the two most-used MLAs, namely neural networks and decision trees. The result of the approach used in the study is a routing protocol that help the networks effectively manage the routing of the data packets. The results shown in the research display tremendous efforts in terms of improvement in the hop count, successful pack delivery and dropping of packets.

In a study published in 2017 [23], focus was on delay tolerant network routing which used machine learning concepts to make a router an intelligent one by eliminating the need of delivery status packets.

In another study released in 2017 [24], an enhancement of Q-routing algorithm was presented that shows splendid results on implementation. The proposed technique changed the traditional approach of feedback into the network and used basic approach with additional learning rates. Keeping this study [24] as a base paper, a new approach was proposed [25] that suggested a modification and outperforms the previous work on implementation, providing better results for route discovery and route memory under high load conditions.

Routing in networks always looks forward for optimization. In this regard, Reference [26] presented a machine learning based routing model that works on optimization of routing protocol parameter, and results show improvement in power allocation in VANETs and delay prediction of the network.

1.2.3 MACHINE LEARNING FOR INTRUSION DETECTION

Machine learning is being adopted in almost all networks, including wired, wireless and optical networks because of its malleability and applicability in various parts of the operating networks. Another such broad domain of the networking is intrusion detection which will ultimately hamper the performance of the network if neglected. Below are some of the significant studies in this domain.

Cryptographic schemes are the most popular when the topic of network security comes into play. A lot of work has been done on technical enhancements in the communication networks to combat various types of attacks on the crucial and sensitive information of the users transmitting across the network. One such effort has been put forth by a study [27] published in 2018, in which various classification algorithms from machine learning were used to conduct an evaluation of block cipher techniques to get the cipher text. A performance analysis of all the MLAs has also been presented in the results of the study, drawing significant attention towards the Rotation Forest (RoFo) classification algorithm, which has shown the best performance among NB, SVM, MPL, IBL, AdaBoost.M1, C4.5 and Bag. The study claimed that the results were worst for IBL classification technique.

In Reference [28], a study was conducted to compare the decision tree (J48) MLA with neural network and support vector machine techniques of ML for performance parameters such as accuracy and false alarm rate. It showed that the decision tree technique does better than the rest of the algorithms.

Another idea presented in Reference [29] proposed various machine learning approaches for making intrusion detection easier in the communication networks by comparing the machine learning classifiers on the basis of their accuracy and power of specifying various possible attacks. A similar approach of comparison of classifiers has been adopted [30, 31] to get the best suitable machine learning approach to detect intrusion detection. Deep learning with machine learning has been combined [32] to get better detection of intrusion that demonstrates another way to achieve security in networks. A performance measurement approach [33] has been adopted to compare random forest, Naïve Bayes, decision tree and stochastic gradient descent MLAs; it shows random forest as best among the rest. Another comparison system [34] shows that the broad learning system approach outperforms the recurrent neural network technique of machine learning.

Comparison of various MLAs is always a better way to get an approach that will be most suitable according to the persisting need of the network. Another way to get a better approach for intrusion detection is to decode the attacks for experiment. Such an approach is followed in a study [35] that reveals that random forest has better performance while detecting denial-of-service attacks to the network than does MLP (multi-layer perceptron).

1.3   DISCUSSIONS

The studies conducted so far worked well on various machine learning approaches as one way or the other improves the network resource utilization and major issues of the same. Combining machine learning techniques with other domains has also been included in some research [36–44] which proved that a lot of work needs to be done in the hybridization of domains under artificial intelligence and networks.

After about 150 research articles that focus on the key role of MLAs in traffic control, routing and intrusion detection paradigms of networks were analyzed,

TABLE 1.1
Literature Survey

Year of Publication	Papers Published on		
	Traffic Control	Routing	Intrusion Detection
2010–2012	1	1	3
2013–2015	0	0	1
2016–2020	18	4	6

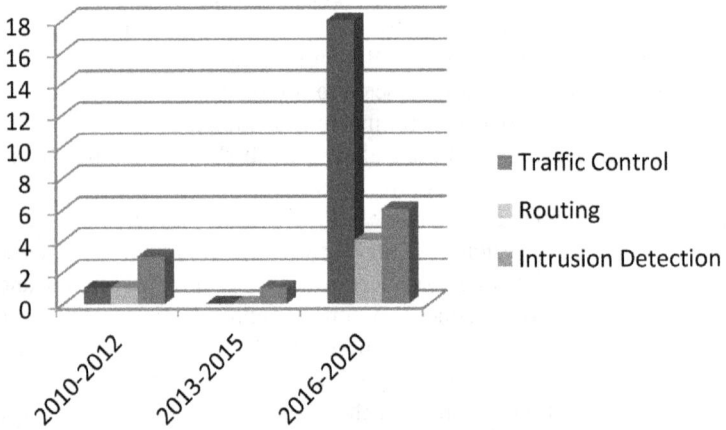

FIGURE 1.1   A chart view of algorithms and techniques.

50 studies were selected to draw conclusions. Table 1.1 summarizes the study conducted from 2010 to 2020 to justify the role of machine learning in communication networks.

The studies reveal that MLAs have been incorporated well in communication networks for traffic management but less focus has been placed on routing algorithms and techniques to make the information transmission timely and fault free. In Figure 1.1 a chart view of the table is shown to give a better view of the scenario presented so far.

1.4   CONCLUSION

The studies conducted so far have focused on the traffic control and management of communication networks, but much more focus needs to be drawn towards the other aspects of networks, such as the network security, efficient routing mechanism and power consumption of the devices in the network, and many more application areas of machine learning are still to be discovered. MLAs such as random forest and neural networks are quite accurate and overwhelmingly used in various network applications. The use of such algorithms along with cryptographic algorithms [45, 46] may bring out more secure network systems; traffic flow of the communication network will get checked if it passes through a series of such algorithmic verifications, and obviously the information flowing through the network will get secured from many existing attacks [47–50]. Dimensionality reduction techniques from machine learning can be used in the networks to get the clustering as well as classification of users, and this may be very useful to scrutinize any intrusion across the network. This chapter will be implemented using Python programming to get the better view of the current scenario, and conclusions can be drawn to get efficient machine learning tools to be worked upon for future prospects.

REFERENCES

Conferences

1. Tomasz Bujlow, Tahir Riaz, and Jens Myrup Pedersen (2012). A Method for Classification of Network Traffic Based on C5.0 Machine Learning Algorithm. Workshop on Computing, Networking and Communications. IEEE.
2. R. Deebalakshmi, DR.V. L. Jyothi. (2016) A Survey of Classification Algorithm for Network Traffic. Second International Conference on Science Technology Engineering and Management (ICONSTEM), IEEE.
3. Zhong Fan and Ran Liu. (2017). Investigation of Machine Learning Based Network Traffic Classification. International Symposium on Wireless Communication Systems (ISWCS), IEEE.
4. Jarrod Bakker, Bryan Ng, Winston K.G. Seah and Adrian Pekar. (2019). Traffic Classification with Machine Learning in a Live Network, IFIP International Symposium on Integrated Network Management (IM2019): Mini-Conference Experiment Session, IEEE.
5. Alba P. Vela, Marc Ruiz and Luis Velasco. (2018). Examples of Machine Learning Algorithms for Optical Network Control and Management, ICTON, IEEE.
6. Shubham Kumar, Gaurang Bansal and Virendra Singh Shekhawat. (2020). A Machine Learning Approach for Traffic Flow Provisioning in Software Defined networks, ICOIN, IEEE.
7. Kaveri Kadam and Navin Srivastava. (2012). Application of Machine Learning (Reinforcement Learning) for Routing in Wireless Sensor Networks (WSNs), 1st International Symposium on Physics and Technology of Sensors (ISPTS-1). IEEE.
8. Yuliya Shilova, Maksim Kavalerov and Igor Bezukladnikov. (2016). Full Echo Q-Routing with Adaptive Learning Rates: a Reinforcement Learning Approach to Network Routing. IEEE NW Russia Young Researchers in Electrical and Electronic Engineering Conference (EIConRusNW). IEEE.
9. Kun Yu, Li-zhuang Tan, Xiao-jin Wu and Zhi-yong Gai. (2019). Machine Learning Driven Network Routing. 6th International Conference on Systems and Informatics (ICSAI), IEEE.
10. Suhaila O. Sharif, L. I. Kuncheva and S. P. Mansoor. (2010). Classifying Encryption Algorithms using Pattern Recognition Techniques. International Conference on Information Theory and Information Security. IEEE.
11. Kamularifin Abd Jalil and Mohamad Noorman Masrek. (2010), Comparison of Machine Learning Algorithms Performance in Detecting Network Intrusion. International Conference on Networking and Information Technology, IEEE.

12. G.V. Nadiammai and M. Hemalatha. (2012). Perspective Analysis of Machine Learning Algorithms for Detecting Network Intrusions. Third International Conference on Computing, Communication and Networking Technologies (ICCCNT'12), IEEE.

13. Sumouli Choudhary and Anirban Bhowal. (2015). Comparative Analysis of Machine Learning Algorithms along with Classifiers for Network Intrusion Detection. International Conference on Smart Technologies and Management for Computing, Communication, Controls, Energy and Materials, IEEE.

14. Fatih Ertam, Ilhan Firat Kilincer, Orhan Yaman. (2017). Intrusion Detection in Computer Networks via Machine Learning Algorithms. International Artificial Intelligence and Data Processing Symposium (IDAP). IEEE.

15. Sasanka Potluri, Navin Francis Henry and Christian Diedrich. (2017). Evaluation of Hybrid Deep Learning Techniques for Ensuring Security in Networked Control Systems. International Conference on Emerging Technologies and Factory Automation. IEEE.

16. Pascal Maniriho and Toahri Ahmad. (2018). Analyzing the Performance of Machine Learning Algorithms in Anomaly Network Intrusion Detection Systems. 4th International Conference on Science and Technology (ICST). IEEE.

17. Zhida Li, Prerna Batta and Ljiljana Trajkovic. (2018). Comparison of Machine Learning Algorithms for Detection of Network Intrusions. International Conference on Systems, Man and Cybernetics. IEEE.

18. Shreekhand Wankhede and Deepak Kshirsagar. (2018). DoS Attack Detection using Machine Learning and Neural Network. Fourth International Conference on Computing Communication Control and Automation (ICCUBEA). IEEE.

19. Tieu Long Mai, Nicolas Navet and Jorn Migge. (2019). A Hybrid Machine Learning and Schedulability Analysis Method for the Verification of TSN Networks. 15th International Workshop on Factory Communication Systems (WFCS). IEEE.

20. Bruno Vicente Alves de Lima and Vinicius Ponte Machado. (2012). Machine Learning Algorithms Applied in Automatic Classification of Social Network Users. Fourth International Conference on Computational Aspects of Social Networks (CASoN). IEEE.

21. Xin Xia, David Lo, Xinyu Wang, Xiaohu Yang and Shanping Li. (2013). A Comparative Study of Supervised Learning Algorithms for Re-opened Bug Prediction. 17th European Conference on Software Maintenance and Reengineering. IEEE.

22. Rachel Dudukovich, Alan Hylton and Christos Papachristou. (2017). A Machine Learning Concept for DTN Routing. International Conference on Wireless for Space and Extreme Environments. IEEE.

23. John C. Porcello. (2017). Designing and Implementing Machine Learning Algorithms for Advanced Communications using FPGAs. IEEE Aerospace Conference, IEEE.

24. Maksim Kavalerov, Yuliya Likhacheva and Yuliya Shilova. (2017). A Reinforcement Learning Approach to Network Routing Based on Adaptive Learning Rates and Route Memory. SoutheastCon 2017, IEEE

25. O. Obulesu, M. Mahendra and M. Thrilok Reddy. (2018). Machine Learning Techniques and Tools: A Survey. Proceedings of the International Conference on Inventive Research in Computing Applications (ICIRCA2018), IEEE.

26. K. Shahina and V. Vaidehi. (2018). Clustering and Data Aggregation in Wireless Sensor Networks Using Machine Learning Algorithms. International Conference on Recent Trends in Advance Computing (ICRTAC), IEEE.

27. Aritra Basu and Budhaditya Bhattacharyya. (2018). A Study on the Integration on Machine Learning in Wireless Communication. International Conference on Communication and Signal Processing. IEEE.

Journals

28. David Cote. (2018). Using Machine Learning in Communication Networks [Invited]. *OSA Journal of Optical Communications and Networking*. IEEE.
29. Min Chen, Yixue Hao, Kai Lin, Zhiyong Yuan and Long Hu. (2018). Label-less Learning for Traffic Control in an Edge Network. *AI For Network Traffic Control*. IEEE.
30. Nikita Lyamin, Denis Kleyko, Quentin Delooz and Alexey Vinel. (2018). AI-Based Malicious Network Traffic Detection in VANETs. *AI For Network Traffic Control*. IEEE.
31. Peng Li, Zhikui Chen, Laurence T. Yang, Jing Gao, Qingchen Zhang and M. Jamal Deen. (2018). An Improved Stacked Auto-Encoder for Network Traffic Flow Classification. *AI For Network Traffic Control*. IEEE.
32. Yibo Zhou, Zubair Md. Fadlullah, Bomin Mao and Nei Kato. (2018). A Deep-Learning-Based Radio Resource Assignment Technique for 5G Ultra Dense Network. *AI For Network Traffic Control*. IEEE.
33. Xiaohong Huang, Tinting Yuan, Guanhua Qiao and Yizhi Ren. (2018). Deep Reinforcement Learning for Multimedia Traffic Control in Software Defined Networking. *AI For Network Traffic Control*. IEEE.
34. Imad Alawe, Adlen Ksentini, Yassine Hadjadj-Aoul and Philippe Bertin. (2018). Improving Traffic Forecasting for 5G Network Scalability: A Machine Learning Approach. *AI For Network Traffic Control*. IEEE.
35. Hao Zhu, Yang Cao, Wei Wang, Tao Jiang and Shi in. (2018). Deep Reinforcement Learning for Mobile Edge Caching: Review, New Features and Open Issues. *AI For Network Traffic Control*. IEEE.
36. Yu Fu, Sen Wang, Cheng-Xiang Wang, Xuemin Hong and Stephen McLaughlin. (2018). Artificial Intelligence to Manage Network Traffic of 5G Wireless Network. *AI For Network Traffic Control*. IEEE.
37. Arish Jindal, Gagangeet Singh Aujla, Neeraj Kumar, Rajat Chaudhary, Mohammad S. Obaidat and Ilsun You. (2018). SeDaTive: SDN-Enabled Deep Learning Architecture for Network Traffic Control in Vehicular Cyber-Physical Systems. *AI For Network Traffic Control*. IEEE.
38. Jian Shen, Tianqi Zhou, Kun Wang, Xin Peng and Li Pan. (2018) Artificial intelligence Inspired Multi-Dimensional Traffic Control for Heterogeneous Network. *AI For Network Traffic Control*. IEEE.
39. Jun Xu and Kaishun Wu. (2018). Living with Artificial Intelligence: A Paradigm Shift toward Future Network Traffic Control. *AI For Network Traffic Control*. IEEE.
40. Alejandro Canovas, Jose Miguel Jimenez, Oscar Romero and Jaime Lloret. (2018). Multimedia Data Flow Traffic Classification Using Intelligent Models Based on Traffic Patterns. *AI For Network Traffic Control*. IEEE.
41. Jie Feng, Xinlei Chen, Rundong Gao, Ming Zeng and Yong Li. (2018). Deep TP: An End-to-End Neural Network for Mobile Cellular Traffic Prediction. *AI For Network Traffic Control*. IEEE.
42. Deepak K. Sharma, Sanjay K. Dhurandher, Rohit K. Srivasta, Anhad Mohananey and Joel J.P.C. Rodrigues. (2016). A Machine Learning-Based Protocol for Efficient Routing in Opportunistic Networks. *IEEE Systems Journal*, IEEE, 2207–2213.
43. Chunxiao Jiang, Haijun Zhang, Yong Ren, Zhu Han, Kwang-Cheng Chen and Lajos Hanzo. (2017). Machine Learning Paradigms for Next-Generation Wireless Networks, *IEEE Wireless Communications*, IEEE.
44. Veerpal Kaur, Kamali Gupta, Vidhu Baggan, Gagandeep Kaur and Amanjyoti. (2019). Role of Cryptographic Algorithms in Mobile Ad Hoc Network Security: An Elucidation. *International Journal of Recent Technology and Engineering (IJRTE)*.

45. Kai Arulkumaran, Marc Peter Deisenroth, Miles Brundage and Anil Anthony Bharath. (2017). Deep Reinforcement Learning-A brief Survey, *IEEE Signal Processing Magazine*, IEEE.

46. Veerpal Kaur and Aman Singh. (2015). Encryption System Based on AES and SHA-512. *International Journal of Applied Engineering Research (IJAER)* ISSN 0973-4562 Vol. 10, Number 10, pp.25207–25218

47. Raouf Boutaba, Mohammad A. Salahuddin, Noura Limam, Sara Ayoubi, Nashid Shahriar, Felipe Estrada-Solano and Oscar M. Caicedo. (2018). A Comprehensive Survey on Machine Learning for Networking: Evaluation, Applications and Research Opportunities. *Journal of Internet Services and Applications*, IEEE.

48. Franceso Musumeci, Cristina Rottondi, Avishek Nag, Irene Macalusa, Darko Zibar, Marco Ruffini and Massimo Tornatore. (2018). An Overview on Application of Machine Learning Techniques in Optical Networks. *IEEE Communications Surveys & Tutorials*, IEEE, pp. 1383–1408.

49. Qian Mao, Fei Hu and Qi Hao. (2018). Deep Learning for Wireless Network: A Comprehensive Survey. *IEEE Communications Surveys and Tutorials*, IEEE.

50. Osvaldo Simeons. (2018). A Very Brief Introduction to Machine Learning with Applications to Communication Systems. *IEEE Transactions on Cognitive Communications and Networking*, IEEE.

2 Evolutionary Optimization Approach for Network Lifetime in Wireless Sensor Networks

Pooja Chaturvedi
School of Management Sciences

Ajai Kumar Daniel
M.M.M. University of Technology

CONTENTS

2.1   Introduction 11
2.2   Related Work 12
    2.2.1   Coverage and PSO 13
2.3   Network Model 14
    2.3.1   Energy Model 14
    2.3.2   Particle Swarm Optimization Approach 15
2.4   Proposed Protocol.. 15
    2.4.1   Performance Metrics 16
    2.4.2   Analytical Modelling... 17
2.5   Simulation Results and Discussion.. 19
2.6   Conclusion and Future Research Directions 25
References.. 27

2.1   INTRODUCTION

Wireless sensor networks (WSN) have been considered a pioneer research area owing to their characteristics: lower cost, maintenance, adaptivity as well as applicability in numerous areas such as environmental monitoring, health care, target tracking, surveillance, etc. Despite the applicability of WSN in diverse fields, several aspects still need to be resolved, such as limited resource capability, localization, security, fault tolerance and coverage. Improvement of coverage is emphasized in WSN because it is often considered a QoS measure which determines how well and for how long the given area is under coverage by a sensor node. A point in a target area lying in a sensor node's communication range is said to be observed. Coverage problem of targets are basically concerned with providing coverage to a predetermined set of points. There are various approaches to overcome the target coverage problem [1–5].

There are several approaches suggested for solving the target coverage problem using scheduling approaches. Node scheduling approaches are based on activation of the subset of nodes to achieve the coverage of the targets and increase the node lifetime. This chapter addresses the target coverage with an objective of coverage duration maximization and node lifetime in energy-efficient coverage protocol (EECP) [6].

This chapter is organized as follows: related work in Section 2.2, network model in Section 2.3, proposed protocol in Section 2.4, results and discussion in Section 2.5, and future research directions in Section 2.6.

2.2 RELATED WORK

The inherent characteristics of sensor nodes, such as limited communication range and energy constraints, cause the network lifetime and coverage problems. There are three viewpoints to define the coverage problems: (1) area coverage problem basically aims to monitor the complete target region. (2) Point/target coverage is to provide coverage to a specific set of observation points. (3) Barrier coverage aims to provide protection against the breaching of the target area.

The problem of coverage and energy conservation objectives can be achieved in several ways, such as adjustment of sensing ranges, node scheduling approaches, and virtual and computational geometry based approaches. One approach is based on the adjustment of the sensing range according to the remaining energy of the nodes. The node scheduling approaches are based on dividing the set of nodes into different subsets which are activated periodically to observe the target area, and the computational geometry based approaches use the Voronoi diagrams and repulsive force based approach to move the sensor nodes near and far from each other to achieve the enhanced coverage [7–9].

In Reference [10], an incremental redeployment approach is shown to increase the coverage for the grid-based deployment. If the coverage rate of a grid is not satisfied, a new node is deployed to improve the coverage. This process is repeated until all the grids have the desired coverage rate.

In Reference [3], the authors have proposed three Voronoi diagram based approaches to remove the coverage holes as follows: Vector force based algorithm (VEC) to push sensors away from a region with high coverage; Voronoi based approach (VOR) to move the sensor towards the farthest vertex in case of coverage hole; mini max approach for moving the sensor towards the farthest vertex.

In Reference [11], authors have proposed a hybrid approach utilizing the static and dynamic nodes. In this approach the static nodes are used in constructing the Voronoi diagram to identify the holes in the coverage; mobile nodes can be used to fill that coverage hole.

In Reference [12], authors have used the concept of exerting the virtual forces due to the potential field due to the nodes. Due to the force of the sensors they are moved towards the unmonitored region to eliminate the coverage hole. Other similar approaches utilizing the node mobility have been proposed in References [13,14]. These approaches are not very efficient for sensor network because node movement requires high energy consumption. The sensor nodes usually exist in battery-operated and limited-energy reservoirs. Reference [15], authors have considered the load balancing parameter too, along with the coverage of the nodes. In Reference [16], authors have proposed the velocity-scheduling approach, which considers energy consumption of the nodes during the movement by considering several metrics, such as road condition, friction, heating, acceleration, etc.

2.2.1 COVERAGE AND PSO

The coverage problem of WSN can also be viewed as a multi-objective that aims to optimize the network performance by limiting the energy consumption, maximizing network lifetime and increasing the coverage level. So PSO has been found to be a good solution for all these objectives. Several approaches have been proposed in the literature that address the coverage problem by utilizing the advantages of PSO technique [16]. In Reference [17], authors describe a PSO-based virtual force based approach to move the sensor to the lesser covered region after the initial random deployment.

In Reference [18], authors have proposed simulated annealing and PSO-based redeployment approach to achieve greater coverage and network lifetime.

In Reference [19], authors propose a sequential PSO-based approach to redeploy the nodes such that the coverage of the network is enhanced. This approach deploys the sensors in a sequential manner (i.e., one at a time until the required coverage is achieved).

In Reference [20], authors have proposed the individual particle optimization technique to address the coverage problem. This approach advantage is increased exploration ability.

The research gap identified from the literature review is summarized below:

1. Most of the existing works on coverage problems are based on increasing the covered area in the context of grid-based deterministic deployment using the PSO-based approaches.
2. The existing works generally consider the Boolean detection model for coverage.
3. The existing approaches are mainly focused on redeployment of nodes, which incurs extra energy consumption due to the mobility.
4. Deterministic deployment, such as grid, triangular etc., is often not efficient in the context of WSN.
5. Existing approaches are based on a deployment-based approach in which all the nodes remain active all the time, which is not efficient because of the limited energy sources in WSN. The sensor nodes may exist in idle, receiving, transmitting and sleep states. It has been studied in the literature that the sensor nodes use high energy consumption in the active state when

they are transmitting the collected data to neighbouring nodes or to the base station and they use minimum energy in the sleep state. So the scheduling approaches which keep the active state a small set of nodes are of great significance.

In Reference [21], authors propose a scheduling approach for nodes to address various QoS parameters such as energy conservation, coverage and network lifetime. The proposed scheduling approach schedules the nodes into several subsets based on the probability of coverage, trust values and contribution of node, and they are activated periodically to observe a specific set of targets. This paper aims to extend the performance of the network by utilizing the evolutionary optimization technique of PSO. The main advantage of using the PSO is as follows:

1. In PSO, parallel computation is possible, which makes it efficient.
2. It is simple to use and understand as it requires fewer parameters.
3. PSO is suitable for the multi-objective problem.
4. The convergence time is low for PSO as it allows quick optimization by considering multiple particles at a time.

2.3 NETWORK MODEL

The target coverage is designed to monitor a predefined group of points with the described coverage level so that the network lifetime is maximized. In a network of x sensor nodes and y targets the proposed protocol aims to maximize the average remaining energy of nodes in all the set covers. The proposed protocol considers following assumptions:

1. The targets to be monitored are distributed in a two-dimensional area.
2. Nodes are homogeneous and static.
3. The range of the node is considered as a circular region, within which the node is situated at the centre.
4. The coverage determination model of the nodes is considered probabilistic, which implies that the node detection probability decreases exponentially with the increasing distance.

2.3.1 ENERGY MODEL

The sensor nodes consume energy during the reception and transmission of data as per the standard energy model. The energy consumed to transmit the l bit data packet at distance d can be determined as follows:

$$E_T(l,d) = \{l \times E\,elec + l \times \varepsilon_{fs} \times d^2, d < d_0$$
$$l \times E\,elec + l \times \varepsilon_{mp} \times d^4, d \geq d_0$$

where E_{elec} is the energy dissipated in the electric circuit, ε_{fs} represents the free space loss and ε_{mp} represents the multipath loss.

2.3.2 PARTICLE SWARM OPTIMIZATION APPROACH

The particle swarm optimization approach is an evolutionary optimization method for the multidimensional problem [22]. PSO considers a swarm of a set of particles in which each particle provides a possible solution to the multi-objective problem of d dimension. In a d dimensional region, every particle p_i is at position x_i and velocity v_i. The position x_i of the particle p_i is represented as follows:

$$x_{i+1} = x_i + v_{i+1}$$

A fitness function is used to determine the quality of the solution provided by a particle p_i. Each particle uses personal best (p_{best}) and global best position (p_g) to update its velocity to reach the best position to obtain the optimal solution. Suppose the initial position and velocity of particle p_i are v_i and x_i respectively. Then the velocity (v_{i+1}) and position (x_{i+1}) of the particle are updated as follows:

$$v_{i+1} = w * v_i + c_1 * r_1 * (p_{best} - x_i) + c_2 * r_2 * (p_g - x_i) \qquad (2.1)$$

$$x_{i+1} = x_i + v_{i+1} \qquad (2.2)$$

where w is the self-adapting parameter and c_1 and c_2 are acceleration coefficients which govern the speed of local and global search. The two random numbers r_1 and r_2 are used to provide the weightage of the local as well as global search. p_{best} and p_g represent the local and global best positions respectively.

2.4 PROPOSED PROTOCOL

The protocol evolves in rounds, and each round performs a steady state phase, a sensing of the environment phase and a data transmission phase. The steady phase is basically concerned with the identification and scheduling of the node subsets. The set covers are determined through contribution node, trust values and coverage detection probability. The sensing phase senses the data from the environment as per the schedule as determined by the sink, and the data communication phase forwards the sensed data to the sink station. The steady phase is further divided into four phases: clustering, trust calculation, set cover determination and set cover optimization. The set covers are determined in the similar way as proposed by Chaturvedi and Daniel [23–25]. The set cover optimization is done using the PSO-based approach as shown in Table 2.1.

To optimize the set cover scheduling, we have considered a Boolean variable a_{ij}, which represents the active status of the nodes and is defined as:

$$a_{ij} = \{1; if\ node\ i\ can\ monitor\ the\ target\ j\ 0; otherwise\} \qquad (2.3)$$

TABLE 2.1
Particle Swarm Optimization Algorithm

PSO Algorithm

Initialize c1, c2 and w

Initialize r1 and r2

max_itr= maximum iterations, max_fitness=maximum fitness

1. Initial population of particles is created with randomly selected positions and velocities;

2. Do

{

3. Compute fitness of each particles according to the (4);

4. Update p_{fi} if the current solution improves the fitness p_{fi};

5. Calculate p_{best}: particle position having best fitness value among all of the neighbours as the p_g is selected;

7. For every individual

{

i. Calculate velocity of the particle according to (1);

ii. Change position of the particle according to (2)

}

} until *max_itr* or *max_fitness*.

The aim of the proposed protocol is to maximize the average residual energy, so the fitness function is defined as:

$$fitness\ function\ (f) = \frac{\sum E_i a_{ij}}{\sum a_{ij}} \tag{2.4}$$

We calculate the average remaining energy of the set covers in each iteration. The set covers having the high residual energy are scheduled to observe the set of targets until all the targets are monitored.

2.4.1 PERFORMANCE METRICS

To evaluate the efficiency of the proposed protocol we have considered the few performance metrics:

1. **Network lifetime:** The lifetime of the node is determined as the number of rounds that the nodes are able to observe all the targets with the required coverage level.
2. **Node contribution:** The contribution of node is determined as set of targets a node *i* can monitor.
3. **Node involvement:** The node involvement is determined as the count of the set covers when the node is in communicating state.

4. **Dead node:** A node is said to be dead if its residual energy is zero or below the threshold so it cannot monitor the specified target.

5. **Node consumption factor:** Node consumption factor is the percentage of nodes dead from the total nodes.

2.4.2 ANALYTICAL MODELLING

To perform the analytical study, a network of 8 sensor nodes and 3 targets is taken. The nodes can cover the targets as follows:

$$x1 = \{t1\}, x2 = \{t1, t3\}, x3 = \{t3\}, x4 = \{t2, t3\}, x5 = \{t2\} \text{ and } x6 = \{t1\},$$

Based on the node monitoring status we determine the set covers as follows

$$c1 = \{s1, s3, s5\}, c2 = \{s3, s5, s6\}, c3 = \{s2, s5\}, c4 = \{s1, s4\}, c5 = \{s4, s6\} \text{ and}$$

$$c6 = \{s2, s4\}$$

Let the initial energy of the nodes be 1 J. Then, using the EECP protocol, the set covers are activated sequentially (i.e., first set cover c1 is activated, then c2 and so on) as follows

$$\{c1, c2, c3, c4\}.$$

On activating the sets, the remaining energy of the nodes in each round is determined as in Table 2.2.

The table shows that the remaining energy of the nodes 1, 3 and 5 is 0 after round 4. So the number of rounds the network is operation is determined as 4 and the dead nodes count is 3. But if we consider the average remaining energy of the set covers in every round then the set covers are activated as {c1, c5, c3, c6}.

TABLE 2.2
Remaining Energy of the Nodes after a Round using EECP

Node/Round No.	r1(c1)	r2(c2)	r3(c3)	r4(c4)
1	**0.7**	**0.5**	**0.3**	**0**
2	0.8	0.6	0.3	0.1
3	**0.7**	**0.4**	**0.3**	**0**
4	0.8	0.6	0.4	0.1
5	**0.7**	**0.4**	**0.1**	**0**
6	0.8	0.5	0.3	0.1
7	0.9	0.8	0.7	0.6
8	0.9	0.8	0.7	0.6

TABLE 2.3

Residual Energy of the Nodes after each Round using the Proposed Approach

Node/Round (Set Cover)	r1(c1)	r2(c5)	r3(c3)	r4(c6)
1	0.7	0.5	0.3	0.1
2	**0.8**	**0.6**	**0.3**	**0**
3	0.7	0.5	0.3	0.1
4	**0.8**	**0.5**	**0.3**	**0**
5	0.7	0.5	0.2	0.1
6	0.8	0.5	0.3	0.1
7	0.9	0.8	0.7	0.6
8	0.9	0.8	0.7	0.6

After the activation of sets, the remaining energy of nodes in a round is in Table 2.3.

It can be observed from the table that the residual energy is 0 for nodes 2 and 4 after 4 rounds. So the number of dead nodes is reduced to 2.

It can be analyzed that out of 8 nodes, 3 nodes run out of energy and cannot provide coverage to the targets, but by using the proposed approach only 2 nodes are dead. Hence, the proposed approach reduces the node consumption from a factor of 38% to 25%.

The comparison of node involvement in the set covers in both the approaches is in Table 2.4.

The comparison of the node involvement in the set covers is shown in Figure 2.1. Nodes 4 and 5 are involved in 3 set covers using EECP, whereas using the proposed approach the maximum node involvement is 2 for the nodes 2, 4 and 5. As the node involvement is less using the proposed protocol, the nodes may exhaust less energy compared to EECP, hence they improve the network lifetime.

The comparison of average remaining energy in every round is shown in Table 2.5.

Figure 2.2 shows the comparative analysis of average remaining energy. It can be seen from the figure that the average remaining energy is decreased with regard to

TABLE 2.4

Node Involvement Comparison

Node	EECP	Proposed
1	2	1
2	2	2
3	2	1
4	3	2
5	3	2
6	1	1

Comparsion of node involment in set covers

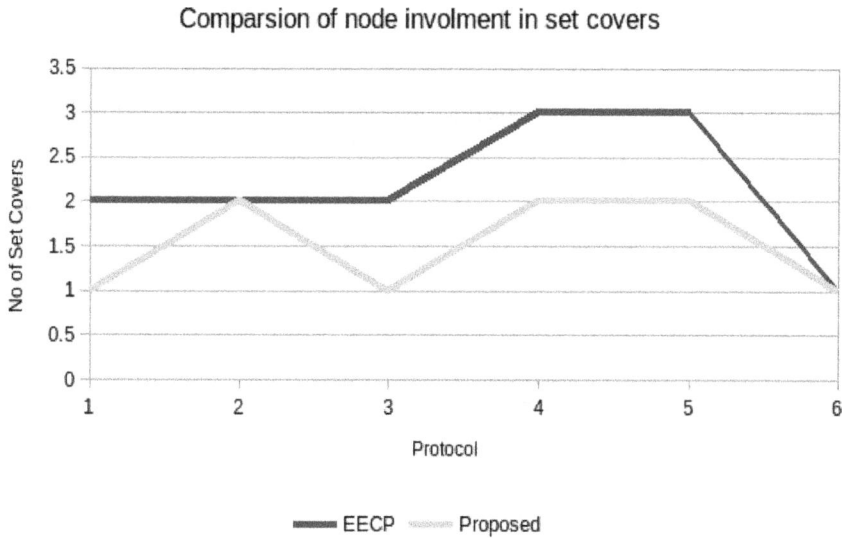

FIGURE 2.1 Comparison of node involvement in the set covers.

the number of rounds in both the approaches. It can also be observed that the average remaining energy is higher, so it will improve the network performance.

2.5 SIMULATION RESULTS AND DISCUSSION

The proposed approach is studied in two scenarios in a network having 10 nodes monitoring 5 targets in a dimension 150 × 150. In the first case we have considered a swarm of 5 particles, and in the second case we have considered a swarm of 10 particles. The remaining simulation settings are shown in Table 2.6:

Case 1

In the first scenario, we have considered the starting population of 5 swarms. The particles are initialized with the parameter values as shown in Table 2.7.

We have considered the same random numbers for the particles in each iteration. The random numbers considered for each particle are as shown in Table 2.8.

The best fitness and position values for the particles in every iteration are shown in Table 2.9.

TABLE 2.5

Comparison of Average Residual Energy

Round	EECP	Proposed
r1	0.78	0.78
r2	0.56	0.58
r3	0.38	0.38
r4	0.18	0.20

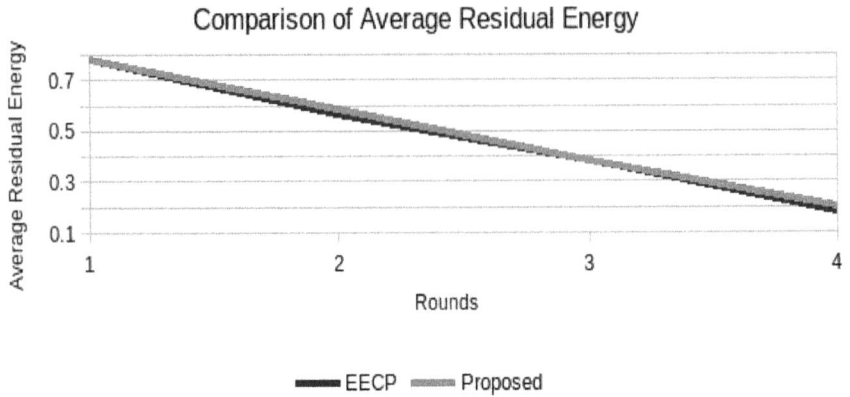

FIGURE 2.2   Comparison of average residual energy.

TABLE 2.6
Simulation Settings

Parameters	Value
No. of Nodes	10
No. of Targets	5
Initial energy	1 J
Simulation Time	1000 sec
No. of Communication	100
Packet size	500 bytes
Required coverage level for each target	0.5
Self-adapting parameter (w)	0.7
Acceleration coefficients	c1=c2=2
Swarm size	5,10

TABLE 2.7
Initial Parameter Values for the Particles

Particle	Position	Velocity
1	2	10
2	9	6
3	8	2
4	3	4
5	5	7

TABLE 2.8
Random Numbers for Each Particle

Particle	R1	R2
1	0.23	0.32
2	0.54	0.10
3	0.36	0.39
4	0.60	0.15
5	0.62	0.06

TABLE 2.9
The Local and Global Parameter Values

Iteration	Local	Global
1	CP(1)= [2 9 8 3 5] CF(1)= [0.2 0.9 0.8 0.3 0.5] V(1)= [10 6 4 2 7] LBF(1)= [0.2 0.9 0.8 0.3 0.5] LBP(1)= 9	GBF(1)= 0.9 GBP(1)= 9 particle= 2 velocity=6
2	CP(2)= [16.7 13.2 12.3 13.4 15.34] CF(2)= [1.67 1.32 1.23 1.34 1.534] V(2)= [14.7 4.2 4.3 10.4 10.34] LBF(2)= [1.67 1.32 1.23 1.34 1.534] LBP(2)= 16.7	GBF(2)= 1.67 GBP(2)= 16.7 particle= 1 velocity= 14.7
3	CP(3)=[26.99 32.64 30.26 33.98 24.42] CF(3)= [2.6 3.2 3.0 3.3 2.4] V(3)= [10.29 19.444 17.96 20.58 9.08] LBF(3)= [2.6 3.2 3.0 3.3 2.4] LBP(3)= 33.98	GBF(3)= 3.3 GBP(3)= 33.98 particle= 4 velocity= 20.58
4	CP(4)= [58.76 50.26 57.922 48.38 35.62] CF(4)= [5.8 5.0 5.7 4.8 3.5] V(4)= [31.97 17.62 27.66 14.40 11.2] LBF(4)= [5.8 5.0 5.7 4.8 3.5] LBP(4)= 58.76	GBF(4)= 5.8 GBP(4)= 58.76 particle= 1 velocity= 31.97
5	CP(5)= [81.13 115.76 78.54 74.03 74.92] CF(5)= [8.1 1.1 7.8 7.4 7.4] V(5)= [22.37 65.50 20.62 25.65 39.3] LBF(5)= [8.1 5.0 7.8 7.4 7.4] LBP(5)= 81.13	GBF(5)= 8.1 GBP(5)= 81.13 particle= 1 velocity=22.37
6	CP(6)= [96.78 126.49 119.57 102.63 138.65] CF(6)= [9.6 12.6 11.9 10.2 13.8] V(6)= [15.65 10.73 41.03 28.60 35.95] LBF(6)= [9.6 12.6 11.9 10.2 13.8] LBP(6)= 138.65	GBF(6)= 13.8 GBP(6)= 138.65 particle= 5 velocity=35.95

Fitness vs Iteration

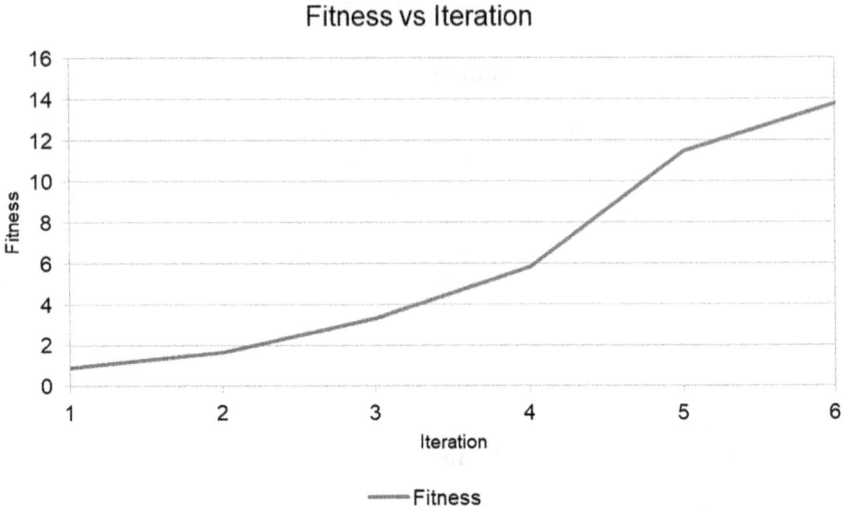

FIGURE 2.3   Fitness vs. iteration.

The fitness value with respect to each iteration is shown in Figure 2.3. It is seen in the figure that the fitness value is increased with the number of iterations.

The maximum and minimum position values for each particle are shown in Figure 2.4.

The comparison of fitness values in each iteration for the particles is shown in Figure 2.5. It shows the increasing fitness value with regard to the iterations.

Case 2

In the next set of experiments we have considered an initial population of 10 swarms. The initial parameter values for individual particles are in Table 2.10.

The random numbers considered for the particles are shown in Table 2.11.

Position vs Iteration

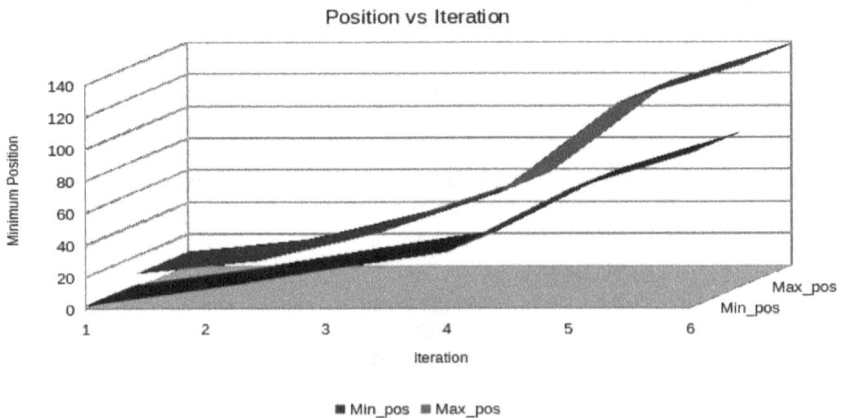

FIGURE 2.4   Performance of position and iteration for each particle.

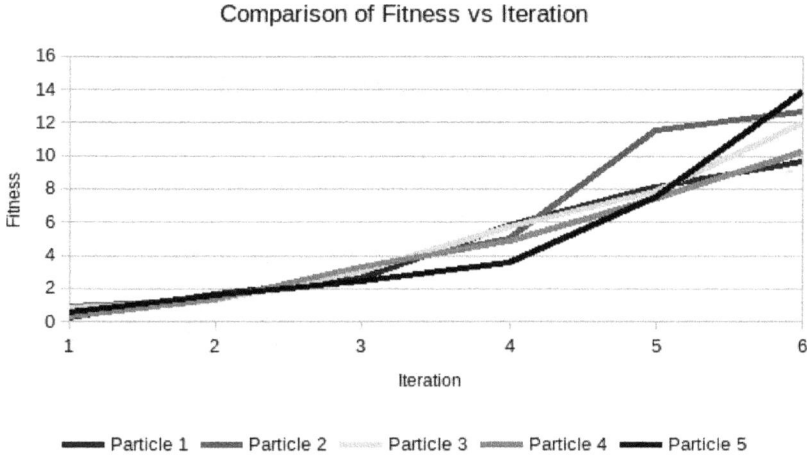

FIGURE 2.5   Performance of fitness and iteration for each particle.

The results of the computation of position and fitness values are summarized in Table 2.12.

The fitness value with respect to each iteration is as shown in Figure 2.6.

The maximum and minimum position values for each particle are shown in Figure 2.7.

The comparison of fitness values in each iteration for the particles is shown in Figure 2.8. It shows the increase in the fitness value with regard to the iteration.

The comparison for the fitness values for both the cases is shown in Figure 2.9. It can be observed from the figure that the fitness value is higher for the higher number of particles, i.e., for the case 2, where the number of particles is higher at 10 compared to the swarm of 5 particles.

TABLE 2.10
Initial Values

Particle	Position	Velocity
1	2	3
2	3	2
3	7	1
4	9	4
5	10	6
6	4	7
7	1	5
8	6	4
9	5	9
10	8	10

TABLE 2.11
Random Numbers for the Particles

Particle	r1	r2
1	0.23	0.32
2	0.54	0.10
3	0.36	0.39
4	0.60	0.15
5	0.62	0.06
6	0.42	0.23
7	0.67	0.42
8	0.8	0.8
9	0.2	0.6
10	0.1	0.9

TABLE 2.12
The Local and Global Parameter Values

Iteration	Local	Global
1	CP(1)=[2 3 7 9 10 4 1 6 5 8] CF(1)=[0.2 0.3 0.7 0.9 1 0.4 0.1 0.6 0.5 0.8] V(1)=[3 2 1 4 6 7 5 4 9 10] LBF(1)=[0.2 0.3 0.7 0.9 1 0.4 0.1 0.6 0.5 0.8] LBP(1)=10	GBF(1)= 1 GBP(1)=10 particle= 5 velocity=6
2	CP(2)=[12.9 13.6 12.2 13.3 14.2 16.7 24.12 21.6 19.3 19] CF(2)=[1.2 1.3 1.2 1.3 1.4 1.6 2.4 2.1 1.9 1,9] V(2)=[10.9 10.36 5.2 4.3 4.2 12.7 23.12 15.6 14.3 11] LBF(2)=[1.2 1.3 1.2. 1.3 1.4 1.6 2.4 2.1 1.9 1.9] LBP(2)=24.12	GBF(2)=2.4 GBP(2)=24.12 particle= 7 velocity= 23.12
3	CP(3)=[45.89 34.38 33.71 32.77 30.63 35.23 40.3 40.58 37.02 56.23] CF(3)=[4.5 3.4 3.3 3.2 3.0 3.5 4.0 4.0 3.7 5.6] V(3)=[32.99 21.02 21.51 19.47 16.43 18.53 16.18 18.98 17.72 37.23] LBF(3)=[4.5 3.4 3.3 3.2 3.0 3.5 4.0 4.0 3.7 5.6] LBP(3)= 56.23	GBF(3)= 5.6 GBP(3)= 56.23 particle= 10 velocity= 37.23
4	CP(4)=[80.34 77.06 82.53 84.79 74.03 75.5 86.17 103.94 80.16 79.29] CF(4)=[8.0 7.7 8.2 8.4 7.4 7.5 8.6 10.3 8.0 7.9] V(4)=[34.45 42.68 48.82 52.02 43.4 40.27 45.87 63.36 43.14 26.06] LBF(4)=[8.0 7.7 8.2 8.4 7.4 7.5 8.6 10.3 8.0 7.9] LBP(4)=103.94	GBF(4)= 10.3 GBP(4)= 103.94 particle= 8 velocity=63.36
5	CP(5)=[130.41 141.33 148.8 149.92 145.08 140.64 157 148.29 134.13 146.83] CF(5)=[13.0 14.1 14.8 14.9 14.5 14.0 15.7 14.8 13.4 14.6] V(5)=[50.07 64.27 66.27 65.13 71.05 65.14 70.83 44.35 53.97 67.54] LBF(5)=[13.0 14.1 14.8 14.9 14.5 14.0 15.7 14.8 13.4 14.6] LBP(5)=157	GBF(5)=15.7 GBP(5)=157 particle=7 velocity=70.83

Iteration vs Fitness

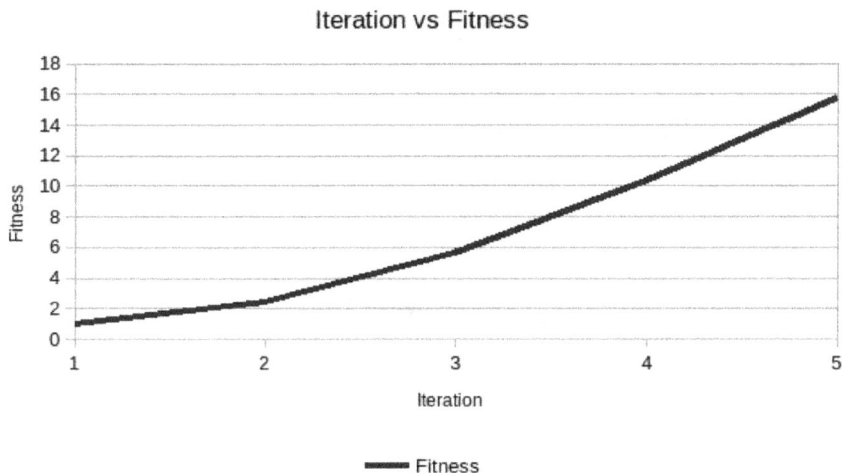

FIGURE 2.6   Fitness value vs. iteration.

2.6   CONCLUSION AND FUTURE RESEARCH DIRECTIONS

The paper proposed a EECP extension protocol using PSO. In this work the residual energy of the sensor nodes is considered to determine the node activation schedule. The results show that the node consumption is reduced from 38% to 25% by using the proposed protocol. The results achieve higher fitness value of 15.7 from 13.8 for more particles for the same number of iterations. So the network efficiency is increased with regard to the node utilization and energy. The analysis of the proposed protocol for different variants of PSO is our future scope.

Position vs Iteration

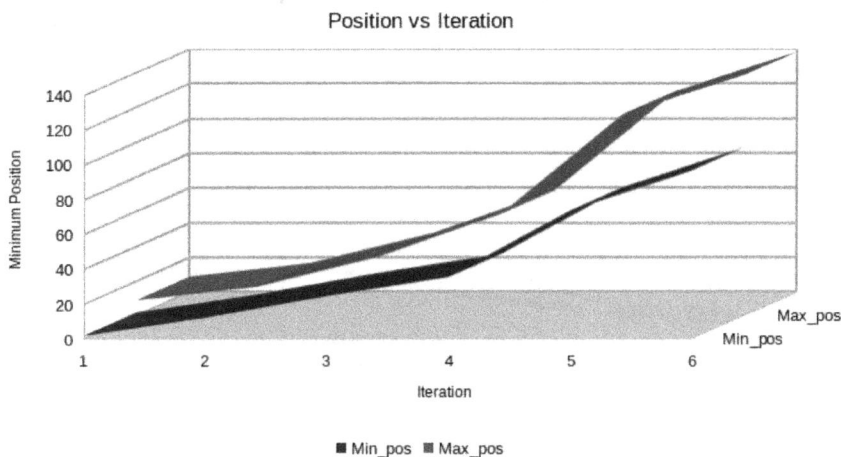

FIGURE 2.7   Performance of position and iteration for each particle.

Comparison of Fitness vs Iteration

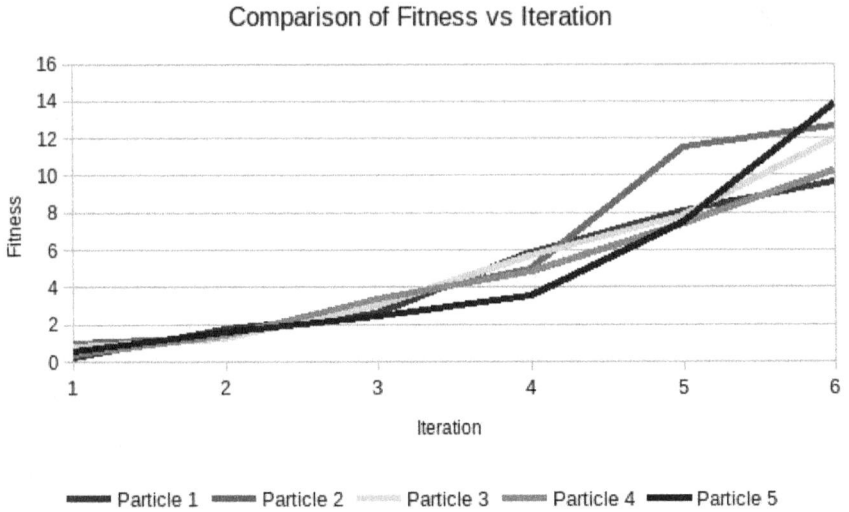

FIGURE 2.8   Performance of fitness and iteration for each particle.

Comparison of Fitness Value vs Iteration

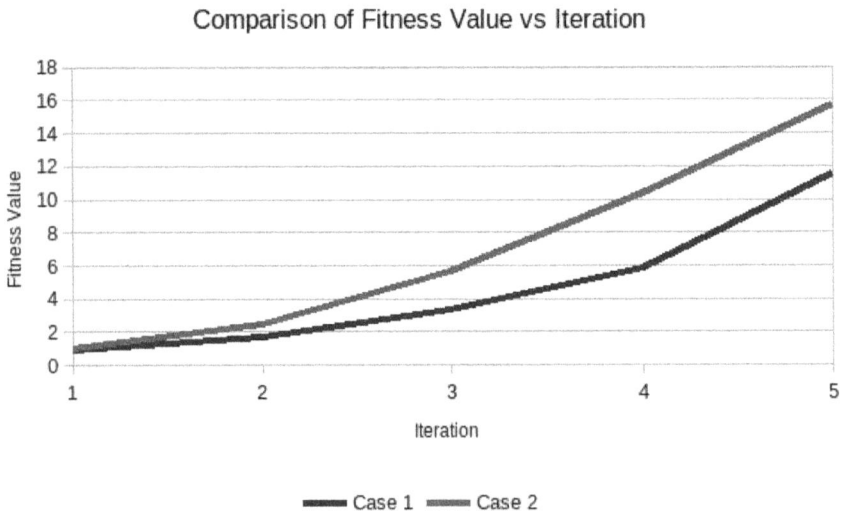

FIGURE 2.9   Comparison of fitness values vs. iteration.

REFERENCES

1. Zhao, J., Wen, Y., Shang, R. & Wang, G. (2004). Optimizing Sensor Node Distribution with Genetic Algorithm in Wireless Sensor Network. In *Advance in Neural Network*, 242–247.
2. Dantu, K., Rahimi, M., Shah, H., Babel, S., Dhariwal, A. & Sukhatme, G. (2005). Robomote: Enabling Mobility In Sensor Networks. In IEEE/ACM 4th International Symposium on Information Processing in Sensor Networks, 404–409.
3. Howard, A. & Poduri, S. (2005). Potential Field Methods for Mobile-Sensor-Network Deployment. In Bulusu, N. and Jha, S. (Eds.), *Wireless Sensor Networks A System Perspective*, London, Artech House, 21–33.
4. Kwok, K.S., Driessen, B.J., Phillips, C.A. & Tovey, C.A. (1997). Analyzing the Multiple-Target-Multiple-Agent Scenario Using Optimal Assignment Algorithms. In Proceedings of SPIE Vol. 3209.
5. Wang, G., Cao, G. & Porta, T.L. (2003). A Bidding Protocol for Deploying Mobile Sensors In Proceedings of 11th IEEE International Conference on Network Protocols, 315–324.
6. Chaturvedi P. & Daniel A. K. (2017). A Hybrid Scheduling Protocol for Target Coverage Based on Trust Evaluation for Wireless Sensor Networks. *IAENG International Journal of Computer Science*, 44(1), 87–104.
7. Rahimi, M., Shah, H., Sukhatme, G.S., Heideman, J. & Estrin, D. (2003). Studying the Feasibility of Energy Harvesting in a Mobile Sensor Network. In Proceedings of the IEEE International Conference on Robotics and Automation, 19–24.
8. Vass, D., Vincze, Z., Vida, R. & Vidács, A. (2005). Energy Efficiency in Wireless Sensor Networks Using Mobile Base Station. In EUNICE 2005: Networks and Applications Towards a Ubiquitously Connected World, 173–186.
9. Wang, G., Cao, G. & Porta, T.L. (2004). Movement-Assisted Sensor Deployment, in 23rd Conference of the IEEE Communication Society (INFOCOM), 2469–2479.
10. Wang, G., Cao, G. & Porta, T.L. (2003). A Bidding Protocol for Deploying Mobile Sensors. In Proceedings of 11th IEEE International Conference on Network Protocols. 315–324.
11. Dantu, K., Rahimi, M., Shah, H., Babel, S., Dhariwal, A. & Sukhatme, G. (2005). Robomote: Enabling Mobility In Sensor Networks. In IEEE/ACM 4th International Symposium on Information Processing in Sensor Networks, 404–409.
12. Zou, Y. & Chakrabarty, K. (2003). Sensor Deployment And Target Localization Based On Virtual Forces. In Twenty-Second Annual Joint Conference of the IEEE Computer and Communications Societies Vol. 2, 1293–1303.
13. Chellapan, S., Gu, W., Bai, X., Xuan, D., Ma, B. & Zhang, K. (2007). Deploying Wireless Sensor Networks under Limited Mobility Constraints. *In IEEE Transactions on Mobile Computing*, 6(10), 1142–1157.
14. Wu, J. & Yang, S. (2005). SMART: A Scan-Based Movement-Assisted Sensor Deployment Method in Wireless Sensor Networks. In Proceedings IEEE 24th Annual Joint Conference of the IEEE Computer and Communications Societies, 2313–2324.
15. Wang, G., Irwin, M.J., Berman, P., Fu, H. & Porta, T.L. (2005). Optimizing Sensor Movement Planning for Energy Efficiency. In Proceedings of the International Symposium on Low Power Electronics and Design, 215–220.
16. Wu, X., Shu, L., Yang, J., Xu, H., Cho, J. & Lee, S. (2005). Swarm Based Sensor Deployment Optimization in Ad hoc Sensor Networks. In Second International Conference Embedded Software and Systems, 533–541.
17. Wang, X., Wang, S. & Ma, J.J. (2007). An Improved Co-evolutionary Particle Swarm Optimization for Wireless Sensor Networks with Dynamic Deployment. *Sensors.* 7(3), 354–370.

18. Wang, X., Ma, J.J., Wang, S., & Bi, D.W. (2007). Distributed Particle Swarm Optimization and Simulated Annealing for Energy-efficient Coverage in Wireless Sensor Networks. *Sensors*. 7(5), 628–648.
19. Ngatchou, P.N., Fox, W.L.J. & El-Sharkawi, M.A. (2005). Distributed Sensor Placement with Sequential Particle Swarm Optimization. In Proceedings of the Swarm Intelligence Symposium, 385–388.
20. Salehizadeh, S.M.A, Dirafzoon, A. & Menhaj, M.B. (2010). Coverage in Wireless Sensor Networks Based on Individual Particle Optimization. International Conference on Networking, Sensing and Control, 501–506.
21. Chaturvedi P. & Daniel A. K. (2017). A Hybrid Scheduling Protocol for Target Coverage Based on Trust Evaluation for Wireless Sensor Networks. *IAENG International Journal of Computer Science*, 44(1), 87–104.
22. Kennedy, J. & Eberhart, R. (1995). Particle Swarm Optimization. Proceeding of IEEE International Conference on Neural Networks. Piscataway: IEEE Service Center, 1942–1948.
23. Chaturvedi, P. & Daniel, A.K. (2014). Recovery of Holes Problem in Wireless Sensor Networks. IEEE 4th International Conference on Information Communication & Embedded Systems (ICICES-2014), 1–6.
24. Chaturvedi, P. & Daniel, A.K. (2015). An Energy Efficient Node Scheduling Protocol for Target Coverage in Wireless Sensor Networks. In the 5th International Conference on Communication System and Network Technologies (CSNT-2015), 138–142.
25. Chaturvedi, P. & Daniel, A.K. (2015). Lifetime Optimization for Target Coverage in Wireless Sensor Networks. In 8th Annual ACM India Conference Compute 2015, 47–53.

3 Characterization of Reverberation Chamber

Abhishek Kadri
Dassault Systèmes

Devendra Chandra Pande
Establishment (LRDE), DRDO

Abhilasha Mishra
Maharashtra Institute of Technology

CONTENTS

3.1	Introduction	30
	3.1.1 EMI Model in Its Basic Form	30
	3.1.2 Reverberation Chamber	31
	3.1.3 Problem Definition	32
3.2	Literature Review	32
	3.2.1 History of Reverberation Chambers	32
	3.2.2 Application-Specific Studies	32
	3.2.3 Statistical Field Uniformity	33
	3.2.4 Stirring Techniques	33
	3.2.4.1 Mechanical Mode Stirring	33
	3.2.4.2 Electronic Mode Stirring	33
	3.2.5 Eigenmodes and Lowest Usable Frequency	34
	3.2.6 Quality Factor	35
	3.2.7 Field Uniformity	36
3.3	System Design	36
	3.3.1 Description of Chamber	36
	3.3.2 Measurements in the Chamber	38
	3.3.3 Modelling and Simulation	39
	3.3.4 Eigenmodes and Lowest Usable Frequency	40
3.4	Results and Analysis	42
	3.4.1 Practical and Simulated Results	42
	3.4.2 Analysis of Results	42
3.5	Conclusion	43
	3.5.1 Advantages	43
	3.5.2 Disadvantages	44

3.5.3   Conclusion 44

3.5.4   Future Scope...44

References..45

3.1 INTRODUCTION

Electromagnetic interference (EMI) is a phenomenon – in the form of radiations and conducted emissions and their effect on the performance of the system – while electromagnetic compatibility (EMC) is a characteristic or property of equipment to behave faithfully in the presence of EMI environment. Parts of electronics and telecommunication systems or subsystems often experience interference within or outside system and subsystems due to their electromagnetic properties. An electromagnetic field (EM) in one system produces an EMI in the form of mutual or induced voltage or current that propagates in the nearby system, subsystems or equipment. The effect of EMI is in the form an unwanted energy or force on the charge carriers in the equipment, resulting in development of unwanted voltage and current causing interference. To have all the equipment, systems and subsystems coexist in harmony, they should be immune to EMI. But the reality is different. EMI-free systems and applications still have a long way to go. However, one can limit the EMI to a specific level in the electronic system.

3.1.1 EMI MODEL IN ITS BASIC FORM

The presence of EMI problems motivates us to address the EMC issue. The International Electrotechnical Commission (IEC) has defined EMI as "degradation of the performance of a device, equipment or system by an electromagnetic disturbance" [1]. Thus, we can characterize the EMI phenomenon in three parts as

- The source of disturbance
- The device interfered with
- The coupling path

Practically, there may be multiple sources of EMI affecting one or more equipment simultaneously. The basic model representing EMI is as shown in Figure 3.1.

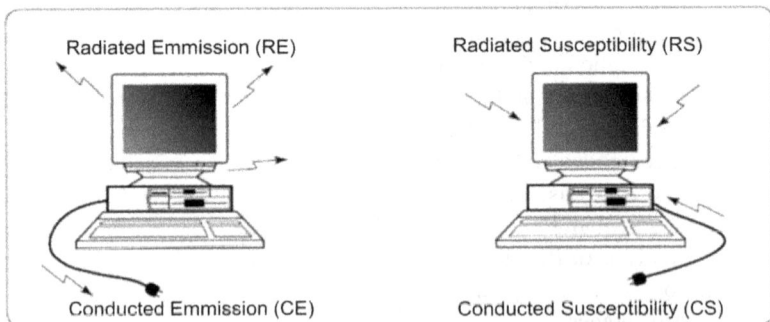

FIGURE 3.1   Interference problem in its basic form [1].

In telecommunication systems, we often come across the term *radio-frequency interference* (RFI). RFI is a subset of EMI in which at least one telecommunication device such as TV, FM/AM receiver, mobile signal and handset walkie-talkie, etc. experience the effect of EMI.

Faraday's cage can be used to demonstrate and minimize EMI practically. But for day-to-day used applications, e.g., watching a TV, Faraday's cage becomes infeasible. To address the issue of EMI, there should be more practical and balanced solution.

3.1.2 REVERBERATION CHAMBER

A reverberation chamber (RC) is an indoor facility with a shielded enclosure and metallic walls with one or more movable objects called *stirrers* or *tuners* placed within the chamber to facilitate variable boundary conditions. The stirrer is generally of a rotating type or laterally moving [2, 3]. An electromagnetic (EM) field is produced from a transmitting antenna, which produces standing waves (SW) due to reflection within the chamber. The SW get reflected of the walls and the stirrer, resulting in a standing wave field pattern which consists of hot spots and cold spots, i.e., places in the chamber having high- or low-intensity fields. The position of the hot spot and cold spots, their directions and polarizations of reflected waves are changed due to the movement of stirrer in the chamber. This arrangement exposes the equipment under test (EUT) to EM waves and polarization with varying field intensity from all the directions in the chamber without rotating or changing the polarization of the transmitting antenna or EUT. Figure 3.2 shows the external view of a mechanically stirred RC.

FIGURE 3.2   A 1.5 m × 1.2 m × 0.9 m reverberation chamber at the Otto-von-Guericke University Magdeburg, Germany [4].

3.1.3 PROBLEM DEFINITION

- To study the effectiveness of the RC using standard performance indicators.
- To perform electromagnetic simulation on a 3D model of a RC and compare practical results with simulated results.

3.2 LITERATURE REVIEW

Testing of EUT requires a predictable EMC testing environment, where the EUT behaviour is not undesirably affected by the test environment properties. The behaviour of the EUT should also be unaffected by the location it is placed within the chamber and it must not favour certain frequencies over others. In other words, there should not be resonances. One way to avoid resonances is lining the walls of test chambers with radiation absorbing material (anechoic chamber). This is an expensive option and requires a large power input to establish high-intensity fields for immunity testing. Another option is to maximize reflections in such a manner that we obtain statistically uniform fields (RC). Engineering ingenuity has resulted in a number of alternative mechanisms to maximize reflections and thus randomize fields. This section deals with the contribution of researchers worldwide in the field of RCs.

3.2.1 HISTORY OF REVERBERATION CHAMBERS

The most widely recognized and accepted RC for EMC testing within the EMC community was developed and reported in 1976 by P. Corona and G. Latmiral at the Istituto Universitario Navale (IUN) in Naples, Italy. Corona and Latmiral included a wider spectrum of EMC tests, including shielding effectiveness, immunity and emission [5, 6].

Since then a lot of innovation has taken place and researchers have developed a variety of methods to produce a reverberant environment. Many of the chief practitioners in the field of EMC, such as R. Serra et al. [4], recently contributed a comprehensive description of the classification of RCs and the prevalent methods that are currently being used in the EMC community.

3.2.2 APPLICATION-SPECIFIC STUDIES

One of the earliest applications of RCs measurement is reported for shielding effectiveness of cables and connectors [7]. RCs have been used to calculate antenna parameters such as total radiated power and radiation efficiency [8]. Niklas et al. have demonstrated the measurement of electromagnetic absorption in RC for two radio-frequency identification (RFID) tags to identify enhanced electromagnetic coupling frequencies to electronic devices by the use of a mode stirred RC [9]. In Reference [10], the application of ZigBee networks to highly reverberant environments was investigated using a RC.

In semi-reverberant environments such as passenger transit vehicles like planes, trains, buses and cars, elevators, factory halls, cargo-container workplaces, etc. people present the most common loading that influences the electromagnetic field

(EMF) characteristics within observed volume. Damir Senic studied the human body absorption characteristics in terms of absorption effectiveness and absorption cross section (ACS) in a RC [11]. The ACS has been further used to calculate the body mass index (BMI), body surface area (BSA) and body fat percentage (BFP) of human subjects [12]. Abdou Khadir Fall et al. have used the RC for conducting experimental dosimetric measurements in the 60 GHz frequency band [13].

3.2.3 STATISTICAL FIELD UNIFORMITY

The concept of statistical field uniformity for RCs is understood differently to the concept of field uniformity normally used in other test environments such as open area test sites, anechoic chambers or gigahertz transverse electromagnetic (GTEM) cells [14]. In those environments, the field is uniform when it is "the same everywhere at any moment", while a statistically uniform field in a RC is found if "on average, and within an acceptable uncertainty, the stirred vector field (and its spatial orientation, magnitude and intensity) is the same at different spatial locations". Even though the fields at sufficiently separated locations within the working volume and different orientations will be widely different, for any given stir state (e.g., a fixed angular position of the stirrer), its statistical parameters (e.g., mean, standard deviation, etc.) are the same everywhere, after completion of a full stirring cycle.

The statistical uniformity does not happen anywhere inside the RC but rather in the working volume alone, which is often defined as the volume whose boundaries are sufficiently far away (typically a quarter of a wavelength at the lowest frequency of operation) from the cavity walls, antennas, stirrer(s) and from any other electromagnetically relevant object inside the RC [15].

3.2.4 STIRRING TECHNIQUES

The stirring techniques have been divided into two main groups:

1. Mechanical mode-stirring techniques
2. Electronic mode-stirring techniques

3.2.4.1 Mechanical Mode Stirring

Mechanical mode stirring denominates to the techniques that make use of movements, i.e., translations, rotations, vibrations, etc. of a RC constituent like a large metallic scatterer or a wall.

Two subclasses of mechanical stirring can be identified: one that makes use of an internal stirrer, acting as a complex scatterer inside the volume confined within the cavity, named in this paper as *rotating paddle*, and another subclass that, conversely, uses changes in the wall(s) of a chamber to allow for mode stirring.

3.2.4.2 Electronic Mode Stirring

Electronic mode stirring came out as a leading technique among other options in which the time-changing boundary conditions are achieved by electronic mechanism. Unlike mechanical stirring, they do not comprise of any movable parts.

Electronic stirring techniques cover a broad spectrum of strategies, from shifting the frequency of excitation to using multiple sources.

3.2.5 EIGENMODES AND LOWEST USABLE FREQUENCY

An empty chamber resembles a rectangular waveguide that is closed on both ends, which implies that waves in a RC do not propagate in a particular direction but instead form a standing wave pattern. Also, there is no preferred axis of propagation as is the case in waveguides. Due to this, all cavity modes are written either as transverse electric (TE) or transverse magnetic (TM) with respect to any of the three coordinate axes [16]. When the cavity is excited with a signal, a finite number of modes significantly contribute to the total field inside the chamber [17]. The approximate number of distinct modes that get excited at a given frequency can be calculated by [18] the following equation:

$$N(f) = \frac{8\pi}{3} abd \left(\frac{f}{c} \right)^3 - (a+b+d)\frac{f}{c} + \frac{1}{2} \ldots \tag{3.1}$$

where 'c' is speed of light (in m/s), 'a' is length of the chamber (in m), 'b' is width of the chamber, 'd' is height of the chamber and 'f' is operating frequency. Figure 3.3 is a schematic representation of a mechanically stirred RC.

The shielded enclosure acts like a high Q multi-frequency resonator. The rational resonance frequencies are [18] as follows:

$$f_{mnp} = \frac{c}{2} \sqrt{\left(\frac{m}{a} \right)^2 + \left(\frac{n}{b} \right)^2 + \left(\frac{p}{d} \right)^2} \ldots \tag{3.2}$$

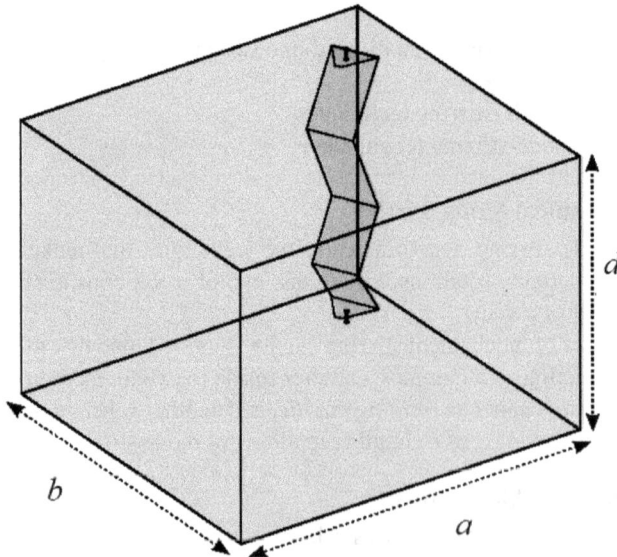

FIGURE 3.3 Generalized schematic of a reverberation chamber.

where, m, n and p are integer numbers, viz., 0, 1, 2, 3, ..., but not more than one of the three may be equal to zero for any one mode.

3.2.6 QUALITY FACTOR

The quality factor (Q-factor) is a parameter that indicates the cavity's ability to store energy. Q-factor plays a vital role in design and measurement of a RC, since several parameters are related to it, such as decay time, Q-bandwidth and the RC's total losses. It is basically the ratio of energy stored in a resonator to the energy dissipated per cycle.

$$Q = 2\pi \frac{energy\ stored}{energy\ dissipated\ per\ cycle} = 2\pi f_r \frac{energy\ stored}{power\ dissipated} \cdots \qquad (3.3)$$

Eq. 3.3 gives the general formula for Q-factor calculation when the stored and dissipated energy are known. Q-factor is also proportional to the average electric field strength inside the chamber. Since there are many eigenmodes in an unperturbed chamber, with each mode having its own Q value, it is not always necessary to define a quality factor for the chamber as a whole. It is desirable to use a composite quality factor (Qc) for a chamber within a specified frequency range. Considering losses due to wall conductivity Qc can be given by [18] this equation:

$$Q_c \approx \frac{3}{2} \frac{V}{S\delta_s\mu_r} \frac{1}{1+\dfrac{3\lambda}{16}\left(\dfrac{1}{a}+\dfrac{1}{b}+\dfrac{1}{d}\right)} \cdots \qquad (3.4)$$

where, V is volume of chamber, S is internal surface area of chamber, σ is wall conductivity, μ is wall permeability, δ_s is skin depth $= \sqrt{2/(\omega\mu\sigma)}$. Qc is equal to the inverse of the mean of the $1/Q$ values of all the possible modes within a small frequency interval around the frequency of interest. The Qc calculated by Eq. 3.4 is considered an upper limit, because it does not take into account other losses such as antenna support structures, RF loading, leakage through gaps and absorption in metallic objects. Hence the general design criteria for RCs are to make the volume as large as possible so that it can operate with an extended lower frequency limit with a large number of eigenmodes within the specified frequency range. The actual average Qc of the chamber, taking into account the total chamber loss, may be estimated by an empirical formula [19]:

$$Q = \frac{1}{2}\varepsilon\omega V|\underline{E}|^2 \cdots \qquad (3.5)$$

where ε is wall permittivity, V is volume of chamber, $\underline{E}$ is the average of normalized electric field over all stirrer positions and sampling points.

3.2.7 Field Uniformity

Field uniformity is one of the most widely used performance indicators used to measure effectiveness of an RC and it is given by:

$$\sigma_{total,J} = \sqrt{\frac{\sum_{i=(x,y,z)}\sum_{k=P_1}^{P_N}(\{\bar{E}_{k\,i\,j}\}-\{\bar{E}_{total,j}\})^2}{3N-1}} \quad \cdots \tag{3.6}$$

$$\text{where, } \bar{E}_{kij} = \frac{E_{max\,(k,i,j)}}{\sqrt{P_{aveinput\,k,i,j}}} \quad \cdots \tag{3.7}$$

$$\text{and } \{\bar{E}_{total,j}\} = \frac{\sum_{i=(x,y,z)}\sum_{k=P_1}^{P_N}\bar{E}_{k\,i\,j}}{3N} \quad \cdots \tag{3.8}$$

where, $E_{max\,(k,i,j)}$ is the absolute value of maximum electric field component obtained in the chamber at position 'k', which ranges from P_1 to P_N, at a frequency 'j' whose polarization is denoted by 'i', which can take values x, y or z. $P_{aveinput\,k,i,j}$ is the average input power for respective position, frequency and polarization.

The IEC limit for field uniformity has been set at 3 dB, but for frequencies from 400 MHz downwards the limit linearly increases to 4 dB at 100 MHz [20]. This is because at lower frequencies, the number of cavity modes that are excited are insufficient. Also, the stirrer fails to effectively distribute all the energy into the modes existing in the chamber, due to which the uncertainty or error in the field uniformity increases.

3.3 SYSTEM DESIGN

3.3.1 Description of Chamber

The RC as shown in Figure 3.4 is a metallic shielded enclosure having length, width and height as 3.7, 2.89 and 2.5 m, respectively. The material used in the construction of the RC is galvanized steel. Generally copper and aluminium are preferred over steel if higher Q-factor is desired, but this increases the cost of construction. A higher Q-factor implies narrow bandwidth, as Q-factor for a given frequency is inversely proportional to the bandwidth of the EM signal within the chamber [7].

Thus the selection of material depends upon the desired Q-factor, budget considerations and required bandwidth of the electromagnetic signal. Special care is taken to ensure that there are no gaps and seams between the metallic sheets that make up the wall as that could lead to unwanted leakage.

The chamber consists of a rotating stirrer made of the same metal as the chamber walls and is rotated along the vertical axis. It is placed in one of the corners of the chamber so that there is enough space in the chamber to place the EUT. While addressing the size of the stirrer, the cylindrical volume of the rotating stirrer is considered, and the height and the diameter of this volume are of importance in the designing stage instead of the actual dimensions of the stirrer. The stirrer measures

FIGURE 3.4   Reverberation chamber at LRDE, DRDO, Bengaluru.

2.7 m in height and 1.35 m in width. The diameter of the cylindrical volume of stirrer is 1.68 m. Figure 3.5 shows the stirrer used at LRDE, which is attached to the metallic extension of a motor axis. The shape of the stirrer is selected such that the electromagnetic waves that are incident on the surface of the stirrer see a change in the

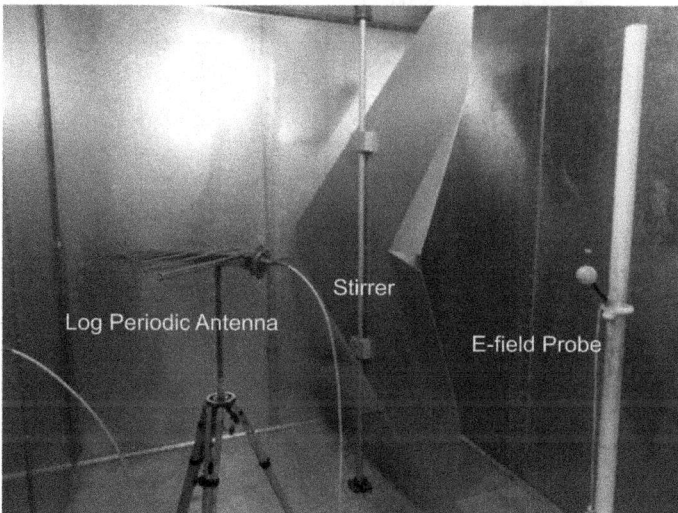

FIGURE 3.5   Structure of the stirrer at LRDE, DRDO, Bengaluru.

angle of incidence, in all three axes, over one complete rotation of the stirrer. This is desired, because for effective reverberation, a random averaged field is desired, in other words, a field which has no dominant polarization.

3.3.2 MEASUREMENTS IN THE CHAMBER

Figure 3.6 shows the block diagram of a RC test setup. A signal generator is used to generate a sine wave signal at the desired test frequency which is amplified using a standard RF Amplifier. This signal is radiated through a directional antenna such as a horn antenna or a log periodic antenna. An e-field probe is used to measure the field level in the chamber at N spatial points in the chamber. The field probe is connected to a field monitor that displays the absolute field intensity as well as the field intensity in x, y and z directions. The stirrer is rotated and M readings of electric field are taken throughout one complete rotation at each of the N points in the chamber. This process is repeated at each test frequency. If T is the number of test frequencies,

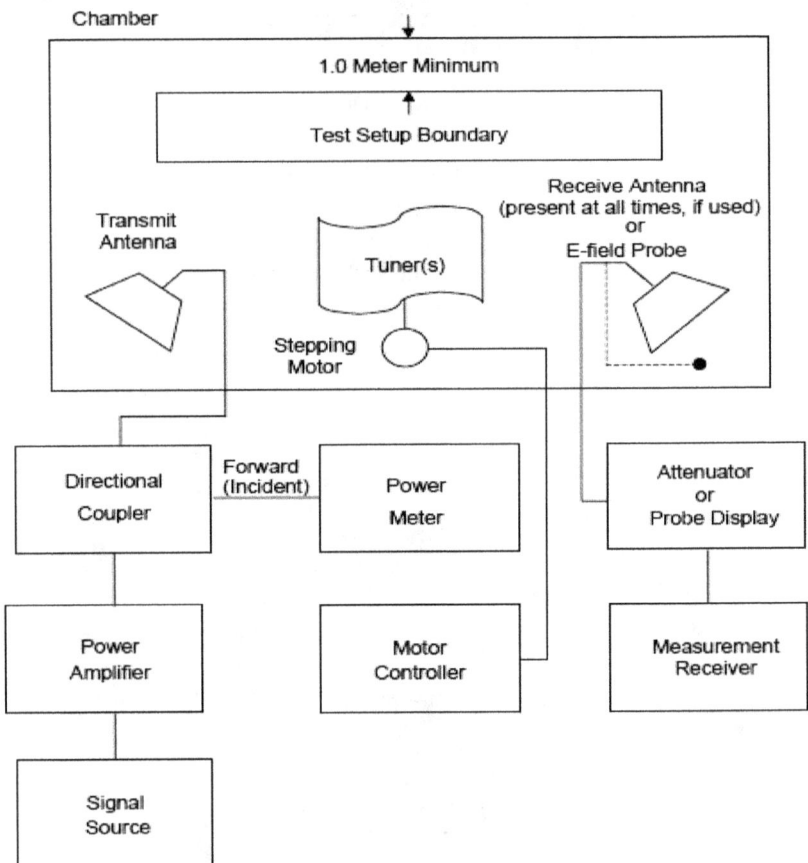

FIGURE 3.6   Block diagram of a reverberation chamber test setup [21].

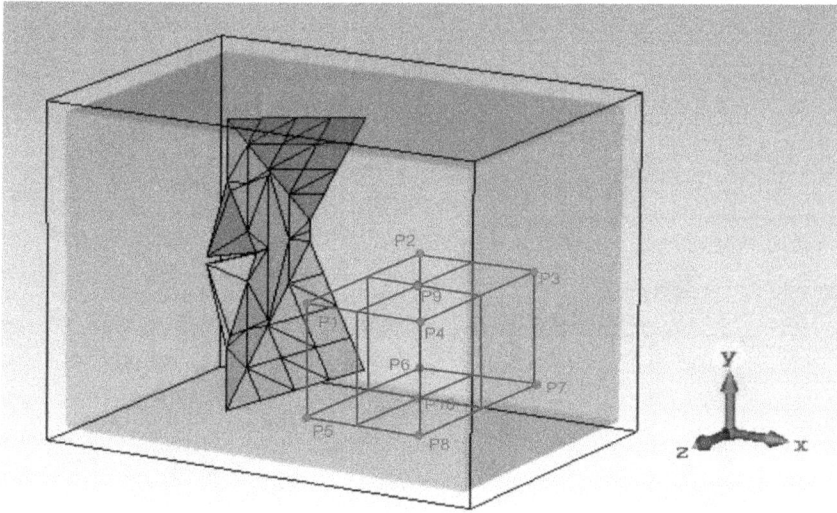

FIGURE 3.7   Probe positions in the chamber.

then the total number of e-field readings that would have to be taken are $T \times M \times N$. A larger value of M increases the testing time but enhances accuracy. According to the application of the chamber, its efficiency is measured over discrete frequencies, which are distributed over the dynamic range of the chamber.

The spatial points selected for placement of the e-field probe are generally the vertices of the working volume as shown in Figure 3.7. According to MIL standard 461G [21] at least eight positions in the chamber must be selected. Hence 8 of 10 positions (P1 to P8) from Figure 3.5 were selected during the practical measurements. The transmitting and receiving antennas, the stirrer or any other electromagnetic equipment such as control panel, etc. are kept outside the working volume.

3.3.3   MODELLING AND SIMULATION

Numerical simulations have been performed using the CST Microwave Studio software, which uses the Finite Integration Technique (FIT) for solving Maxwell's equations in integral form in the time domain. An approximate model of the RC is designed and the e-field monitor is selected. Figure 3.8 shows the 3D model of the RC with the original stirrer. The stirrer is rotated in 50 steps, and the field measurements have been carried out at 10 different positions in the working volume (P1 to P10), as shown in Figure 3.7. A Gaussian beam was used to excite the chamber with a signal of desired frequency. Input power was set to 1 W. Figure 3.9 shows the electric field simulated in CST Studio at the test frequency of 1 GHz. A preliminary material study helped us identify the difference in the results when the chamber walls were simulated with lossy and lossless metals. Lossless materials were used for all following simulations as they saved 45% of simulation time while negligibly affecting result accuracy.

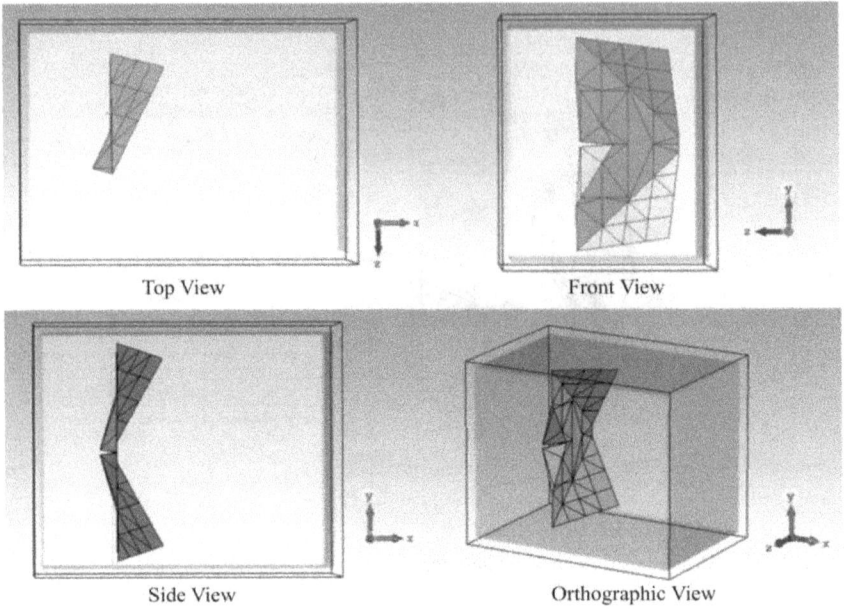

Top View   Front View

Side View   Orthographic View

FIGURE 3.8   3D model of the reverberation chamber in CST.

3.3.4   EIGENMODES AND LOWEST USABLE FREQUENCY

Table 3.1 shows the approximate number of eigenmodes existing in a shielded enclosure (LRDE) that are excited by the respective resonant frequencies. These resonant frequencies can be obtained from Eq. 3.2, while the numbers of distinct eigenmodes are obtained by substituting the value of these frequencies into Eq. 3.1.

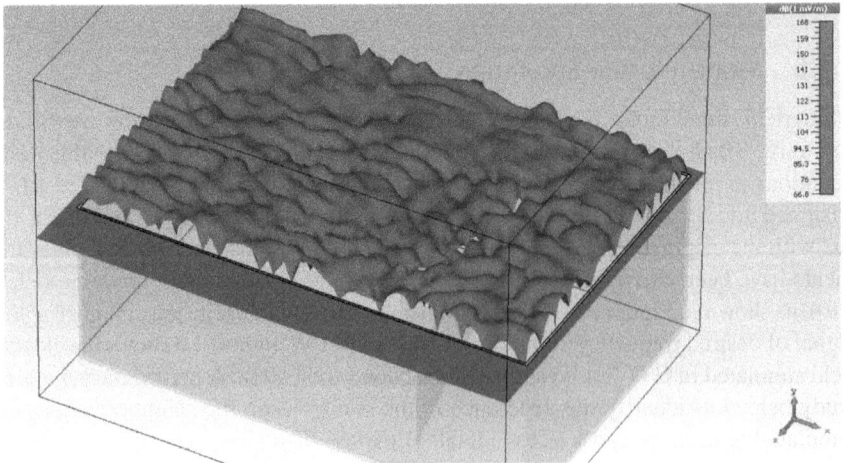

FIGURE 3.9   E-field simulated at 1 GHz in the RC model using CST Studio Suite.

TABLE 3.1

Distinct Frequencies of Modes in 2.89 m Wide by 3.7 m Long by 2.5 m High Shielded Chamber Situated at LRDE, Bengaluru

Distinct Modes	Mode #	Frequency (MHz)	Distinct Modes	Mode #	Frequency (MHz)
1	110	65.9	31	310	160.9
2	011	72.4	32	212	163.8
3	101	79.3	34	301	166.9
4	111	89.1	37	140	170.3
5	120	96.3	37	231	170.8
6	021	100.9	37	032	170.9
9	210	111.4	38	311	171.7
10	121	113.4	39	041	172.9
12	201	119.9	41	320	175.6
14	211	126.6	43	222	178.2
14	012	126.7	43	132	178.6
16	102	130.7	44	141	180.5
16	220	131.7	48	013	184.5
16	130	132.2	48	321	185.5
18	031	135.6	50	103	187.3
18	112	136.9	54	113	191.7
22	221	144.7	54	240	192.5
22	022	144.8	58	302	196.6
22	131	145.2	59	023	197.4
27	122	153.8	59	330	197.6
29	202	158.7	61	232	199.9
30	230	159.9			

The lowest usable frequency (LUF) is defined as the frequency above which the chamber meets operational requirements. The frequency in Eq. 3.2 respective to 60 distinct modes can be assumed to be the LUF [19]. The LUF generally lies between 3 and 6 times of the first resonant frequency [22]. Table 3.2 lists the distinct modes at every resonant frequency for the given RC based on Eqs. 3.1 and 3.2.

TABLE 3.2

Required Number of Tuner Positions for a Reverberation Chamber

Frequency Range (MHz)	Stirrer Positions
200–300	50
300–400	20
400–600	16
Above 600	12

The LUF is theoretically found to be 200 MHz. In practice, the LUF is settled by the chamber mode density, the effectiveness of the stirrer and the quality factor. Therefore, experimental LUF can be higher than the one estimated theoretically [23].

3.4   RESULTS AND ANALYSIS

3.4.1   PRACTICAL AND SIMULATED RESULTS

Using Eqs. 3.4 and Eq. 3.5, the theoretical and practical Q values have been found out. Additionally, Eq. 3.5 is also substituted with simulated e-field readings to find out a simulated Q-factor. Figure 3.10 shows a graph of all three Q values.

The field uniformity is calculated for simulated results. Figure 3.11 shows the graph of practical and simulated field uniformity. The red line indicates the field uniformity limit specified by IEC for effective reverberation.

3.4.2   ANALYSIS OF RESULTS

The difference in the practical and simulated readings can be attributed to various reasons. First, the simulations are performed at 10 probe positions, whereas practical measurements are carried out at eight probe positions, because increasing the number of probe positions in simulation does not increase simulation time, but while conducting practical measurements, the time required increases with increase in number of probe positions, as the probe has to be manually placed at every position sequentially. The number of stirrer positions, on the other hand, has a different relationship with the measurement and simulation times. While conducting simulations at lower frequencies, computational time is low; therefore a large number of stirrer positions can be considered for better accuracy. But at higher frequencies, the computational time increases manifold; hence the minimum required number of stirrer positions has to be considered, as specified by MIL standard 461G [21].

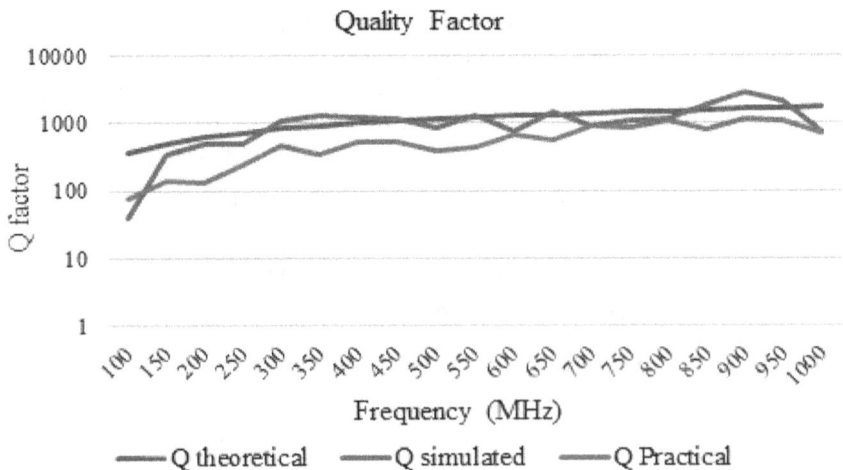

FIGURE 3.10   Quality factor for the reverberation chamber at LRDE.

Field Uniformity

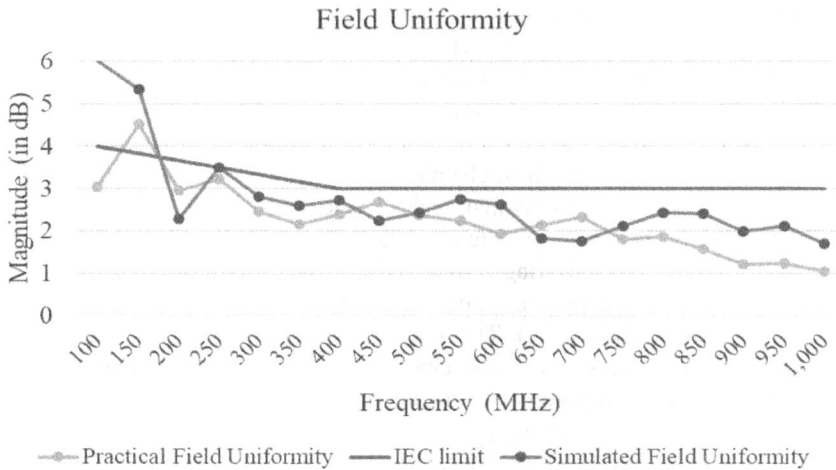

FIGURE 3.11   Graph of practical and simulated field uniformity.

On the other hand, the time required to conduct practical measurements in the chamber is not a function of frequency. The number of required stirrer positions is less at higher frequencies because at higher frequencies, the wavelengths are much smaller compared to the dimensions of the stirrer, and thus even a small change in the position of the stirrer causes the field levels to change drastically. Hence, there is little or no correlation between readings taken at adjacent stirrer positions. This makes the field more random in nature at higher frequencies.

3.5   CONCLUSION

3.5.1   ADVANTAGES

The RC has a number of advantages over the traditional anechoic chamber, which is more popular than its former counterpart.

1. The construction of RC is easier and cheaper because all the surfaces need to be made up of metal. In case of an anechoic chamber, the construction requires several pyramidal structures made up of a dielectric material, such as foam or polyethylene impregnated with graphite, which covers the entire chamber from inside, including the floor. A narrow walkway has to be constructed to allow the user to move inside the chamber which should be made up of a dielectric material and should occupy minimum total surface area to make sure absorption is above a certain limit. No such challenges exist in the designing of a RC. Therefore, comparatively fewer number of variables have to be taken into consideration while physically designing a RC.
2. Since the RC operates on the principle of stochastic electromagnetic theory, a tight shielding is not necessary. On the other hand, the anechoic chamber requires a good amount of shielding effectiveness, i.e., it cannot permit

leakage of EM radiation, and thus special care is needed in designing the air vents doors and cable connectors.

3. There is no need to move or rotate the equipment under test while testing. Since the field in the reverberation is polarization-less, and is incident in all the directions, neither the EUT nor the transmitting antenna is required to be moved. In an anechoic chamber, either the EUT or the transmitting antenna is needed to be rotated, and the antenna has to be placed in both planes to see the effects of different polarizations. This becomes very critical, especially when dealing with large EUTs such as aircraft.

4. Large electromagnetic fields can be generated over the test volume using comparatively lower power. This is due to the constructive interference of the SW within the RC. Anechoic chambers have been seen to require about 10 times the input power to generate the same level of electromagnetic fields for the same volume [7].

3.5.2 DISADVANTAGES

1. Since the stirrer is movable, the boundary conditions are constantly changing, thus making it difficult to solve Maxwell's equations. Therefore, it is difficult to mathematically model the chamber.

2. A minor change in the placement of the EUT or transmitting antenna can change the entire field, which causes non-repeatability of measurements.

3.5.3 CONCLUSION

Because the RC functions efficiently from 200 MHz onwards, the LUF can be said to be experimentally validated to be 200 MHz. This verifies that the LUF theoretically predicted in Section 3.3.4 is correct. The same can be said for Q-factor.

RCs have time and again proved to be the best candidate for radiated immunity tests.

The resemblance in practical and simulated results for Q-factor and field uniformity highlights accuracy of simulation definition and encourages engineers to incorporate 3D modelling and simulation of RCs not only as a prerequisite step in the manufacturing process of an RC, but also possibly to conduct radiated emissions and immunity tests virtually, thus cutting down costs manifold. As far as time is concerned, virtual tests can promise faster computation of results with minimal compromise in accuracy, provided a system with good configuration is used.

3.5.4 FUTURE SCOPE

There are two main areas in the RC field which require the attention of researchers – RC testing and improved RC performance. The testing of the RCs using standard performance indicators is a very time-consuming process. This is because the field levels in all three polarizations at multiple points must be measured at every stir state throughout the frequency range. Even if the entire process is automated, it still takes hours to analyse the performance of a stirrer.

The effectiveness of a stirrer alone cannot be virtually analysed. It has to be accompanied with the chamber, and only then can its effectiveness be measured. Research is needed to find a new method of analysing the performance of a RC which consumes less time and efforts without reducing accuracy.

As far as the RC performance is concerned, one of the most common improvements is lowering of the LUF. As mentioned before, the LUF is inversely proportional to the number of contributing eigenmodes, which in turn depends upon the volume. Therefore to lower the LUF further, the volume of the chamber must be increased. This will require a larger stirrer which in turn affects feasibility of construction and increases load on the motor used to rotate the stirrer. The same problem also arises while testing large subjects such as aircrafts or satellites.

Another challenge is the placement of the stirrer in the chamber, since a huge portion of the chamber is permanently occupied by the stirrer. Moreover, many RCs employ more than one stirrer to meet required performance values. These disadvantages can be avoided by using an electronic stirring technique such as multiple antenna stirring [24] or source position stirring [25] which eliminate the need for a physical stirrer and increase the working volume, while making the RC more cost-effective.

Some of these alternatives may prove to be advantageous only if the RC is sufficiently large enough to accommodate the required extra equipment such as multiple antennas. Otherwise the same stirring techniques might end up occupying a considerable portion of the chamber. Similarly, the requirements of all the dependent parameters such as LUF, quality factor, maximum e-field strength, dimensions of equipment under test, etc. must be taken into account before the selection of stirring technique.

REFERENCES

1. Goedbloed J.J. and Mart C. (1993). Electromagnetic compatibility. MYbusinessmedia.
2. Nikla W, Olof L. and Mats B. (2007). Experimental investigation and mathematical modeling of design parameters for efficient stirrers in mode-stirred reverberation chambers. *IEEE Transactions on Electromagnetic Compatibility*, Vol. 49, pp. 94–103.
3. Serra R., Marvin A.C., Moglie F., Mariani V., Cozza A., et.al. (2017). Reverberation chambers an overview of the different mode-stirring techniques. *IEEE Electromagnetic Compatibility Magazine*, Vol. 6, Quarter 1, pp. 63–78.
4. Wikimedia Commons. (2020). Small mode swirling chamber at the OVGU Magdeburg. jpg. https://commons.wikimedia.org/wiki/File:Kleine_Modenverwirbelungskammer_an_der_OvGU_Magdeburg.jpg
5. Corona P. and Latmiral G. (1976). "Valutazione ed impegno normativo della camera reverberante dell'Istituto Universitario Navale" in 'Atti I Riunione Nazionale di Elettromagnetismo Applicato', L'Aquila, Rome, Italy, pp. 103–108, 1976
6. Corona P., Latmiral G., Paolini E. and Piccioli L. (1976). Use of reverberating enclosure for measurement of radiated power in the microwave range. *IEEE Transactions on EMC*, Vol. 18, No. 2, pp. 54–59.
7. Crawford M.L., Ladbury J.M. (1988). Mode-stirred chamber for measuring shielding effectiveness of cables and connectors: An assessment of MIL-STD-1344A method 3008. IEEE International symposium on Electromagnetic Compatibility, pp. 30–36.
8. Christopher L.H., Haider A.S., et.al. (2012). Reverberation chamber techniques for determining the radiations and total efficiency of antennas. *IEEE Transactions on Antennas and Propagations*, Vol. 60, No.4, pp. 1758–1770. Doi: 10.1109/TAP.2012.21862663.

9. Niklas E., Mattias E. and Tomas H. (2017). Finding frequencies of enhanced electro-magnetic coupling to electronic devices by the use of mode stirred reverberation chambers. Proc. Of the 2017Proc. of the 2017, International Symposium on Electromagnetic Compatibility – EMC Europe, Angers, France, pp. 1–6.

10. David H., John D., et.al. (2008). Assessing the performance of ZigBee in the reverberant environment using a mode stirred chamber. International Symposium on Electromagnetic Compatibility, pp. 1–6.

11. Damir Senic, A.S. and Zbigniew M.J. (2013). Preliminary results of human body average absorption cross section measurements in reverberation chamber. Proceedings of the International Symposium on Electromagnetic Compatibility (EMC Europe 2013), Bruges, Belgium.

12. Martin P.R., Xiaotian Z. and Ian D.F. (2017). Time domain technique for rapid broadband measurement for human absorption cross section in reverberation chamber. *32nd URSI Gass, Montreal*, pp. 19–26.

13. Abdou Khadir, F., Philippe, B., et.al. (2016). An experimental dosimetry in a mode-stirred reverberation chamber in the 60 GHz band. *IEEE Transactions on Electromagnetic Compatibility*, Vol. 58, No. 4.

14. Karima E.F., Christop L., et.al. (2011). Statistical uncertainties due to the directivity of an EUT when testing its immunity in an anechoic chamber. Proceedings of the 10th International Symposium on Electromagnetic Compatibility (EMC Europe 2011), York, UK, pp. 153–158.

15. Wojciech J.K., Robert B. and Bartosz B. (2010). Design considerations of nested reverberation chambers for shielding effectiveness testing. Proceedings of the ICECOM, 20'th International Conference on Applied Electromagnetic Communications.

16. David A.H. (1998). Electromagnetic theory of reverberation chambers. National Bureau of Standards Technical Note 1506.

17. Florian M., Andrea C. (2014). Average number of significant modes excited in a mode-stirred reverberation chambers. *IEEE Transaction on Electromagnetic Compatibility*, Vol. 56, No. 2, pp. 259–265.

18. Pande D.C. (1997). Measurement using mode stirred chamber (MSC) and EMI specifications. Proceedings of the Seminar on EMI Problems and Design for EMC, ER and DC.

19. Song W., Zhancheng W., et.al.,(2013). A new method of estimating reverberation chamber Q-factor with experimental validation. *Progress in Electromagnetic Research Letters*, Vol. 36, pp. 103–112.

20. Ramiro S. (2017). Reverberation chamber through the magnifying glass: An overview and classification of performance indicators. *IEEE Electromagnetic Compatibility Magazine*, Vol. 6.

21. Department of Defense Interface standards, USA, Draft (2015). MIL Standard 461G.

22. Abdulkhadir F., Philippe B., et.al. (2015). Design and experimental validation of mode-stirred reverberation chamber at millimeter waves. *IEEE Transaction on Electromagnetic Compatibility*, Vol. 57, No. 1, pp. 12–21.

23. Aziz A., Guillaume A. and Alain R. (2013). *Determination of the Quasi-Ideal* reverberation chamber minimum frequency according to the mode of the stirrer geometry. Proceedings of the 2013 International Symposium on Electromagnetic Compatibility, EMC-Europe., pp. 437–442.

24. Cozza A., Koh W.J. et.al. (2012). Controlling the state of the a reverberation chamber by means of a random multiple-antenna stirring. Asis-Pacefic Symposium on Electromagnetic Compatibility, Singapore, pp. 765–768.

25. Kunthong J. and Bunting C. (2009). Source-stirring and mechanical-stirring reverberation chamber measurement comparison for 900 Mhz and 1800 MHz. IEEE International Symposium on Electromagnetic Compatibility, pp. 193–196.

4 Intelligent Intrusion Detection System

Anurag Singh Tomar
University of Petroleum and Energy Studies

Aditya Bakshi
Shri Mata Vaishno Devi University

CONTENTS

4.1   Introduction 48
4.2   Traditional Intrusion Detection Approaches 48
    4.2.1   Signature-Based NIDS 48
    4.2.2   Pattern Matching 49
    4.2.3   Rule-Based Techniques 49
    4.2.4   State-Based Techniques 51
4.3   Techniques Based on Data Mining..52
    4.3.1   Anomaly-Based NIDS 53
    4.3.2   Advanced Statistical Models 54
    4.3.3   Haystack..54
    4.3.4   Nides 55
    4.3.5   Emerald...55
4.4   Rule-Based Techniques 55
    4.4.1   Wisdom & Sense 55
    4.4.2   Network Security Monitor 56
    4.4.3   Time-based Inductive Machine (TIM)..56
    4.4.4   NADIR...56
4.5   Machine-Learning Techniques for Intrusion Detection Systems.................57
    4.5.1   K-Nearest Neighbour 57
    4.5.2   Support Vector Machine...58
    4.5.3   Decision Tree 59
    4.5.4   Random Forest 59
    4.5.5   Neural Network 59
4.6   Conclusion 61
References..61

4.1   INTRODUCTION

In the era of technology, people use telecommunication on a daily basis to send email, documents, video chats, video conference, online transactions etc. On the other hand, cyberattacks are increasing exponentially day by day, which causes companies and businesses huge financial loss, misuse of information, theft of identity and data leakage. Data is fuel for any business; by analyzing the data a company can develop the strategy for business, target more customers and understand customer patterns, so loss or theft of data can be harmful. Yahoo, one of the industry giants, claimed that about 3 billion accounts were hacked in 2016; Uber reported the attackers have stolen the information of 57 million riders and drivers; and these attacks cost organizations in terms of financial loss and company credibility. According to a report by NITI Aayog, Government of India, attackers sent phishing email to an employee of the Union Bank of India and accessed their credentials, trying to transfer $171 million, but due to immediate action from the bank, the attacker was not able to execute it. The food delivery company Zomato faced the problem of data theft by unethical hackers. Due to such real-life incidents almost every organization should have a system that can provide the security against such attacks, detect the attacks and send an alert message to company administration. An intrusion detection system (IDS) can fulfil these requirements, detecting threats, providing security and monitoring network traffic activities. As the attackers are continually changing their pattern of attack, it's really difficult for an IDS system to detect the threats. To avoid such kind of attacks, a traditional approach will not work; hence intelligent intrusion detection — which can learn the pattern of attack and understand the variation of attacks — is required. An IDS system can be made intelligent using machine-learning algorithms to learn the trends and pattern of threats.

4.2   TRADITIONAL INTRUSION DETECTION APPROACHES

In this section, various traditional approaches used in IDS, such as signature-based detection and rule-based intrusion detection, are explained. As these approaches are lagging to detect the unknown threats, may not be efficient in today's context.

Conventionally, we can divide the network-based intrusion detection systems into two categories [1].

4.2.1   SIGNATURE-BASED NIDS

In real life, signature is defined as a unique property corresponding to a person or entity. It is widely used to identify the various attacks or also known as misuse detection. A signature-based NIDS [2] is basically used to define a known attack. In this, the signatures of incoming packet payloads are compared with signatures that are stored in the database for detection of an attack. Thus, for comparing systems a signature database has to be maintained.

Figure 4.1 describes a signature-based detection model. Collection of network traffic data, building or defining the normal behaviour or profile of the system, developing intrusion detection and action that corresponds to protection against the malicious activity all play a vital role to secure the network. There are numerous sources

Add new rules   Match Rule?

System profile ⟹ Misuse Detection ⟹ Response

Timing
Information   Modify existing profile   Data Collection

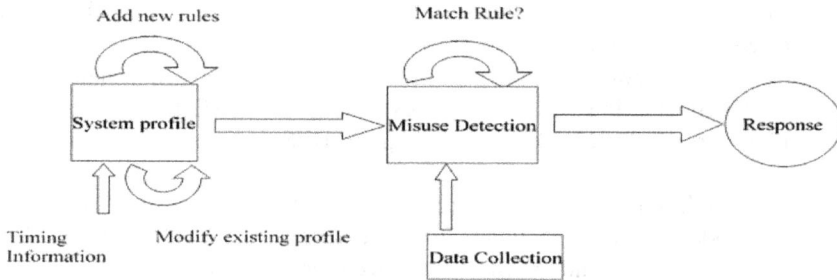

FIGURE 4.1   Misuse detection.

of data, such as audit trails, network traffic and system call trace, from which data can be collected. To make the data clear to other components of the system, data has to be transferred into the corresponding format. For characterizing usual and irregular behaviours, a system profile is used. The basic use of profiles is to tell what operations should be performed on the objects by the subjects and what should be the behaviour of the normal subject. If any deviation in the profiles in the actual system is reported, then an imposter has been accessing the system.

To implement misuse detection, the common techniques that are used, namely

1. Pattern matching
2. Rule-based techniques
3. State-based techniques
4. Data mining

4.2.2   PATTERN MATCHING

Enhancements in technology, infrastructure, processing speed, hardware and software have caused an increase in cyberattacks. Pattern-matching algorithms are widely used in intrusion detection systems to identify the pattern of attacks, and in future these patterns will be used to detect the same kind of malicious attacks. In this, packet head and packet content are matched and identified for various attack patterns to identify interruption in the system. The pattern-matching approach is more costly in terms of the searching for patterns in a database and more time-consuming as new and different forms of attack are continuously emerging. Moreover, the solution is proposed by Abbes et al., which is a novel technique to increase the speed to match the pattern from a database, decrease the overhead in terms of search time and identify the attack signatures by analyzing the working of protocol [3]. For constructing a decision tree, checking the pattern on either the header or payload portion of packet is implemented.

4.2.3   RULE-BASED TECHNIQUES

Rule-based techniques are probably the most common for signature detection. These frameworks encode meddlesome situations as a lot of principles, which are coordinated against review or system traffic information. Any abnormality in the standard coordinating procedure is accounted for as an interruption. Instances of standard-based frameworks incorporate Multics Interruption Detection and Alerting System (MIDAS) [4], intrusion detection expert framework (IDES) [5], and Next-Generation Intrusion Detection Expert System (NIDES) [6, 7].

a. **MIDAS**: It is a rule-based technique used to detect malicious activity over a network. National Computer Security Center (NCSC) built the framework to screen intrusion into NCSC's centralized server, Dockmaster. It collects information on the log activity of the user and investigates it with the help of statistical analysis and master framework innovation. MIDAS utilizes the Production-Based Expert System Toolset (P-BEST) [8], carved in LISP language for segregating and actualizing the rule base.

  The two-layer P-BEST principle base is used for incorporating rules. The minor layer is used to coordinate specific kinds of occasions, for example, measuring the activity of the user, and afterward includes new occasions by setting up a specific limit of doubt. Rules in the advanced layer implement these doubts and raise a caution by choosing a specific framework.

b. **IDES**: IDES is one of the initial models for intrusion discovery. The IDES model is proposed in Denning's original paper [9], which gives a scientific systematization of intrusion-location components. The model depends on the assumption that typical cooperation between subjects (for example, clients) and articles (for example, records, projects or gadgets) can be portrayed, and furthermore that clients consistently carry on in a reliable way when they perform tasks on the PC framework. These uses can be portrayed by finding different insights and connecting with established profiles of ordinary practices. New review accounts are confirmed by coordinating recognized profiles for the two subjects and their compared gatherings. Standard user profiles are created based on the activity log, and further variations in the profiles are considered as suspicious activity in the network. The identification rate can be improved by screening the action of IDES of clients as clients exercises this technique on an on or off day. For instance, exercises for ordinary clients on a regular day might be anomalous on a non-regular day. IDES utilizes P-BEST to depict its rule base comprising two kinds of principles: conventional guidelines and explicit rules. Non-exclusive guidelines can be used for various objective frameworks, and explicit rules are carefully subject to the working framework and the relating execution. IDES design comprises the profiles record and the framework safety official (SSO) user interface (UI).

c. **NIDES**: It is a combination of intrusion discovery frameworks comprising a framework based on signature, just as a discovery segment is dependent on measurable methodologies. The signature-based framework develops an ancient IDES form by converting progressively known intrusion situations and refreshing the P-BEST form used. The discovery segment dependent on measurable methodologies depends on abnormality location. In these methodologies more than 30 criteria are used to build up ordinary client profiles including use of CPU and I/O, order utilized, nearby system action, etc. The NIDES framework is exceptionally modularized with well-characterized interfaces between segments.

4.2.4 STATE-BASED TECHNIQUES

State-based methods differentiate the well-known threats by using articulations of the organization state and state changes. It improves detail of examples for well-known threats and can be used to portray outbreak situations more easily than rule-based languages, for example, P-BEST. In state-based methods, exercises adding to interruption situations are characterized as advances between organization states, and in this way interruption situations are characterized as state transition charts.

Figure 4.2 depicts a general state diagram; system state is represented by node, and an action is represented by an arc. The condition of the system depends on the clients or procedures. Intrusions characterized by the state transition diagram incorporate states, which are of three kinds: starting state, transition state and compromised state. An underlying state alludes to the start of the attack, while conceded state represents the effective finishing of the attack. Change states compare to the transition states happening between a starting state and a compromised state. An interruption is compromised if and only if it comes last. In other sections, two instances of state-based methods are explained, specifically a mechanism proposed by Ilgun and partners [10, 11] in which state transition analysis tool are used and the IDIOT (Intrusion Detection In Our Time) framework by Kumar et al. [12–15] where coloured petri nets are proposed.

UNIX STATE TRANSITION ANALYSIS TOOL (USTAT): The UNIX State Transition Analysis Tool (USTAT) depends on the supposition that an attack starts from an original state by all attackers to get to an objective framework by some approved restrictions, and afterward finishes a few activities on the objective system; they procure some already unapproved capacities. USTAT is an experienced model usage of the state change investigation system for interruption identification. It screens the framework state progress from safe to hazardous by speaking to all known susceptibilities or intrusion situations as a state change graph. In USTAT, more than 200 review occasions are spoken to by ten USTAT activities, for example, read file, alter owner file, where the parameter record represents the name of specific documents. The attacks are displayed as a change in sequence of states that leads an advancement from an underlying restricted approval state to a completely control

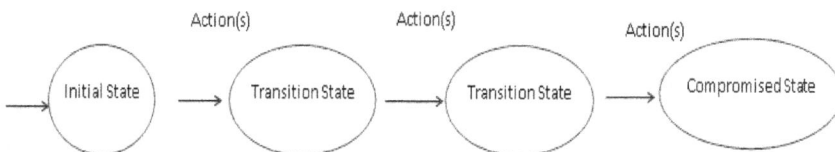

FIGURE 4.2 General state diagram.

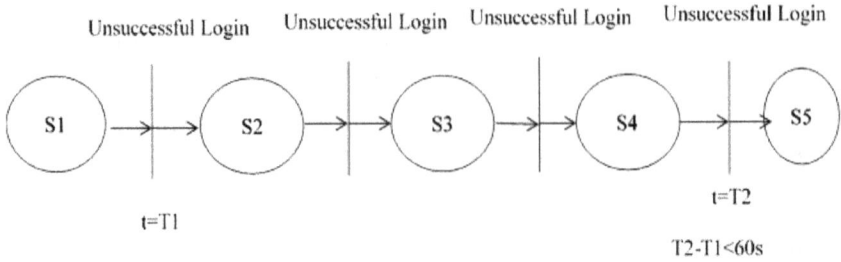

FIGURE 4.3   Example of failed login attempts four times in one minute that illustrate CPA.

under the attacker's hand. Any derivation in USTAT maintains the list of activities and decides if the present activity is the source of a movement of state from present to its next state by coordinating with the state transition table.

COLOURED PETRI NETS: Petri nets are a mathematical modelling approach to model the behaviour of attacker, containing parameters such as state, transition, arcs. States can be modelled as user or their states or nodes; transition shows the relationships between these places. IDIOT, the abbreviation of Intrusion Detection in Our Time, is a technique that uses a pattern-matching model using coloured Petri nets; it is also called a state-based signature detection that depends on the coloured Petri nets (CPN) model. Patterns in IDIOT are formed into various situations for intrusion detection. The patterns coming from the network are checked by matching them with known patterns. In the CPN model, IDIOT is implemented as a safeguard against false positive and false negative intruder signatures, and the nodes represent one possible state of system. The Petri nets that are selected in CPN model are called as coloured Petri automata (CPA). The CPA characterizes a severe definitive particular of intrusions and indicates which examples should be coordinated rather than the most effective method to coordinate them.

Figure 4.3 illustrates a CPA example by labelling the following intrusion situation: The intrusion can be reported if the quantity of ineffective login endeavour surpasses four inside one minute. The mixture of arrows and vertical bar represents system states which are divided into different transitions. For instance, the change from states S1 to S2 happens when the token reaches to S1; this represents a failed login endeavour. The hour of first ineffective login endeavour is spared in the token variable T1. When the token is in S4, there is a transition from S4 to S5.

The difference between the failed login attempts and transition state is more than one minute in duration. An alarm will be generated when the system state is moved to the last state S5.

4.3   TECHNIQUES BASED ON DATA MINING

In the modern era, methods of data mining are significantly used in detecting the intrusion in a network [16–18]. In this case, the user's normal or intrusive behaviour is automatically discovered by incorporating data mining techniques. Basically, for data mining there are three distinct algorithms: sequence analysis, classification and link analysis.

The algorithm that implements classification produces sufficient data for suspicious and non- suspicious activity that works on a decision tree paradigm and requires different learning methods. Link analysis is used to control the relation between normal profiles and audit database records that are usually derived from these relations. For auditing of data, sequential patterns can be discovered, and these patterns are implanted in intrusion detection systems. This process is called sequence analysis.

4.3.1 ANOMALY-BASED NIDS

Anomaly-based NIDS can also be classified as host intrusion detection systems (HIDSs). In this, the events have been checked by the deviances between pre-defined normal profiles and observed events. The network connection shows the normal behaviour by presenting the normal profiles of the events. Monitoring of suspicious activities is the main function of normal profiles, which can occur over a period of time. Therefore, to check the network connection and its normal behaviour, anomaly-based NIDSs plays a very important role.

Figure 4.4 A typical anomaly detection model. The four components that make up the model are collection of network traffic data, definition of the normal behaviour of system, detection of abnormal behaviour or intruder and action corresponding to abnormal behaviour. Collection of network traffic data is used to obtain the normal user activities or traffic data. Normal behaviour of system is identified by specific modelling techniques. Anomaly detection serves two purposes; the first is to determine what percentage of activity should be labelled as abnormal, and the second is to determine how much the current activities deviate from normal profiles. Finally, intrusion is reported by the response component, with equivalent timing information.

Misuse detection is the major bane for anomaly detection (i.e., capability to find unknown threats), but, due to some underlying rules of anomaly detection, the false alarm rate is very high.

The main limitation for high false alarm rate is as follows:

1. Intrusive activities missed during the period of normal operations are considered normal behaviour when data is collected by the user's normal behaviour model.

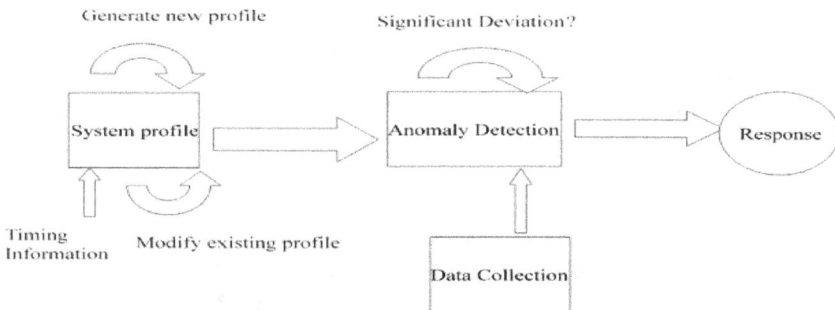

FIGURE 4.4   Anomaly detection model.

2. The second major drawback of anomaly detection techniques is ineffectiveness against stealthy attacks. The reason for this is the hidden nature of the attack as large quantities of data are bypassed as normal data. Furthermore, security experts finalize the list of attributes or parameters that have to be considered as inputs for normal behaviour of the system. Any error in defining these parameters will affect the efficiency of the anomaly detection system and will escalate the fake alarm rate. The key problem in anomaly detection is monitoring and selection of network features and designing of detection models. In literature, several anomaly detection techniques have been proposed for intrusion detection purposes. The classification of anomaly detection techniques is very difficult. So, after going through an extensive survey on anomaly detection systems [19–23] we have divided them in two categories, i.e., statistical models and rule-based models.

4.3.2 ADVANCED STATISTICAL MODELS

The theoretical anomaly detection approach is discussed in a seminal paper by Denning [9]. Her detection framework comprises nine components that are defined on the basis of statistical analysis. The components are profiles, statistical models, objects, subjects, statistical metrics, audit records, profile templates, anomaly records and activity rules. A user, process or a system can be treated as a subject. The files, programmes and messages can be defined as objects as they receive the actions and entities from subjects. The set of actions executed on objects by subjects is handled by audit records. Once audit records are created by an event generator, the appropriate profile matching is done by statistical model. The statistical model also makes decisions for checking abnormal behaviour, reporting detected anomalies and profile updating. The activity profile is used for determining the performance of the system by incorporating statistical metrics. The Haystack [24], NIDES [6, 7] and EMERALD [25] explained by Denning's model use a basic idea with improvement in many intrusion detection systems [26]. In the next section, overview of these system are explained in detail.

4.3.3 HAYSTACK

The Haystack framework was structured and executed for the identification of intrusions in a multi-client air force PC framework [25]. Statistical methods are used to distinguish anomalies in exercises. A lot of structures, for example, use a measurement of I/O and CPU; a number of document are watched that are passed off, and afterward the typical scope of qualities for these highlights is characterized. Actions come under these scope are accounted for as intrusions. Haystack is one of the initial abnormality location frameworks dependent on statistical models. It uses an exceptionally straightforward factual model in which each component is allotted a load by the SSO. The primary shortcoming of Haystack is that no test is directed to check if the weight is sensitive to interruption designs, and moreover no clarification is offered about how the weight worth is dealt out.

4.3.4 NIDES

As demonstrated in the previous sections, Next-Generation Intrusion Detection Expert System (NIDES) incorporates a statistical irregularity identifier. The review data gathered comprise client names, client CPU time and names of records, absolute number of documents unlocked, secondary storage pages, number of clients logged in through different machines and so on. Insights are processed from the gathered data. NIDESs copy only statistics identified with frequencies, means, differences or covariance of measures rather than all-out review information.

4.3.5 EMERALD

The Event Monitoring Enabling Responses to Anomalous Live Disturbances (EMERALD) [26] is a mechanism for identification of anomaly and misuse detection. It comprises a sign investigation part and a segment of statistical profile based anomaly discovery. The anomaly identifier depends on the numerical methodology, with other occasions named as intrusive as they are generally vitiated using normal conduct. In EMERALD more than 30 distinct criteria, including CPU and I/O utilization, directions utilized, neighbours action, and framework mistakes, are used to fabricate normal client profiles. An EMERALD screen can be uninvolved through perusing movement records or system bundles. The expository outcomes can be traded non-concurrently between various client monitors worked on various layers (for example, space layer, venture layer and so on.). Besides, each screen has an example of the EMERALD resolver, a countermeasure choice motor combining the cautions from its related investigation motors and counter intruded exercises with the help of response handlers.

4.4 RULE-BASED TECHNIQUES

Different rule-based techniques that yield several anomaly detection models have been developed in the past. In this section, a few rule-based models such as Wisdom & sense (W&S), NSM, TIM and NADIR are discussed.

4.4.1 WISDOM & SENSE

W&S is a special way to deal with inconsistency location, which comprises, as the name shows, two parts: Wisdom and sense [27]. The intelligence segment comprises a lot of standards portraying typical practices of the framework dependent on historical review information. About 10 000 records for every client are read from the document to develop the principles. The sense part is a specialist framework dependent on the previous rules; it confirms whether the resulting review information damages the rule base or not. The framework alerts the user by sending an alarm message when the sense recognizes malicious activity. In the execution of W&S, an anomalousness score for each string yields the figure of merit (FOM) metric. The FOMs assessments for a few occasions are added and looked at against an edge. At the certain point, when the aggregate parameter value is above the threshold value, activity is classified as abnormal by the framework.

4.4.2 NETWORK SECURITY MONITOR

Network security monitor (NSM) is a standard framework for recognition of intruder activities such as unauthorized access of resources, manipulation of secret information or access of password files in network. It collects the traffic packets that pass through the network or system then analyze the traffic to gather information about the traffic, such as source IP address, protocol, number of packets sent, etc. Analyzing this information and patterns in traffic will help to identify the suspicious activity in network [28]. Network traffic profiles are made by creating host vectors from the system information. Host vectors and association vectors are the principle contributors to the master framework in NSM. The expert system framework shows which frameworks are relied on to convey the determinations of application layer conventions at a more significant level. Intrusions are chosen by examination effects of expert system. The conclusive outcome answered to the SSO comprises an association vector and comparing suspicious level. The suspicious level estimates the probability that a specific association speaks to some intrusion behaviour.

4.4.3 TIME-BASED INDUCTIVE MACHINE (TIM)

the Time-Based Inductive Machine (TIM) models the client's ordinary standards of conduct by progressively producing movement principles using inductive speculation [29]. TIM finds typical rules for the sequence of events rather than a solitary event. From these sequences of events the following event can anticipate some rule that depicts some events. On the off chance that the client's conduct matches with past event groupings and the genuine event is excluded in the anticipated event sets, the client's conduct is reflected as intrusive. The drawbacks of TIM is that the principles model only nearby events. When several submissions are implemented simultaneously, events in one submission might be incorporated with events in different applications, in which case the standards produced by TIM may not precisely distinguish the client's typical practices. Also, events that can't coordinate with any past event successions in the preparation informational index consistently fire an intrusion caution, which in all likelihood may relate to some false alarm.

4.4.4 NADIR

Network Anomaly Detection and Intrusion Reporter (NADIR) was developed at Los Alamos National Laboratory to monitor the activities of interior PC systems [30]. It gathers review data from three types of administration hubs: security controller, basic document framework and security assurance machine. The system security controller gives client validation and access control; normal document framework stores the information; and security confirmation machine records endeavours to corrupt the security level of the information. The review data gathered for every event incorporate a one-of-a-kind client ID related with the subject, the date and time of the event, an accounting parameter, the wrong code, a flag showing whether the comparing activity is fruitful or not, and the depiction of the event. The individual client profile is processed on a week-after-week basis from the review database. The client profiles are contrasted, and a lot of expert system framework administration is done so as to recognize potential deviations in clients' practices. The expert rules are generated based on the past activity logs and security-related attacks and their prevention strategies. Each rule is assigned some weightage based on security threats; if the particular activity matches with n rules and the sum of these weights corresponding to n rule is more than the threshold limit, then an alarm is raised that is a signal to alert the user or administrator.

4.5 MACHINE-LEARNING TECHNIQUES FOR INTRUSION DETECTION SYSTEMS

Various machine-learning techniques [31, 32] are used in intrusion detection systems to make them efficient or intelligent to detect threats. Machine-learning algorithms can be supervised or unsupervised; in supervised learning, labelled data is required, while in unsupervised, we can make a cluster of data based on their similarity with respect to other data points.

In supervised learning we required some mapping function $f : X \rightarrow Y$ to map input samples X to target output, where X can be the n dimensional feature vector and Y can be column vector. Some machine-learning algorithms are discussed below.

4.5.1 K-NEAREST NEIGHBOUR

The k-nearest neighbour algorithm can be used as multi-class classifier to detect unknown threats. It is based on the distance of the test instance of network traffic from all instances of network traffic present in the training dataset, then selecting the k-nearest instances from training dataset; if the majority of neighbours out of k belong to some pattern of attacks, then it will be detected as threat.

Let's discuss the dataset of network traffic that contains the parameters shown in below Table 4.1. It can be one instance of any network traffic, so the intrusion detection system has to take the decision whether it's normal traffic or malicious activity.

Suppose each network traffic instance is denoted by X_i that can be n dimensional. (In this case it contains the fifth parameter). In training the dataset for each X_i, we have value of Y_i (that is, a single column vector may have a value such as threat, normal, risky, less risky or may be type of attack). As shown in Figure 4.5, the test instance will calculate the distance from all training instances of network traffic and select the k-nearest instances from the training set. If most of the instances

TABLE 4.1

Sample Dataset Attributes

Source IP Address	Source Port Number	Protocol	Length of data	Type of Traffic

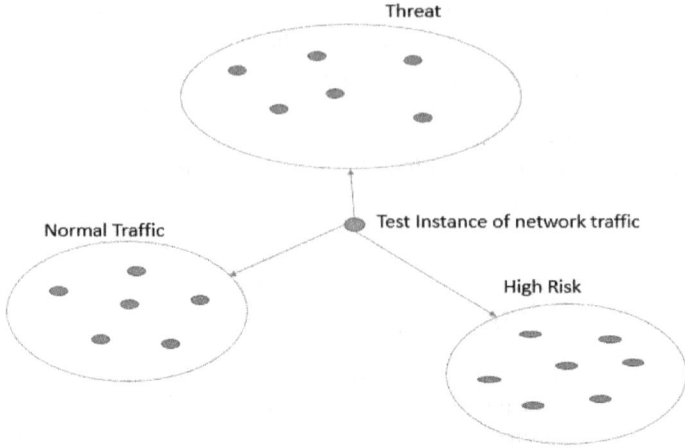

FIGURE 4.5 KNN implementation to classify the network traffic.

belong to the class of threat, risk or normal traffic, the test instance will be classified accordingly.

4.5.2 SUPPORT VECTOR MACHINE

A support vector machine is another supervised machine-learning algorithm to classify the network traffic into multiple class such as threat, normal and risky traffic. In this algorithm we need to find the hyperplane, which can be linear or non-linear, to classify the network traffic.

Figure 4.6 demonstrates the linear classifier identifying the normal traffic or threat, but the linear classifier may not be able to classify the network traffic every

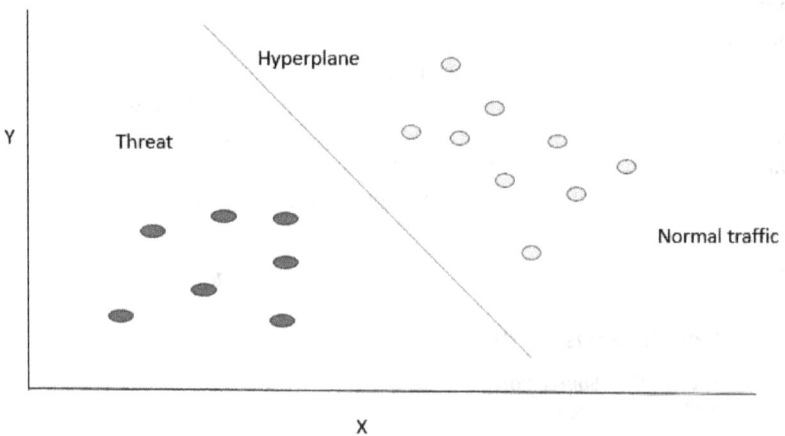

FIGURE 4.6 Binary classification using SVM to classify the network traffic.

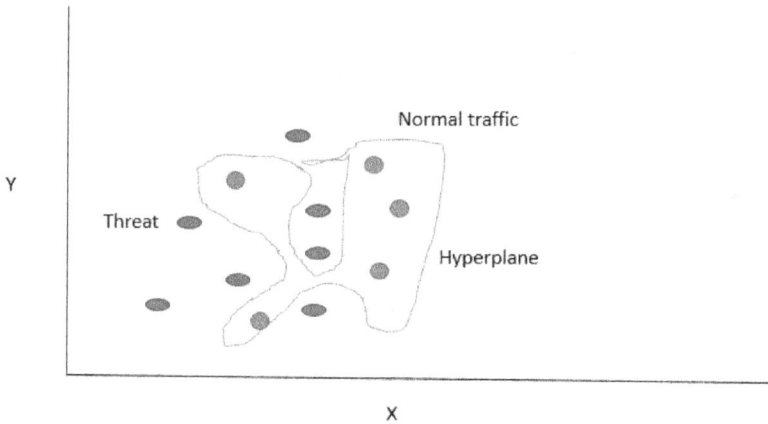

FIGURE 4.7   Multi-class classification using SVM to classify the network traffic.

time. In that case, a non-linear classifier, as shown in Figure 4.7, will be used to classify the traffic.

4.5.3   DECISION TREE

A supervised machine-learning algorithm contains the number of nodes, edges and leaves. Each node of tree is subdivided based on the value of attribute that it's representing. Leaves of tree represent classes such as normal, risky, threat. Decision trees have three types of nodes: chance node, decision node and end node or leave. Decision nodes represent that a decision has to be made; chance nodes shown the possible of value of the attribute; and leave or end nodes represent the final outcome, such as threat, normal or risky traffic, as shown in Figure 4.8.

4.5.4   RANDOM FOREST

Random forest is a supervised machine-learning algorithm that can be used for regression as well as classification. In the case of cyberattacks, we need to classify the traffic pattern (threat, normal or risky traffic). A more generalized form of decision tree or multiple decision tree is constructed to make the decision about the traffic pattern as shown in Figure 4.9. A test instance of traffic is passed through the multiple decision trees, and an individual tree gives some result about the class of the traffic. After that these decisions are passed through the majority voting mechanism to decide the final decision about the test instance.

4.5.5   NEURAL NETWORK

A neural network is a supervised machine-learning model that contains the input layer, hidden layer and output layer, each layer having set of neurons. The input layer contains the features extracted from the dataset, such as port number, IP address, protocol, length of data. For each neuron in hidden layer, get the input as the sum

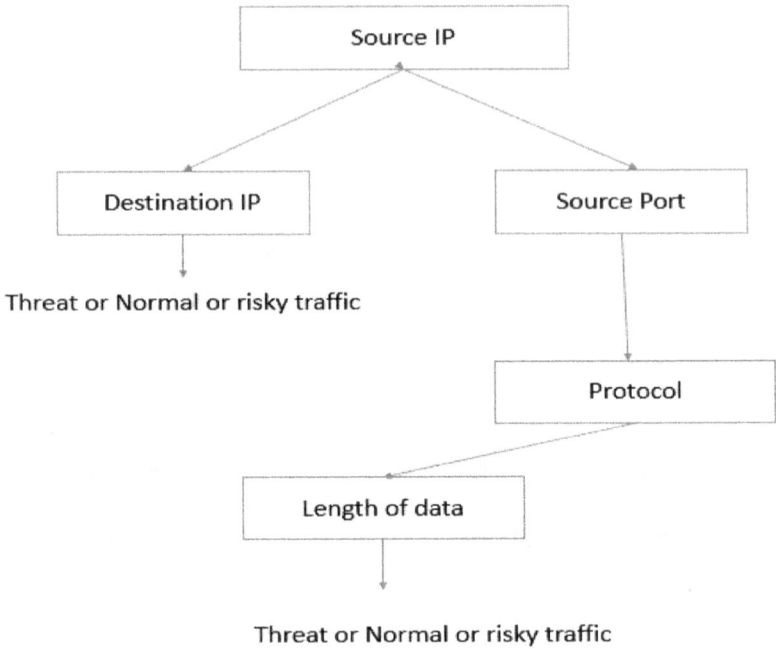

FIGURE 4.8 Classification of network traffic using a decision tree.

of product of feature and weights; then each neuron will fire with some activation function on the received input vector and the same process will be repeated in the output layer. In the case of classification, the output layer will produce the output as 0 (normal), 1 (threat) or 2 (risky traffic) as shown in Figure 4.10.

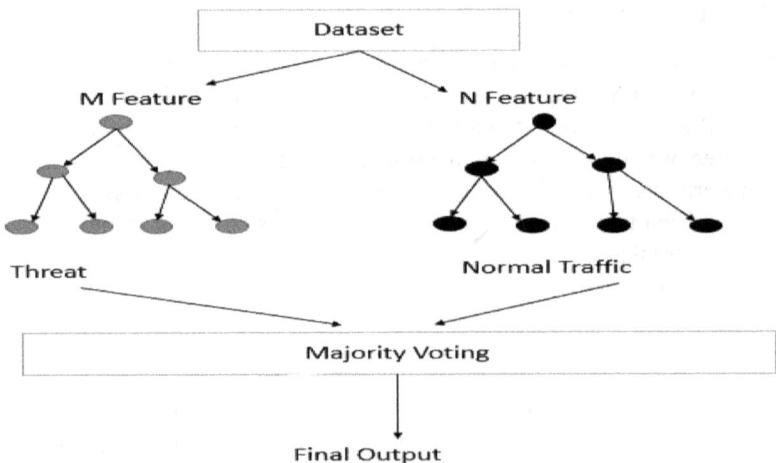

FIGURE 4.9 Classification of network traffic using random forest.

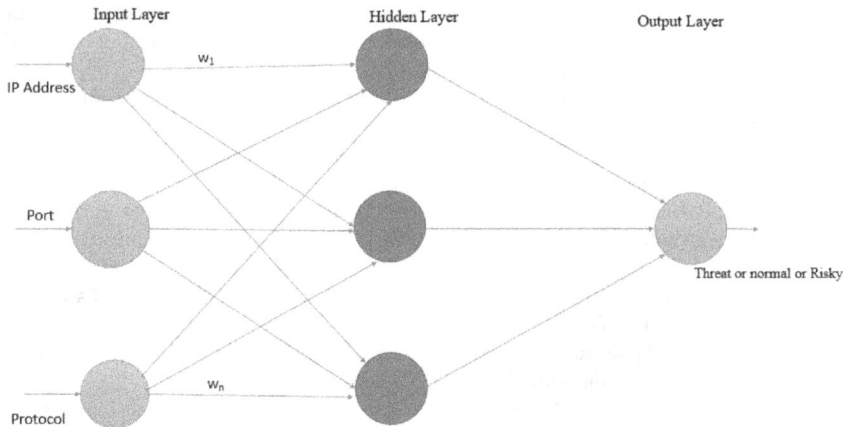

FIGURE 4.10   Classification of network traffic using a neural network.

4.6   CONCLUSION

Intrusion detection techniques have become a lot more advance in recent years, but there are also lots of problems that have arisen and need to be solved. Huge datasets with a variety of attacks are required for training and testing algorithms. IDSs should look into the real-time traffic and filter the traffic based on intruder history and self-learning of IDS to make an intelligent traffic classifier. Various machine-learning, deep-learning and clustering algorithms can be used to refine the data to make the intrusion detection system more intelligent.

REFERENCES

1. Scarfone, Karen, and Peter Mell. *Guide to intrusion detection and prevention systems (idps).* No. NIST Special Publication (SP) 800-94 Rev. 1 (Draft). National Institute of Standards and Technology, 2012
2. Roesch, Martin. "Snort: Lightweight intrusion detection for networks." *Lisa.* Vol. 99. No. 1. 1999
3. T. Abbes, A. Bouhoula & M. Rusinowitch (2004). Protocol analysis in intrusion detection using decision tree. Proceedings of International Conference on Information Technology: Coding and Computing (ITCC).
4. M. Sebring, E. Shellhouse, M. Hanna & R. Whitehurst (1988). Expert systems in intrusion detection: A case study. Proceedings of the 11th National Computer Security Conference. 74–81.
5. T. Lunt, R. Jagannathan, R. Lee, S. Listgarten, D. Eclwards, P. Neumann, H. Javitz & A. Valdes (1988). IDES: The Enhanced Prototype. A Real-Time Intrusion Detection System, Tech. report, Technical Report SRI Project.
6. D. Anderson, T. Frivold & A. Valdes (1995). Next-generation intrusion detection expert system (NIDES): A summary. International, Computer Science Laboratory.
7. D. Anderson, T.F. Lunt, H. Javitz, A. Tamaru & A. Valdes (1995). Detecting unusual program behaviour using the statistical component of the Next-Generation Intrusion Detection Expert System (NIDES). SRI International Computer Science Laboratory.

8. U. Lindqvist & P.A. Porras (1999). Detecting computer and network misuse through the production-based expert system toolset (P-BEST). Proceedings of the IEEE Symposium on Security and Privacy, 146–161.

9. D.E. Denning (1987). An intrusion-detection model. *IEEE Transactions on Software Engineering*, 13(2), 222–232.

10. K. Ilgun (1993). USTAT: A real-time intrusion detection system for UNIX. Proceedings of the IEEE Symposium on Security and Privacy, 16–28.

11. K. Ilgun, R.A. Kemmerer & P.A. Porras (1995). State transition analysis: A rule-based intrusion detection approach. *IEEE Transactions on Software Engineering*, 21(3), 181–199.

12. S. Kumar (1996). Classification and detection of computer intrusions. Ph.D. Dissertation. Purdue University, USA.

13. S. Kumar & E. Spafford (1994). A pattern matching model for misuse intrusion detection. Proceedings of the 17th National Computer Security Conference.

14. S. Kumar & E. Spafford, A software architecture to support misuse intrusion detection. Proceedings of the 18th National Information Security Conference.

15. S. Kumar & Eugene Spafford (1994). An application of pattern matching in intrusion detection. Tech. Report, Purdue University, Department of Computer Sciences.

16. W. Lee, S. J. Stolfo & K. W. Mok (1999). A data mining framework for building intrusion detection models. IEEE Symposium on Security and Privacy, 120–132.

17. W. Lee & S.J. Stolfo (1998). Data mining approaches for intrusion detection. Proceedings of the 7th USENIX Security Symposium.

18. W. Lee, S.J. Stolfo & K.W. Mok (1998). Mining audit data to build intrusion detection models. Proceedings of the 4th International Conference on Knowledge Discovery and Data Mining, AAAI Press, 66–72.

19. S. Axelsson (2000). Intrusion detection systems: A survey and taxonomy, Tech. Report. Chalmers University of Technology, Department of Computer Engineering.

20. A. Khraisat, I. Gondal, P. Vamplew et al (2019). Survey of intrusion detection systems: techniques, datasets and challenges. *Cybersecurity*, 2(20), 1–22.

21. A. Jones & R. Sielken (2000). Computer system intrusion detection: A survey. Tech. report, Department of Computer Science, University of Virginia, Thornton Hall, Charlottesville.

22. J. McHugh (2001). Intrusion and intrusion detection. *International Journal of Information Security*, 1(1), 14–35.

23. M. Dacier, H. Debar & A. Wespi (1999). A revised taxonomy for intrusion-detection systems. Tech. report, IBM Research Report.

24. Smaha, Stephen E. "Haystack: An intrusion detection system." *Fourth Aerospace Computer Security Applications Conference*. Vol. 44. 1988.

25. A. Ph. Porras, Phillip A., and Peter G. Neumann. "EMERALD: Event monitoring enabling response to anomalous live disturbances." *Proceedings of the 20th national information systems security conference*. Vol. 3. 1997.

26. R. Mitra, S. Mazumder, T. Sharma, N. Sengupta & J. Sil (2012). Dynamic network traffic data classification for intrusion detection using genetic algorithm. International Conference on Swarm, Evolutionary, and Memetic Computing (SEMCCO), Lecture Notes in Computer Science, Springer, Berlin, Heidelberg, 509–518.

27. H.S. Vaccaro & G.E. Liepins (1989). Detection of anomalous computer session activity. Proceedings of the Symposium on Research in Security and Privacy (Oakland, CA), May 1989, pp. 280–289.

28. L.T. Heberlein, G.V. Dias, K.N. Levitt, B. Mukherjee, J. Wood & D. Wolber (1990). A network security monitor. IEEE Computer Society Symposium on Research in Security and Privacy, Oakland, CA, USA, 296–304.

29. H.S. Teng, K. Chen & S.C. Lu (1990). Adaptive real-time anomaly detection using inductively generated sequential patterns. Proceedings of the Symposium on Research in Security and Privacy (Oakland, CA), 278–284.
30. K.A. Jackson, D.H. DuBois & C.A. Stallings (1991). An expert system application for network intrusion detection. Proceedings of the National Computer Security Conference.
31. P. Sadotra & C. Sharma (2016). A survey: Intelligent intrusion detection system in computer security. *International Journal of Computer Applications*, 151(3).
32. M. Almi'ani, A. A. Ghazleh, A. Al-Rahayfeh & A. Razaque (2018). Intelligent intrusion detection system using clustered self-organized map. Fifth International Conference on Software Defined Systems (SDS).

5 A Morphological Filtering-Based Image-Enhancement Method for Citrus Plant Diseases

Bobbinpreet Kaur, Tripti Sharma and Bhawna Goyal
Chandigarh University

Ayush Dogra
Ronin Institute

CONTENTS

5.1 Introduction 65
5.2 Related Work 68
5.3 Image Sharpening.. 69
5.4 High-Boost Filtering.. 70
5.5 Morphological Filters 71
5.6 Proposed Methodology... 73
5.7 Experimental Setup 75
    5.7.1 Performance Metrics 75
    5.7.2 Results and Discussion 75
5.8 Conclusion 80
References.. 80

5.1 INTRODUCTION

Image enhancement is the broad subfield of image processing that aims at improvisation of the image quality qualitatively and quantitatively. It caters to the applications pertaining to image processing in almost every domain. The images acquired through the acquisition process suffer from certain artefacts due to illumination conditions while capturing the image. Images acquired through acquisition devices have many ambiguities. Due to the inherent properties of sensors and the acquisition device, the images need to be passed through certain processing algorithms for quality improvement. The trend is going toward achieving an image that is more accurate, on one hand, and on the other hand it needs to be visually pleasing to the end user. The images thus acquired need to be preprocessed in terms of contrast enhancement and noise removal. As the demand for high-quality colour images is growing day by day, this arouses the need for efficient and less complex methodologies for making images visually better. In almost every application listed in literature, some kind of preprocessing is required. This creates a continuous thrust to develop an algorithm for improving the preprocessing of images.

The choice of preprocessing technique will be dependent solely on the performance and type of application under study. Various filtering approaches have been developed to remove ambiguities from the images. An appropriate filter is chosen according to the noise characteristics. In image processing, image enhancement may be required to overcome physical limitations such as camera resolution and degradation. Enhancement deals with improving the image for better visual satisfaction, while restoration works by changing the statistical properties of the image to make it visually pleasing as well as free of ambiguities and abnormalities.

Low-contrast images reduce the ability of the user to analyze and interpret the information hidden in the image [1]. Factors affecting the contrast of image include low-light environment, a faulty imaging device, poor environmental conditions, etc. [2]. Such effects can be countered by deployment of various types of enhancement techniques. The main aim of contrast enhancement is to show up the information buried in the lower frequency pixels of the image and to make it visually more pleasing.

Overall improvement is achieved in terms of the quality of an image and brings out the latent information buried in the degraded image to make it more interpretable in terms of image data and information for further analysis. The preprocessed (enhanced) image can drive the systems in many kinds of applications. One of the applications is disease detection in plants, wherein efficient enhancement can lead to a more accurate classification of the disease. Most of the image-enhancement techniques are based on histogram-based methods. Histogram-based techniques can lead to improper contrast enhancement, which eventually results in making the processed image appear unnatural with visual artefacts [3]. The local histogram equalization (HE) methods often lead to overenhancement of noise and artefacts in the image [4]. The process of quality enhancement is described in Figure 5.1. Overall quality enhancement requires both the removal of noise and contrast enhancement. Nevertheless, the noise eventually degrades the images produced by modern cameras, contributing to declining visual image quality. Therefore there is a need to reduce noise without losing object characteristics. An in-depth study of various approaches produces various assumptions, pros and cons for different sets of applications. Quite often the properties of noise are known well in advance, and certain types of noises destroy all the pixels equally. The filtering approach used for denoising will be considered both in the spatial and frequency domains.

As there is an explosive increase in the number of images generated under improper conditions, such as poor illumination and atmospheric turbulence, the requirement of tools that can cancel out the effect of degradations is on the rise. Improvements done on an image are governed by predefined criteria dictated by the kind of application under consideration. An image acquired through an acquisition device gives a two-dimensional (2D) representation of a three-dimensional (3D) scene. But many images represent the 3D scene in an unsatisfactory manner. Thus there is a need for improvement of image properties and information by looking for

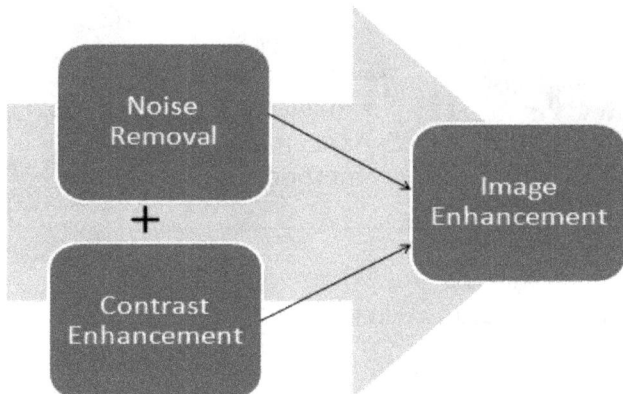

FIGURE 5.1 Process of image enhancement.

statistical improvement and applying certain mathematical tools, because the 2D representation of an image can be easily considered to be a matrix. All the principles of matrix mathematics can be applied on images to make them visually better and describe more information content. Because the physical imaging systems are not perfect and the conditions under which the images are captured are never ideal, an image captured often gives a degraded version of the ideal 2D mapping of the scene. The capturing system, the atmospheric factors, and the recording medium all act as sources of degradation in the captured image, which results in inefficiency of the image to represent the 3D scene adequately. Image restoration aims at addressing the problem of the unsatisfactory mapping of 3D to 2D representation. The main aim of image enhancement is to make an image capable of closely depicting the scene that it actually aims to represent. Figure 5.2 lists the various image-enhancement methods.

Other contrast-enhancement methods include adjacent-blocks-based modification for HE [4], entropy scaling in wavelet domain [3], optimization-based methods [5–10], spatial filtering approach[11, 12], morphological filters [13–19], frequency domain filters [20–24], thresholding based [25–27] and contrast stretching [28, 29]. All these methods aim for the reduction of illumination and improving the contrast of the image, which results in overall quality enhancement for visual perception.

This article proposes a hybrid filtering approach for contrast enhancement using the high-boost filtering of red, green and blue (RGB) subimages followed by top-hat and bottom-hat filtering. High-boost filtering sharpens the image components, and

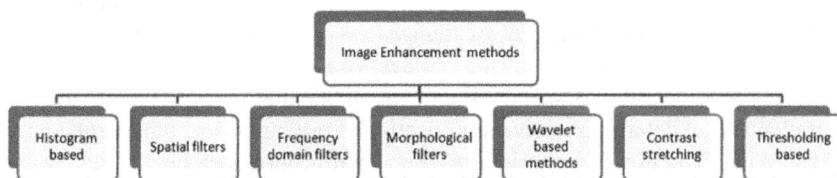

FIGURE 5.2 Types of image-enhancement methods.

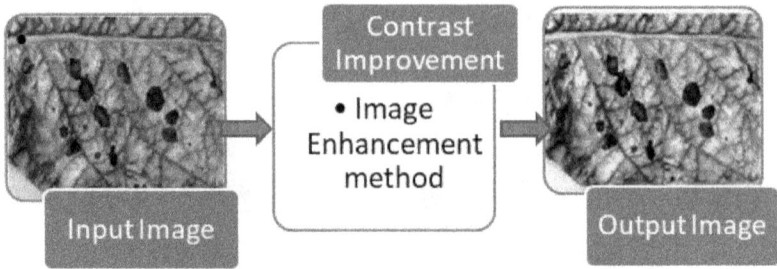

FIGURE 5.3   Example of contrast enhancement.

top-hat and bottom-hat filtering preserves both maximas and minimas and enhances them. The experimental results, considering both the subjective and objective metrics, have verified that the proposed hybrid approach results in contrast enhancement while preserving the signal-to-noise ratio.

5.2   RELATED WORK

Contrast enhancement refers to change in intensity value, which will result in enhanced contrast. This improvement in contrast makes it easier to interpret and analyze the image. The features of the image are highlighted and shown more accurately. The choice of enhancement method depends upon the type of application [30]. Figure 5.3 shows an example of contrast enhanced image.

Contrast stretching is a technique of linear contrast enhancement. The original pixel intensity values are mapped onto new values linearly such that the linear operator is selected in a way to enhance the image. This is more appropriate when the histogram values are flooded around the central value, making it look like a Gaussian curve. There are three methods for linear contrast enhancement: maxima-minima linear constant search, percentage linear stretch and piecewise linear stretch [31]. Image contrast is the difference in intensity value of the brightest spot and the darkest spot in images [31]. Histogram equalization is the technique applied to spread all the pixels over the entire intensity range equally. The modified histogram will result in improved contrast of the image. This can be a general purpose or matching with the required specification. This will make areas of lower contrast to be highlighted. Certain adaptive histogram equalization methods were developed in order to remove the constraints of global equalization. The transformation function in the case of histogram equalization is directly dependent upon the frequency of occurrence of grey level in the image, which leads to overenhancement of high-frequency grey values and less enhancement of low-frequency grey values. The image-enhancement method suffers from certain drawbacks such as amplification or enhancement of noise artefacts, saturation loss of details, excessive modification of brightness and unnatural contrast enhancement. The non-parametric histogram modification method adequately controls the histogram peaks and minimizes the distortion present in smoother regions without the pragmatic change of parameters [32].

In recent times, a variety of nature-inspired approaches have been used to achieve the best possible outcomes. In the bat algorithm, the neural network and bat algorithm are blended, in which the bat methodology has been used to refine the parameters of the modified neural model to optimize two competitive image reliability indices of contrast enhancement factor and mean opinion score [33]. In the modified cuckoo algorithm, optimum wavelet the optimum scale value of the wavelet is used to perform medical image enhancement. The improved algorithm will redevelop the worst nests dynamically. For all iterations, the fitness of each nest is calculated and the threshold value is set based on the fitness value [34]. An improved immune algorithm is proposed in Reference [35]. This method possesses three stages. First, the unique coding strategy of the MRI brain image is structured instead of simple binary coding. Second, the mutation distance is connected to the mutation variable to better control the advancement of the mutation and to eliminate any small local optimization. Finally, in the Gauss distribution, uniform distribution and chaotic distribution, both the clone selection and mutation are configured together rather than in the Gauss distribution alone. The articles discussed above are all intended to increase the visual quality of the medical images, but the quantitative metrics for evaluation of image quality have not been considered. In Reference [36] a novel approach for improving the visible appearance of images captured in dusty conditions is proposed. The fuzzy operator is applied in RGB decomposed images individually, thus attaining overall improvement in the image quality. The technique is fine-tuned by taking different values of tuning parameter zeta.

5.3   IMAGE SHARPENING

In many digital image applications, such as disease detection and lesion detection, there is a need to enhance or make the edges crisp and highlight the higher-frequency pixels while preserving the low-frequency components. Traditional methods for image sharpening include Unsharp Filtering. Figure 5.4 explains the process of sharpening an image. Unsharp Filtering is one such method that finds it application in the area of photographic and printing industries for edge crisping. This operator enhances the high-frequency components on one hand, and on the other hand it sharpens the edges. Since noise also belongs to high-frequency components, its cancellation is required [37].

$$I_{out}(m,n) = I(m,n) - I_{unsh} \tag{5.1}$$

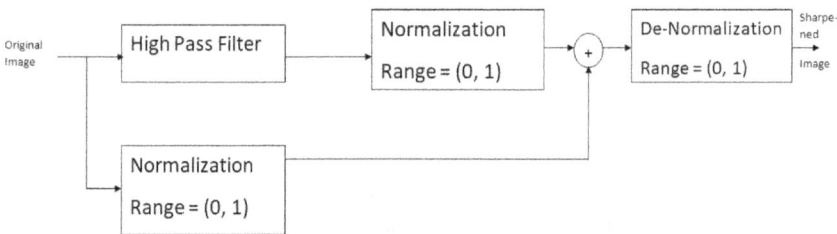

FIGURE 5.4   Image sharpening.

where I_{out} is the sharpened image and I_{unsh} is the blurred counterpart of original image. The procedure followed for this is as follows:

1. The input image is smoothed by applying a low-pass filter.

$$I_{lpf}(m,n) = lpf(I(m,n)) \qquad (5.2)$$

2. An edge description and other desired high-frequency components of an image are obtained by performing subtraction of the smoothed image and the original image.

$$I_e(m,n) = I(m,n) - I_{lpf}(m,n) \qquad (5.3)$$

3. The image thus obtained sharpens the edges by performing the addition of the original image and the image obtained in step 2.

$$I_{sh}(m,n) = I(m,n) + k * I_e(m,n) \qquad (5.4)$$

5.4 HIGH-BOOST FILTERING

For many applications in image processing, it is often desired that images must be enhanced before passing on to subsequent stages. The process for enhancement focuses on higher-frequency pixels and low-frequency components. The broad details of the scene and objects in the image are represented by high-frequency pixel components, and fine details such as lines and small points are represented by low-frequency pixel components. To sharpen the image, more emphasis needs to be placed on boosting the high-frequency components on one hand and on the other hand preserving the low-frequency components. One filter suited for this kind of operation is the high-boost filter. It is simply a sharpening operator that is applied on the image to get sharp edges. It is used for boosting high-frequency components present in the image. The sharpening of high-frequency components is achieved by subtracting a smoothed version, i.e., a low-pass filtered version of the image from the original one. This enhances the relative importance of details depicted by high-frequency components by increasing the relative weight of high-frequency components. This filter is particularly useful for the dark images.

$$I_{hb}(m,n) = A * I(m,n) - I_{lpf}(m,n) \qquad (5.5)$$

$$I_{hb}(m,n) = (A-1) * I(m,n) + I(m,n) - I_{lpf}(m,n) \qquad (5.6)$$

$$I_{hb}(m,n) = (A-1) * I(m,n) + I_{hpf}(m,n) \qquad (5.7)$$

$$I_{hb}(m,n) = (A-1) * I(m,n) + I(m,n) * h_{hpf}(m,n) \qquad (5.8)$$

where A is a factor controlling weights and is known as amplification factor, I(m,n) is the original image, $I_{lp}(m,n)$ is the low-pass filtered image and $I_{hpf}(m,n)$ is the high-pass filtered image.

0	-1	0
-1	w	-1
0	-1	0

-1	-1	-1
-1	w_1	-1
-1	-1	-1

FIGURE 5.5 High-boost filtering mask.

The implementation of filter on an image is achieved by applying a suitable mask, or kernel, and performing the convolution operation. The high-boost filtered image can be obtained by convolution with masking kernels H_B represented as follows:
Where $w = A + 4$ and $w_1 = A + 8$.

Figure 5.5 shows two types of mask to achieve high-boost filtering.

The constraint while designing high-boost filtering is appropriate selection of the amplification factor in order to achieve simultaneous edge sharpening and noise removal. The objective function aims at achieving maximum sharpened edges and minimum noise levels [38].

The operation of high-boost filtering can be represented in spatial domain as

$$I(m,n) = I(m,n) - (I(m,n)**H_B) \tag{5.9}$$

where symbol '**' represents a 2D convolution operation of the original image and high-boost kernel H_B. The main drawback pertaining to this technique is amplification of noise pixels along with the information pixels, depending on appropriate choice of amplification factor. So a suitable noise-eliminating filter needs to be deployed in order to cancel the effect of noise amplification.

5.5 MORPHOLOGICAL FILTERS

In image processing, morphological operators deal with characterization of shapes and regions. Morphology is particularly used for boundary extraction and skeletonization [44]. It initiates by performing a comparison of each pixel in the image with its neighbours in various ways, and a decision is made depending on the comparison – either adding or removing the particular pixel and brightening or darkening of the pixel. The overall operation of morphological filters is defined by the application of a structuring element (SE) which classifies the neighbourhood of pixels. Different shapes and sizes of kernels or SEs are deployed in order to ensure optimum performance of the filter [39].

The basic morphological operators are described as follows [40]:

1. Erosion: It removes or erodes the pixels present at the object boundaries, depending upon the size and shape of SE.

$$I(m,n) \ominus SE = Z \,|\, (SE)_z \subseteq I(m,n) \tag{5.10}$$

where I(m,n) is the original image and SE is the particular structuring element.

Erosion of the image and structuring element can be termed as the set of all points of z, such that SE translated by Z is contained in I(m,n).

2. Dilation: It adds pixels to the edges and boundaries. The number of pixels added will depend primarily upon the shape and size of structuring element.

$$I(m,n) \oplus SE = Z\llbracket SE_{\hat{z}}^{\wedge} \cap I(m,n)\rrbracket \subseteq I(m,n) \tag{5.11}$$

The reflection of set SE about its origin is shifted by dilation of the image, and SE is the set of all displacements such that SE^ overlaps I(m,n) by at least one element.

3. Opening: This is a composite operation in which first erosion is performed and then dilation is performed by using the same structuring element. It smooths out the contours and break down sharp peaks.

$$I(m,n) \circ SE = (I(m,n) \ominus SE) \oplus SE \tag{5.12}$$

4. Closing: This is also a composite operation in which first dilation is performed and then erosion is performed. It smooths out the contours and eliminates small holes.

$$I(m,n) \cdot SE = (I(m,n) \oplus SE) \ominus SE \tag{5.13}$$

These basic morphological operators are combined in different manners for achieving the requirements of a particular application.

For image contrast enhancement, two filters – top-hat filter and bottom-hat filter – find applications for improving contrast of especially darkened images which are devised using image subtraction with opening and closing operation. Figure 5.6 illustrates the morphological filter module for image contrast enhancement. The top-hat filtered image is obtained by subtracting the opening operation of image and structuring element from the original image. This can lead to removal of background illumination problems. It enhances the bright objects on a dark background. The bottom-hat filter or black top-hat filter performs the reverse operation of top-hat

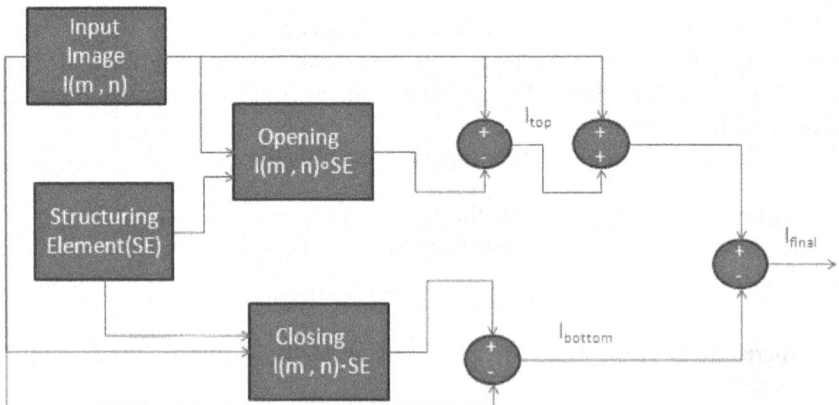

FIGURE 5.6 Morphological filter module.

filtering. The bottom-hat filter image is obtained by subtracting the original image from the closing operation of image and structuring element. In this article the citrus disease images are taken as input and passed through subsequent stages for overall enhancement. The enhanced image will be used as input for subsequent stages.

$$I_{top} = I(m,n) - (I(m,n) \circ SE) \qquad (5.14)$$

$$I_{bottom} = (I(m,n) \cdot SE) - I(m,n) \qquad (5.15)$$

$$I_{final} = I(m,n) + I_{top} - I_{bottom} \qquad (5.16)$$

The I_{final} image possesses the advantage of both top-hat filtering and bottom-hat filtering. The bright areas are added, and dark regions are subtracted from the original image, which results in overall contrast enhancement. The performance of these filters is dependent on the appropriate qualities of SE as per the application. There are many different shapes of structuring elements quoted in the literature, and an extensive research has been carried in order to select a particular structuring element. A few examples of structuring elements shapes include disk, diamond, octagon, periodic line, rectangle, etc. The disk-shaped SE exhibits rotation invariance, so it finds its appropriate place in many image processing applications.

5.6 PROPOSED METHODOLOGY

This chapter proposes a hybrid filtering approach towards image contrast enhancement based upon spatial and morphological filters. Image sharpening is achieved by deploying high-boost spatial filtering, and contrast enhancement is achieved by deploying suitable morphological filtering. High-boost filtering also aims at noise reduction so that the performance of morphological filtering reaches an optimum level.

Figure 5.7 gives the detailed description of various stages of proposed algorithm. The input image is a coloured image, so a detailed description can be extracted

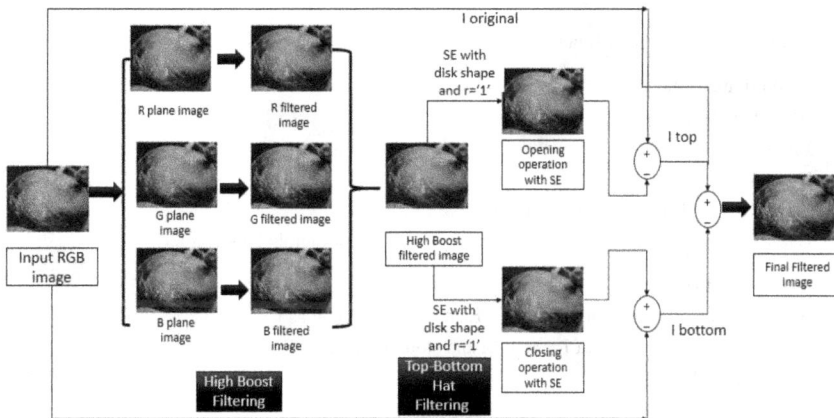

FIGURE 5.7 Proposed methodology.

by decomposition into RGB planes. These decomposed images are sharpened by application of high-boost filtering. The application of high-boost filtering leads to noise reduction so the further stages of our proposed methodology can work more efficiently. These decomposed, and thus filtered, images are recombined to form a single filtered image. This image is then processed by application of morphological filtering. The morphological filter is applied on the filtered image by a suitable structuring element. The top-hat and bottom-hat filter is deployed in order to enhance both maximas and minimas of the image, thus preserving the finest details and achieving overall contrast enhancement. We tested our algorithm with various sizes and shapes of the structuring element, and the resultant image was both subjectively and objectively evaluated for quality assurance.

1   **Begin**
2   Input the input image.
3   Resize the input image to particular dimensions.
4   Decompose the image into subsequent RGB planes, i.e., I_R, I_G, I_B.
5   Initialize the value of amplification factor(A) and weight(w).
6   Generate filter mask or kernel for high-boost filtering, i.e., M_{HBF}
7   **for** R plane image
8   **while** stopping condition is not met (i.e., all pixels are mapped onto centre of kernel)

9   **for** optimum value of A.
10 $HB_R = I_R ** M_{HBF}$.
11 **End for.**
12 **End while**
13 **for** G plane image
14 **while** stopping condition is not met (i.e., all pixels are mapped onto centre of kernel)

15 **for** optimum value of A.
16 $HB_G = I_G ** M_{HBF}$
17 **End for.**
18 **End while**
19 **for** B plane image
20 **while** stopping condition is not met (i.e., all pixels are mapped onto centre of kernel)

21 **for** optimum value of A.
22 $HB_B = I_B ** M_{HBF}$
23 **End for.**
24 **End while.**
25 Concatenate I_R, I_G, I_B and form a filtered image I_{HB}.
26 Chose different values of radii (r) and shapes (s) of SE for morphological filter

27 **for** r=1 and shape=disk.
28 Perform Opening operation of I_{HB} and SE and store resultant as I_o.
29 Perform closing operation of I_{HB} and SE and store resultant as I_C.
30 Calculate $I_{top} = I_{HB} - I_o$.
31 Calculate $I_{bottom} = -I_C - I_{HB}$.
32 Calculate $I_{final} = I_{HB} + I_{top} - I_{bottom}$.

33   **End for.**
34   Repeat the process for different values of radii and shapes.
35   Calculate the image quality parameters.
36   **End.**

5.7   EXPERIMENTAL SETUP

The implementation of the proposed algorithm is carried out in MATLAB R2018a by an Intel Core i5 @2.10 GHz processor and having 4.00 GB RAM. For implementation and evaluation of proposed algorithm a total of five images are taken from the citrus disease image gallery belonging to four different types of plant diseases, namely black spot, greening, canker, scab and healthy class [41]. All the images taken are of 256 * 256 dimensions with 72 dpi resolution. The dataset in total contains 759 images belonging to each class of disease. For high-boost filtering the mask is chosen with weight w=A+4; for implementation of a morphological filter, the different shapes considered for SEs are disk, octagon and diamond. The values of radii chosen for disk-shaped SEs are r=1,2,3,4,; for octagon-shaped SEs they are 3,6,9; and for diamond-shaped SEs they are r=1,2,3.

5.7.1   PERFORMANCE METRICS

The proposed method was tested objectively using certain image-enhancement quality metrics [42].The value of these metrics validated the overall quality of the image and the information content present in the image. The metrics used for qualitative evaluation were MSE (mean squared error), PSNR (peak signal-to-noise ratio), SC (structural content), AD (average difference), MD (maximum difference), NAE (normalized absolute error) and NCC (normalized cross correlation)

5.7.2   RESULTS AND DISCUSSION

The performance evaluation of the proposed methodology was done on a citrus disease dataset [43]. The system performance was verified on five images of black spot, greening, canker, scab and healthy class. The experiment was conducted with different values of radii and shapes of structuring element, i.e., r=1,2,3 for disk- and diamond-shaped SE and r=3,6,9 for octagon-shaped SE. On conducting the experiment it was found that the best results were achieved through r=1 for both disk- and diamond-shaped SE. The diamond- and disk-shaped kernels perform equally well but we prefer disk-shaped kernels due to the rotation invariant nature. The weights of a disk operator are same in all the directions. This will help in yielding same result irrespective of orientation angle with respect to sampling grid of an object present in a scenic image. As we scan from the tabular values, our proposed method in which disk operator for radii value '1' is giving the best values for objective metric. Figure 5.8 shows a plot of values of PSNR obtained with different techniques.

With an increase in the value of radii, the performance starts degrading. Also the permissible value of radii is greater than or equal to 1 and for octagon-shaped kernels;

PSNR Analysis

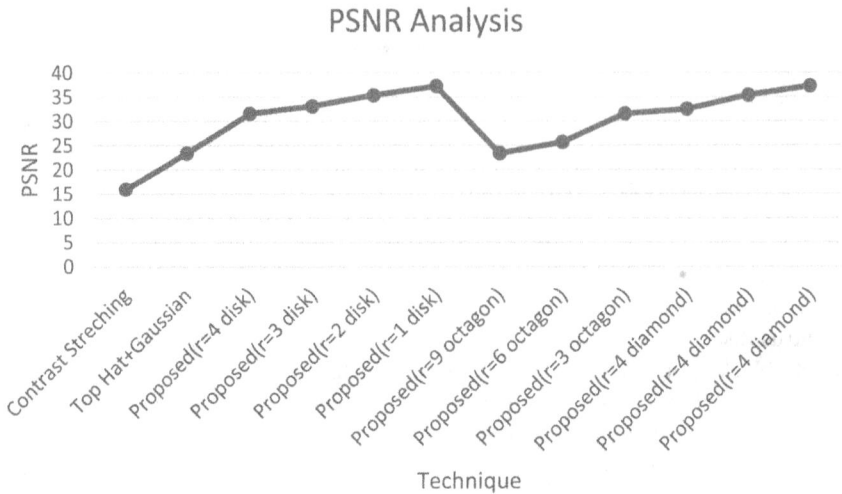

FIGURE 5.8   PSNR values for scab image with different techniques.

the permissible values are multiples of 3. The image required for further stages, such as segmentation, classification, etc., requires the objects to be distinctively sharpened so that the region of interest can be extracted accurately and crisply. For considering disease detection, the diseased area must have sharp edges. The application of high-boost filtering resulted in sharp object boundaries and noise removal. The deployment of morphological filtering resulted in overall contrast enhancement. The result of proposed algorithm shows a high degree of improvement in the value of peak signal-to-noise ratio which depicts overall quality improvement and reduction of noise. The other quality metrics show considerable improvement by deployment of proposed algorithm. The image thus obtained possesses both visual and objective enhancement. The edges are sharp, and crisp gradient values are obtained. The diseased area is embodied. The diseased area is perceptible with high accuracy and precision. The fruit and leaf texture is also visibly enhanced. Most of the methods failed at detecting finer lesion details.

Table 5.1 shows the values of parameters MSE, PSNR, normalized cross correlation, average difference, structural content, maximum difference and normalized absolute error obtained experimentally for different techniques. The proposed method was applied with 10 combinations of radii and shapes, and the values were recorded. For the sake of simplicity we have shown the results in the form of images for scab disease taken from the dataset. There is considerable visual enhancement and the parameters show the objective improvement in the image. Figure 5.9 shows the detailed step-by-step outputs obtained using the proposed algorithm. It can be seen from the Figure 5.9 that there is visual quality enhancement in input and output image. We have chosen the output in Figure 5.9 i to be best in both objective and subjective terms.

TABLE 5.1
Experimental Results Obtained for Different Techniques

	Contrast Stretching	Top-Hat + Gaussian Filter	Proposed									
			Disk				Octagon			Diamond		
			r=4	r=3	r=2	r=1	r=9	r=6	r=3	r=3	r=2	r=1
Image 1_black spot												
MSE	390.7551	400.3487	179.0599	149.8145	115.9602	96.6878	694.4293	430.5477	179.0599	156.9167	115.9602	96.6878
PSNR	22.2118	22.1064	25.6008	26.3753	27.4877	28.2771	19.7145	21.7906	25.6008	26.1741	27.4877	28.2771
Normalized cross correlation	0.87	0.9969	1.024	1.0245	1.0243	1.0231	1.0228	1.0238	1.024	1.0247	1.0243	1.0231
Average difference	20.4503	-16.2864	-3.005	-3.2003	-3.3619	-3.4264	0.2038	-1.585	-3.005	-3.1736	-3.3619	-3.4264
Structural content	1.2974	0.6764	0.9441	0.9448	0.9469	0.9502	0.9187	0.9309	0.9441	0.944	0.9469	0.9502
Maximum difference	33	255	105	86	70	60	126	123	105	100	70	60
Normalized absolute error	0.1791	0.6042	0.0612	0.0548	0.0459	0.0382	0.1445	0.105	0.0612	0.0564	0.0459	0.0382
Image 2_canker												
MSE	699.3366	428.7168	127.6075	101.1155	79.8264	64.0648	430.3195	285.8047	127.6075	112.0863	79.8264	64.0648
PSNR	19.6839	21.8091	27.072	28.0826	29.1093	30.0646	21.7929	23.5701	27.072	27.6353	29.1093	30.0646
Normalized cross correlation	1.1359	1.1359	1.0221	1.0219	1.0213	1.0205	1.0276	1.0247	1.0221	1.0224	1.0213	1.0205
Average difference	-13.5541	-13.5541	-3.0519	-3.1376	-3.1595	-3.192	-2.0554	-2.5504	-3.0519	-3.133	-3.1595	-3.192
Structural content	0.7718	0.7718	0.9527	0.954	0.9559	0.9579	0.9322	0.9426	0.9527	0.9528	0.9559	0.9579
Maximum difference	21	21	90	45	43	28	116	116	90	89	43	28
Normalised absolute error	0.1454	0.1454	0.0543	0.0479	0.0407	0.0333	0.1071	0.0846	0.0543	0.0502	0.0407	0.0333
Image 3_greening												
MSE	313.7824	1265.6	618.5369	523.4621	413.0418	334.1313	1354.3	1058.3	618.5369	545.7477	413.0418	334.1313
PSNR	23.1645	17.1079	20.2171	20.942	21.9709	22.8916	16.8136	17.8845	20.2171	20.7609	21.9709	22.8916
Normalized cross correlation	1.1501	1.1305	1.0799	1.0731	1.0628	1.0529	1.1265	1.1094	1.0799	1.0761	1.0628	1.0529

(continued)

TABLE 5.1
Experimental Results Obtained for Different Techniques (Continued)

	Contrast Stretching	Top-Hat + Gaussian Filter	Disk r=4	Disk r=3	Disk r=2	Disk r=1	Octagon r=9	Octagon r=6	Octagon r=3	Diamond r=3	Diamond r=2	Diamond r=1
Image 1_black spot												
Average difference	-9.1619	-47.8307	-7.8341	-7.7154	-7.3531	-7.1752	-7.3125	-7.8424	-7.8341	-7.7915	-7.3531	-7.1752
Structural content	0.7539	0.4436	0.8196	0.8353	0.8576	0.8783	0.7231	0.757	0.8196	0.8296	0.8576	0.8783
Maximum difference	9	236	117	94	78	74	98	105	117	117	78	74
Normalized absolute error	0.1482	0.997	0.1865	0.1668	0.1406	0.1146	0.3132	0.2646	0.1865	0.1721	0.1406	0.1146
Image 4_healthy												
MSE	875.3943	173.8145	53.6304	42.4371	29.3429	21.917	204.4555	138.7442	53.6304	44.3528	29.3429	21.917
PSNR	18.7088	25.7299	30.8367	31.8534	33.4558	34.723	25.0248	26.7087	30.8367	31.6616	33.4558	34.723
Normalised cross correlation	1.0884	0.9432	1.0132	1.0133	1.0128	1.0117	1.0057	1.009	1.0132	1.0134	1.0128	1.0117
Average difference	-8.9984	-3.2343	-1.8441	-1.9151	-1.9264	-1.8837	-0.1542	-0.9196	-1.8441	-1.9078	-1.9264	-1.8837
Structural content	0.843	0.8778	0.9725	0.9726	0.9741	0.9763	0.9816	0.9777	0.9725	0.9725	0.9741	0.9763
Maximum difference	11	173	42	40	32	26	82	77	42	40	32	26
Normalised absolute error	0.09	0.4516	0.0315	0.028	0.0228	0.0185	0.0649	0.0511	0.0315	0.0286	0.0228	0.0185
Image 5_scab												
MSE	1649.8	299.7912	45.3054	31.913	18.9095	12.3249	297.9529	174.5127	45.3054	36.5271	18.9095	12.3249
PSNR	15.9565	23.3626	31.5693	33.0911	35.364	37.223	23.3893	25.7125	31.5693	32.5047	35.364	37.223
Normalized cross correlation	1.328	1.328	1.0197	1.0178	1.0146	1.0112	1.0302	1.0277	1.0197	1.0183	1.0146	1.0112
Average difference	-29.4447	-29.4447	-2.1042	-2.0223	-1.8301	-1.5941	-1.5773	-2.1155	-2.1042	-2.0366	-1.8301	-1.5941
Structural content	0.5653	0.5653	0.9593	0.9636	0.9704	0.9772	0.9246	0.9366	0.9593	0.9623	0.9704	0.9772
Maximum difference	12	12	35	26	20	10	70	64	35	31	20	10
Normalized absolute error	0.3117	0.3117	0.038	0.0317	0.024	0.018	0.1101	0.0805	0.038	0.0341	0.024	0.018

Fig a) Input Image

Fig b) R plane image

Fig c) G plane image

Fig d) B plane image

Fig e) R plane filtered image

Fig f) G plane filtered image

Fig g) B plane filtered image

Fig h) High Boost Filtered image

Fig i) r=1 disk shape SE output

Fig j) r=2 disk shape SE output

Fig k) r=3 disk shape SE output

Fig l) r=4 disk shape SE output

Fig m) r=3 Octagon shape SE output

Fig n) r=6 Octagon shape SE output

Fig o) r=9 Octagon shape SE output

Fig p) r=1 Diamond shape SE output

Fig p) r=2 Diamond shape SE output

Fig p) r=3 Diamond shape SE output

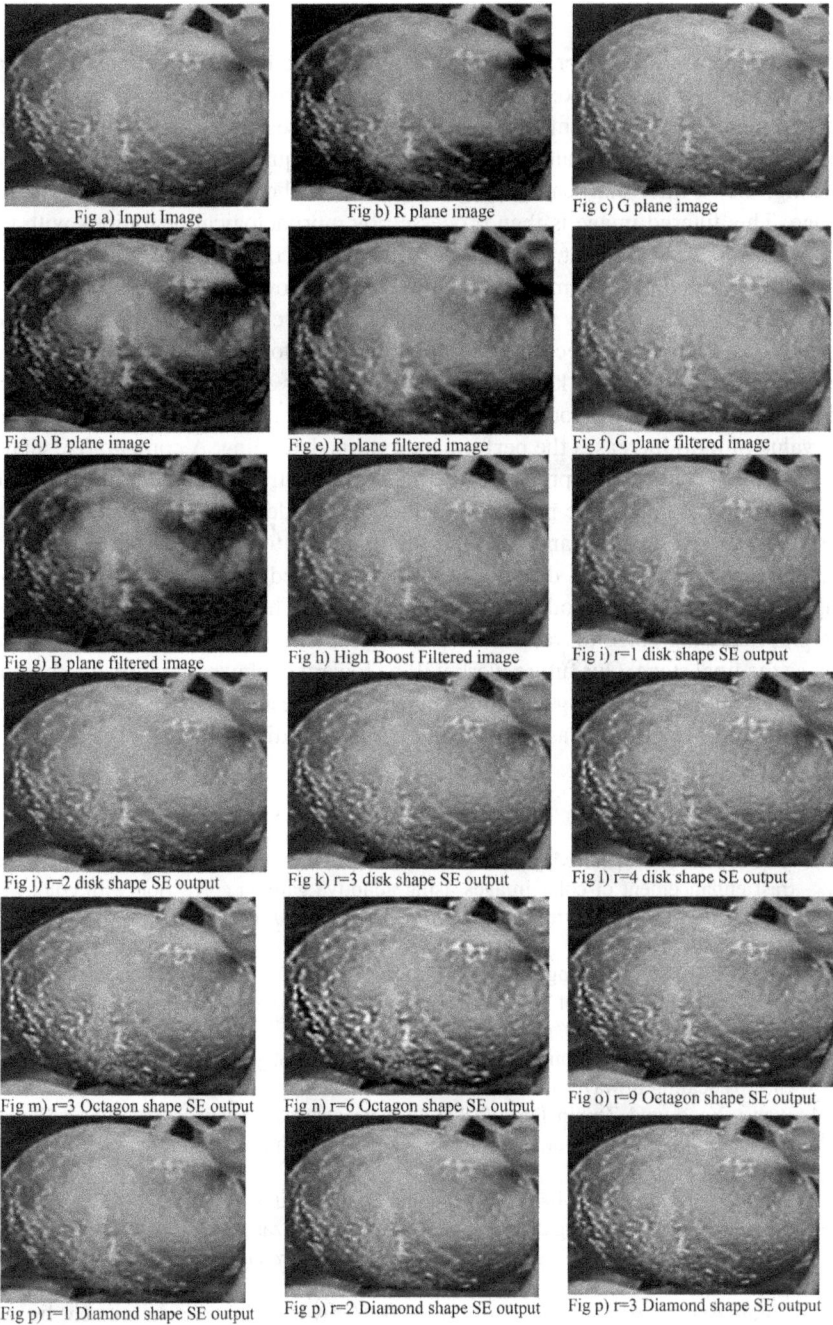

FIGURE 5.9 Experimental results for scab disease image.

5.8   CONCLUSION

This article proposed a hybrid methodology for image enhancement based on spatial domain filtering and morphological filtering in terms of denoising and contrast improvement. The algorithm was tested on citrus disease images, and a considerable improvement in the values of PSNR and other quality metrics was achieved. The high-boost filter sharpens the edges and reduces the noise of the input diseased image. This filtered image is then processed by morphological operations with the use of a structuring element. Various sizes and shapes of structuring elements were applied and tested on five images. Obtained values show that the proposed method improved the image quality significantly and can prove to be a boon for the subsequent stages of disease detection. The value of PSNR for the scab image is found to be 37.2 with a disk-shaped kernel of radii '1'. Also for different sets of radii values and shapes of kernels, it is found that the disk-shaped kernel gives best results, and as the value of radii increases the performance starts degrading. A comparative analysis was conducted for the proposed method with contrast stretching and a hybrid top-hat Gaussian filter. The proposed method has made the information content in the image more enhanced and can lead to a higher information retrieval rate. The comprehensive comparison describes that the proposed method show considerable improvement compared to other existing techniques. This method can serve well in the preprocessing stage of various disease-detection methods involving classification as the final stage. Our future work will be toward deployment of this method for preprocessing and achieving robust segmentation of lesions present in citrus fruits. This can improve overall classifier accuracy for citrus disease detection.

REFERENCES

1. Gupta, Bhupendra, and Mayank Tiwari. (2016). "Minimum mean brightness error contrast enhancement of color images using adaptive gamma correction with color preserving framework." *Optik-International Journal for Light and Electron Optics* 127, no. 4: 1671–1676.
2. Jung, Cheolkon, and Tingting Sun. (2016). "Optimized perceptual tone mapping for contrast enhancement of images." *IEEE Transactions on Circuits and Systems for Video Technology* 27, no. 6: 1161–1170.
3. Kim, Se Eun, Jong Ju Jeon and Il Kyu Eom. (2016). "Image contrast enhancement using entropy scaling in wavelet domain." *Signal Processing* 127: 1–11.
4. Wang, Yang, and Zhibin Pan. (2017). "Image contrast enhancement using adjacent-blocks-based modification for local histogram equalization." *Infrared Physics & Technology* 86: 59–65.
5. Sarangi, P. P., B. S. P. Mishra, Banshidhar Majhi, and S. Dehuri. (2014). "Gray-level image enhancement using differential evolution optimization algorithm." In *2014 international conference on signal processing and integrated networks (SPIN)*, 95–100. IEEE.
6. Gorai, Apurba, and Ashish Ghosh. (2009). "Gray-level image enhancement by particle swarm optimization." In *2009 World Congress on Nature & Biologically Inspired Computing (NaBIC)*, 72–77. IEEE.
7. Draa, Amer, and Amira Bouaziz. (2014). "An artificial bee colony algorithm for image contrast enhancement." *Swarm and Evolutionary computation* 16: 69–84.

8. Benala, Tirimula Rao, Sathya Harish Villa, Sree Durga Jampala, and Bhargavi Konathala. (2009). "A novel approach to image edge enhancement using artificial bee colony optimization algorithm for hybridized smoothening filters." In *2009 World Congress on Nature & Biologically Inspired Computing (NaBIC)*, 1071–1076. IEEE.

9. Hanmandlu, Madasu, John See and Shantaram Vasikarla. (2004). "Fuzzy edge detector using entropy optimization." In *International Conference on Information Technology: Coding and Computing, 2004. Proceedings. ITCC 2004.*, vol. 1: 665–670. IEEE.

10. Hadhoud, Mohiy M. (1999). "Image contrast enhancement using homomorphic processing and adaptive filters." In *Proceedings of the Sixteenth National Radio Science Conference. NRSC'99 (IEEE Cat. No. 99EX249)*, C5–1. IEEE.

11. Sharmila, R., and R. Uma. (2011). "A new approach to image contrast enhancement using weighted threshold histogram equalization with improved switching median filter." *International Journal of Advanced Engineering Sciences and Technologies* 7, no. 2: 208–211.

12. Greenberg, Shlomo, Mayer Aladjem and Daniel Kogan. (2002). "Fingerprint image enhancement using filtering techniques." *Real-Time Imaging* 8, no. 3: 227–236.

13. Jivet, Ioan, Alin Brindusescu and Ivan Bogdanov. (2008). "Image contrast enhancement using morphological decomposition by reconstruction." *WSEAS Trans. Cir. and Sys* 7, no. 8: 822–831.

14. Jamil, Nursuriati, Tengku Mohd Tengku Sembok and Zainab Abu Bakar. (2008). "Noise removal and enhancement of binary images using morphological operations." In *2008 International Symposium on Information Technology*, vol. 4:1–6. IEEE.

15. Srivastava, Akansha, Abhishek Raj and Vikrant Bhateja. (2011). "Combination of wavelet transform and morphological filtering for enhancement of magnetic resonance images." In *International Conference on Digital Information Processing and Communications*, 460–474. Springer, Berlin, Heidelberg.

16. Thriveni, R. (2013). "Satellite image enhancement using discrete wavelet transform and threshold decomposition driven morphological filter." In *2013 International Conference on Computer Communication and Informatics*, 1–4. IEEE.

17. Radha, R. and Bijee Lakshman. (2013). "Retinal image analysis using morphological process and clustering technique." *Signal & Image Processing* 4, no. 6: 55.

18. Ma, Hua, Gerardo Dibildox, Jyotirmoy Banerjee, Wiro Niessen, Carl Schultz, Evelyn Regar and Theo van Walsum. (2015). "Layer separation for vessel enhancement in interventional X-ray angiograms using morphological filtering and robust PCA." In *Workshop on Augmented Environments for Computer-Assisted Interventions*, 104–113. Springer, Cham.

19. Maragos, Petros. (2005) "Morphological filtering for image enhancement and feature detection." *The Image and Video Processing Handbook*, 135–156.

20. Adelmann, Holger G. (1998) "Butterworth equations for homomorphic filtering of images." *Computers in Biology and Medicine* 28, no. 2: 169–181.

21. Mitra, Sanjit K., Hui Li, I-S. Lin and T-H. Yu. (1991). "A new class of nonlinear filters for image enhancement." In *[Proceedings] ICASSP 91: 1991 International Conference on Acoustics, Speech, and Signal Processing*, 2525–2528. IEEE.

22. Cheng, Fan-Chieh, and Shih-Chia Huang. (2013). "Efficient histogram modification using bilateral Bezier curve for the contrast enhancement." *Journal of Display Technology* 9, no. 1: 44–50.

23. Vishwakarma, Virendra P., and Tripti Goel. (2019). "An efficient hybrid DWT-fuzzy filter in DCT domain based illumination normalization for face recognition." *Multimedia Tools and Applications* 78, no. 11: 15213–15233.

24. Chu, Junqiu, Haotong Ma, Ge Ren and Bo Qi. (2019). "A nonlocal filter for local denoising using the Wigner transform." *Optics and Lasers in Engineering* 122: 105–112.

25. Haralick, Robert M., and Linda G. Shapiro. (1992). *Computer and robot vision.* Vol. 1. Reading: Addison-wesley.

26. Kandhway, Pankaj, and Ashish Kumar Bhandari. (2019). "An optimal adaptive thresholding based sub-histogram equalization for brightness preserving image contrast enhancement." *Multidimensional Systems and Signal Processing,* 1–36.

27. Bhakat, Sudeshna, and Sivagami Periannan. (2019). "Brain tumor detection using cuckoo search algorithm and histogram thresholding for MR images." In *Smart Innovations in Communication and Computational Sciences,* 85–95. Springer, Singapore.

28. Trongtirakul, Thaweesak, Denis Ladyzhensky, Werapon Chiracharit and Sos Agaian. (2019). "Non-linear contrast stretching with optimizations." In *Mobile Multimedia/ Image Processing, Security, and Applications 2019,* vol. 10993: 1099303. International Society for Optics and Photonics.

29. Srinivasan, S., and N. Balram. (2006). "Adaptive contrast enhancement using local region stretching." In *Proceedings of the 9th Asian symposium on information display,* 152–155.

30. Shivhare, Priyanka, and Vinay Gupta. (2015)."Review of image segmentation techniques including pre & post processing operations." *International Journal of Engineering and Advanced Technology* 4, no. 3: 153–157.

31. Gu, Ke, Guangtao Zhai, Weisi Lin and Min Liu. (2015). "The analysis of image contrast: From quality assessment to automatic enhancement." *IEEE transactions on cybernetics* 46, no. 1: 284–297.

32. Kim, Hyoung-Joon, Jong-Myung Lee, Jin-Aeon Lee, Sang-Geun Oh and Whoi-Yul Kim. (2006). "Contrast enhancement using adaptively modified histogram equalization." In *Pacific-Rim Symposium on Image and Video Technology,* 1150–1158. Springer, Berlin, Heidelberg.

33. Singh, Munendra, Ashish Verma and Neeraj Sharma. (2017). "Bat optimization based neuron model of stochastic resonance for the enhancement of MR images." *Biocybernetics and Biomedical Engineering* 37, no. 1: 124–134.

34. Daniel, Ebenezer, and J. Anitha. (2016). "Optimum wavelet based masking for the contrast enhancement of medical images using enhanced cuckoo search algorithm." *Computers in biology and medicine* 71: 149–155.

35. Gong, Tao, Tiantian Fan, Lei Pei and Zixing Cai. (2017). "Magnetic resonance imaging-clonal selection algorithm: An intelligent adaptive enhancement of brain image with an improved immune algorithm." *Engineering Applications of Artificial Intelligence* 62: 405–411.

36. Al-Ameen, Zohair. (2016). "Visibility enhancement for images captured in dusty weather via tuned tri-threshold fuzzy intensification operators." *International Journal of Intelligent Systems and Applications* 8, no. 8: 10.

37. Srivastava, Rajeev, J. R. P. Gupta, Harish Parthasarthy and Subodh Srivastava. (2009). "PDE based unsharp masking, crispening and high boost filtering of digital images." In *International Conference on Contemporary Computing,* 8–13. Springer, Berlin, Heidelberg.

38. Anoop, B. N., Justin Joseph, J. Williams, J. Sivaraman Jayaraman, Ansa Maria Sebastian and Praveer Sihota. (2018). "A prospective case study of high boost, high frequency emphasis and two-way diffusion filters on MR images of glioblastomamultiforme." *Australasian physical & engineering sciences in medicine* 41, no. 2: 415–427.

39. Kushol, Rafsanjany, Md Raihan, Md Sirajus Salekin and A. B. M. Rahman. (2019). "Contrast Enhancement of Medical X-Ray Image Using Morphological Operators with Optimal Structuring Element." *arXiv preprint arXiv:1905.08545.*

40. Gonzalez, Rafael C., and Paul Wintz. (1977). "Digital image processing (Book)." *Reading, Mass., Addison-Wesley Publishing Co., Inc.(Applied Mathematics and Computation* 13: 451.

41. Raj, Abhishek, Akansha Srivastava and Vikrant Bhateja. (2011). "Computer aided detection of brain tumor in magnetic resonance images." *International Journal of Engineering and Technology* 3, no. 5: 523

42. Dogra, Ayush, Bhawna Goyal and Sunil Agrawal. (2017). "From multi-scale decomposition to non-multi-scale decomposition methods: A comprehensive survey of image fusion techniques and its applications." *IEEE Access* 5: 16040–16067.

43. Rauf, Hafiz Tayyab, Basharat Ali Saleem, M. Ikram Ullah Lali, Muhammad Attique Khan, Muhammad Sharif and Syed Ahmad Chan Bukhari. (2019). "A citrus fruits and leaves dataset for detection and classification of citrus diseases through machine learning." *Data in Brief* 26: 104340. https://data.mendeley.com/datasets/3f83gxmv57/2. [Accessed on 20.07.2019.]

44. Goyal, Bhawna, Ayush Dogra, Sunil Agrawal, B. S. Sohi and Apoorav Sharma. (2020). "Image denoising review: From classical to state-of-the-art approaches." *Information Fusion* 55: 220–244.

6 MOSFET-Based Low-Power Hardware Design for Autonomous Applications

Suchismita Sengupta
CMR Institute of Technology

Ananya Dastidar
College of Engineering and Technology

CONTENTS

6.1 Introduction 86
6.2 Power Consumption in MOSFET-Based Circuit Technology 87
    6.2.1 Static Power Dissipation 87
    6.2.2 Dynamic Power Consumption 87
    6.2.3 Short-Circuit Power Dissipation 89
    6.2.4 Leakage Power Consumption 89
6.3 Techniques Employed to Implement Low-Power Designs 89
    6.3.1 Voltage Scaling 90
    6.3.2 Capacitances 90
    6.3.3 Frequency of Operation 90
    6.3.4 Short-Circuit and Leakage Current 90
    6.3.5 Transistor Sizing 91
    6.3.6 Signal Gating 91
    6.3.7 Clock Gating 91
    6.3.8 Low-Power Techniques for Memory Banks 91
    6.3.9 Pipelining 92
6.4 MOS-Based Memory Design 93
    6.4.1 SRAM 93
    6.4.2 DRAM 94
6.5 Design Example: Low-Power Memory Components for a Custom Pipelined Processor 95
    6.5.1 Programme Counter (PC) 95
    6.5.2 Register File (RF) 95
    6.5.3 Instruction Memory (IM) 96

      6.5.4   Data Memory (DM)..96
      6.5.5   ALU and Control Unit (CU)...96
6.6   Summary 97
References..98

6.1 INTRODUCTION

Innovation drives the integrated circuit design industry, and in order to develop autonomous and intelligent designs, circuit design engineers are employing cutting-edge technologies such as reconfigurable computing and artificial intelligence. This design engineering has paved the way for autonomous designs which, with the technique of artificial intelligence, enable the devices around us to connect and adapt to the digital ecosystem with ease and efficiency. Design of autonomous systems demands intelligent transformation of the traditional devices by using the concept of parallel design, or co-design. Intelligent hardware must cater to emerging applications that must be expandable as well as easy to use. One important hardware that is omnipresent in any autonomous system is the Memory component is one of the most important part of any autonomous system design

With the increasing demand for storage in electronic devices, owing to the heavy data transfer, including the huge amount of data processing, engineers around the world have been working towards designing scaled-down memory elements capable of increased storage for high-speed applications with reduced area requirements. The trade-off in this type of design is the increase in power consumption, which leads to other issues such as heating effects, increased battery back-up and reduced efficiency. Different Scaling techniques does help conform to Moore's second law, but with it comes the power-hungry devices and gadgets. When the dimensions of metal oxide semiconductor field effect transistor (MOSFET) devices are scaled below 22 nm, we see overshooting of leakage power over dynamic power, which is a design concern for a very large scale integration (VLSI) designer. So, design and optimization of MOS-based devices are of utmost importance.

Intelligent hardware design caters to real-time information acquisition and response and is usually a combination of both software and hardware. It is distinguished from non-intelligent counterparts based on the computational capacity. An intelligent device may be a stand-alone element, such as a medical diagnostic wearable or handheld module or a computer system capable of performing autonomous processing. Various parts of such an intelligent hardware include a sensing component and a controlling element with some inbuilt memory. The important part of the controller is the processor, which is the most complex and power-thirsty component, but in order to cater to the demands of the current age wireless applications, there is a requirement for low-power solutions that will provide long product lifetimes. The commercial success of autonomous systems depends on the duration for which the autonomous operation can run reliably.

To satisfy this need of the industry, VLSI engineers have used many design techniques to implement low-power designs that may or may not have a performance trade-off. One of the techniques of implementation of autonomous application is the use of system on chip (SoC), the fundamental components of which are capable of mixed signal processing with suitable storage and radio frequency (RF) functionalities [1]. In order to satisfy these capabilities, a suitable memory design is necessary for the SoC that can handle the autonomous and intelligent functionalities with low-power consumption.

This chapter will highlight the design and implementation of a low-power memory bank and its application in high-speed integrated circuit design. It starts the various power consumption associated with the MOSFET-based designs and subsequently the various MOSFET-based techniques employed to implement low-power designs. The next section will elaborate on various memory implementations using MOSFET for low-power applications. This would be followed by the design and implementation of a MOSFET-based memory bank using the Cadence design suite (45 nm). The chapter will end by summarizing the limitations and applications of the MOSFET-based low-power designs.

6.2 POWER CONSUMPTION IN MOSFET-BASED CIRCUIT TECHNOLOGY

MOS devices are still ruling the industry today for the design of memory and microcomponents owing to their excellent performance and standardized assembly line and literature. Fundamentals of MOSFET devices and their characteristics can be found in various literature [2]. Complementary MOSFET (CMOS) is a derivative of the MOS device where an n and a p enhancement-mode MOS device are connected in series to achieve excellent device performance with almost negligible static power dissipation. Pass transistor logic (PTL) transmission makes use of the principle that while an n type device allows a strong logic '0' to pass, the p type device allows a strong logic '1' to pass. The respective n and p MOSFETs cannot pass a strong '1' and a strong '0', and this complexity of PTL is overcome by the Complementary Pass Transistor Logic (CPL) by making use of both true and complementary signals as its inputs to get a true and complementary form of the output functions. CPL has reduced parasitic capacitance and higher speed of operation over a complete CMOS-based circuit. Power dissipation can be broadly categorized, in CMOS circuits, as *static* and *dynamic* power dissipation.

6.2.1 STATIC POWER DISSIPATION

Static or Direct Current (DC) power dissipation, which is a measure of battery life of circuits, is the product of the power supply voltage and the amount of current flowing between the power rails during the idle mode of operation.

$$P_{static} = f(I_{static}, V_{DD})$$

In MOS circuits using depletion type device or resistive loads, the P_{static} is non-zero, while for a CMOS-based circuit this component of power consumption has been found to be negligible.

6.2.2 DYNAMIC POWER CONSUMPTION

Dynamic power is the power consumed due to switching activities or when the circuit makes a transition from one state to another; so it is also referred to as switching power dissipation. The leading source of dynamic power dissipation are – charging and discharging of the capacitors associated with the input of the driving gates, interconnect and the output node of the gate. Where oxide capacitance is associated with the driving gates and junction parasitic capacitances are associated with output node of the gate. This power consumption component is dependent on the amount of energy dissipated in charging and discharging the parasitic capacitances and is independent of the rise and fall time of the circuit. The parasitics are unavoidable, and hence proper analysis and optimization are required, failing which would lead to spikes in power and timing considerations of the circuit. This will ultimately degrade the performance metrics of the design.

$$P_{dyn} = \alpha_T \times C \times V_{DD}^2 \times f$$

where,

P_{dyn} = dynamic power consumed,

α_T = node transition factor

C = capacitance,

V_{DD} = supply voltage and

f = frequency of operation.

The frequency of operation of the given circuit is calculated as follows

$$f = \frac{1}{T}$$

where,

f = frequency of operation and

T = Time period of the circuit.

The node transition factor is a measure of the number of power-hungry transitions carried out per clock cycle. The power consumption due to the switching component of the total power can be reduced by reducing the supply voltage which is inherent to low-power designs. But we must keep in mind the delay and signal transmission to peripheral circuits. Another method is to reduce the load capacitances that can be achieved by proper transistor sizing. Reducing this switching component of the dynamic power can also bring down the switching power.

. Power delay product (PDP) is the *energy metric* and is not concern with the speed of computation. On the other hand, energy delay product (EDP) is the product of average energy and the time required for computation. Hence, EDP is the *performance metric* [3].

$$PDP = \frac{Average\ Energy}{No.\ of\ Transitions\ the\ Clock\ makes}$$

or, $PDP = (Average\ Power\ Consumed) \times (Average\ Delay\ of\ the\ circuit)$

$EDP = PDP \times Delay$

6.2.3   SHORT-CIRCUIT POWER DISSIPATION

Apart from static and dynamic power, *short-circuit power dissipation* also plays a role in digital circuits. The short-circuit power dissipation occurs during the switching of the signal, i.e., when the signal itself is going through a transition from one logic state to another or an event, but unlike the switching power component it does not depend on the energy used to charge up or discharge the capacitance but on the rise and fall time of the circuit. During switching activity, when the n net and the p net may conduct simultaneously for some time, this triggers a short-circuit current from the power supply, V_{DD}, to ground (GND) which leads to power dissipation as heat via the MOSFETs involved. This again is unavoidable, as a signal will switch between states. The short-circuit power component is a function of the rise time/fall time and transconductance (k) of the device.

$$P_{sc} = f(\tau, k)$$

Implementation of a capacitive load does lower the short-circuit current, but again the combined effect of short-circuit current and capacitor current has an alarming effect on power dissipation. Again, utilization of a faster signal will not help the cause. Therefore, to balance out short-circuit power dissipation, the designer has to minimize capacitance and use an input signal whose time period is in the range of the transistor's delay.

6.2.4   LEAKAGE POWER CONSUMPTION

Although n type and p type MOS devices are devoid of leakage current, a circuit containing millions of transistors are affected by leakage current and its effects. Leakage current in MOS-based circuits is contributed by a reverse-biased PN junction and sub-threshold conduction, and both have a positive temperature coefficient and may occur even during standby mode. Even though leakage current is a demerit, some researches have used it in power-on reset signal generation.

Because of the unpredictable nature of leakage current, it becomes near impossible to optimize it. One way to limit it may be to reduce the threshold voltage of the device with a speed trade-off. For devices that fall under the 22 nm technology and above, dynamic power is a major concern but for the devices that fall under the 22nm technology and below, the severity of static power dissipation prevalent.. In this category of devices, static power dissipation has an overshoot over dynamic power. Nearly every high-end digital device, nowadays, is below 22 nm technology, for example, the Qualcomm Snapdragon Processors – 845, 855, 855+ and 865, to name a few – do fall in this category.

6.3   TECHNIQUES EMPLOYED TO IMPLEMENT LOW-POWER DESIGNS

In order to curb power dissipation, a designer has to decide whether to conserve or to look for a trade-off technique, keeping in mind the design constraints and specifications of the device given.

6.3.1 VOLTAGE SCALING

Scaling the power supply would significantly reduce the overall power dissipation, which can be best understood from the quadratic dependence of P_{dyn} on the power supply voltage. But excessive scaling of V_{DD} would lead to loss of performance of the device, as the transistors will become slow with reduced noise immunity and signal integrity. Since the supply voltage is inversely proportional to propagation delay, scaling down the V_{DD} will lead to an increase in the delay of the circuit.

This may be compensated by reducing the threshold voltage of the device, but there may be a trade-off with subthreshold leakage current. This can be remedied by the use of variable-threshold CMOS (VTCMOS) circuits, which make use of adjustable substrate bias, or multiple-threshold CMOS (MTCMOS) circuits, which employ the multiple V_t (using a n type and p type device) where a low V_t will allow high switching speed and high V_t device will isolate the device during standby mode thereby reducing leakage power dissipation. Scaling of power supply, V_{DD}, can reduce power dissipation drastically, but after a point it will cease to cast an effect and it will also increase the delay of the total circuit as transistors will fail to respond to scaled V_{DD}. This is because the threshold voltage, V_t, doesn't get scaled down with V_{DD}. But if V_t of the transistor is reduced then it enhances the overall performance of the circuit at hand. Therefore by scaling down both V_t and V_{DD}, power dissipation and delay of the circuit are reduced. Scaling of V_t as well should be limited as it would lead to an exponential rise in subthreshold leakage, which reduces the battery life of the device.

6.3.2 CAPACITANCES

Minimization of not just capacitive loads but also parasitic capacitances reduces overall power dissipation as all the major sources of power dissipation depend on capacitances as discussed in the earlier section. But minimizing capacitances also has a limitation on its effect on power dissipation because, after a certain point, power dissipation cannot be reduced by minimizing capacitances.

6.3.3 FREQUENCY OF OPERATION

The direct dependency of power on frequency has been discussed in the earlier sections. But frequency has interdependency on capacitance too. Therefore minimizing frequency (clock frequency and switching frequency) has limited effects on the reduction of power dissipation, and that may gain affect the frequency-capacitance trade-off.

6.3.4 SHORT-CIRCUIT AND LEAKAGE CURRENT

Leakage currents are unpredictable and have a positive temperature coefficient, which makes it difficult to optimize. Occurrence of leakage power is seen mostly in ultra-low frequency or when the device is in its sleep mode. To prevent leakage power, reduction schemes are applied at the design stage. Memory banks are more prone to leakage power dissipation due to high-density architecture.

6.3.5 TRANSISTOR SIZING

The gate area given by the product of the width and length of the channel of the device is scaled in order to reduce the chip area of a circuit. Keeping the aspect ratio $\left(\frac{W}{L}\right)$ of a transistor close to unity helps to achieve this feature, but there is a trade-off with other design criteria, such as noise margins, current driving capability and switching speed. One of the methods of reducing the chip area can be to replace spiral resistors structures with areas of undoped polysilicon to realize resistors on-chip. Usually transistor sizing is done in circuits where primary emphasis may be on low power but not on the other transistor performance characteristics.

6.3.6 SIGNAL GATING

To put it in simple terms, signal gating is a masking technique that prevents redundant switching activities. As stated earlier, the biggest reason for power dissipation is switching activities. Therefore, if the redundant switching can be masked, then curbing power issues becomes easier. Signals can be masked by implementing AND/OR gates or latches/flip flops or transmission gates/tristate buffers. For these circuits to act as masking devices, a control input must be provided that indicates whether the signal should propagate. Also by adding these circuits to prevent redundant switching, the area-delay trade-off suffers.

6.3.7 CLOCK GATING

In high-end devices, the clock signal is another source of power dissipation. This is because, the clock signal has to be fed to all the sequential components spread across the integrated circuit, and hence, the clock has to drive a large load. Therefore the clock signal alone takes up 40% of the total power dissipation. Similar to the signal gating, the clock signal gets masked by either a NAND or NOR gate. The power dissipation improves but at the cost of area-delay trade-off.

6.3.8 LOW-POWER TECHNIQUES FOR MEMORY BANKS

Low-power techniques involve two of the most important parameters related to memory devices – standby power reduction and operating power reduction – and are totally dependent on the application. Generally, memory units stay in hibernation, and therefore standby power is of minimal importance. One way of dealing with power dissipation is to actively change the threshold voltage of the memory unit – when active, the threshold voltage is reduced for timing considerations and when inactive, threshold voltage is increased to prevent static power overshoot.

Clock Cycle	1	2	3	4	5	6	7
Instruction #				Pipeline Stage			
1	FTH	DEC	EXE	MA	WBK		
2		FTH	DEC	EXE	MA	WBK	
3			FTH	DEC	EXE	MA	WBK
4				FTH	DEC	EXE	MA
5					FTH	DEC	EXE

FIGURE 6.1   Stages of pipelining in a processor.

6.3.9 PIPELINING

The concept of pipelining is used to increase the speed of operation. Pipelining is the process that incorporates parallelism within a processor, which allows faster throughput. Pipelining breaks down the whole process into smaller modules. Multiple tasks that use different resources operate simultaneously. It helps throughput of the entire workload. The slowest stage decides the rate. A pipelined processor doesn't wait for the results from the previous operation. Pipelining has five stages it follows one step per stage. They are instruction fetch (FTH), instruction decoding (DEC) of the fetched instruction, execution (EXE) of the instruction, memory access (MA) to get the operand, and storing the result in a specific register by the process of Write Back (WBK) to the memory. This is illustrated in Figure 6.1.

In Figure 6.1 we see that at the fifth clock cycle, when the first instruction has been executed simultaneously, the second instruction is accessing the operands from the memory, the third instruction is being executed, the fourth instruction is being decoded and the fifth instruction is being fetched. So multiple operation are taking place in the processor in a parallel manner, thereby increasing the speed of completion of an operation. The hardware implementation of a pipelined processor can be best understood from Figure 6.2.

A pipelined processor has two end points (input and output), within which the five stages are interconnected. Intermediate registers (IR) are used to store the immediate results of the stages. These registers are called latches or buffers. All the stages and the intermediate registers are controlled by a single clock. Here IR1 stores the fetched instruction code; IR2 holds any implicit operand after the instruction code

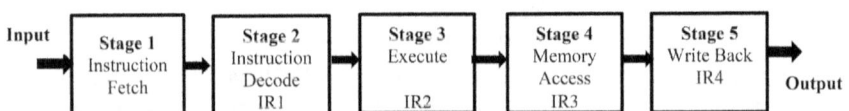

FIGURE 6.2   Stages of pipelining with intermediate registers.

has been decoded; IR3 stores any operand; and IR4 holds the results before being sent for storage in destined registers or memory.

6.4 MOS-BASED MEMORY DESIGN

Memory is the unit that is used for storing data. Memory can classified by its capability of modification. That is, if it has the capability of modifying the stored data, then it's a random-access memory (RAM), and if it doesn't allow modification to the stored data, then it's a read-only memory (ROM). Therefore, RAM has 'read' as well as 'write' modes, but ROM has only a 'read' mode. RAM can be classified as static random-access memory (SRAM) or dynamic random-access memory (DRAM).

6.4.1 SRAM

A SRAM cell is composed of a cross-coupled inverter using four transistors which is used for latch operation, while the rows and column access can be done by two more transistors. So a six-transistor memory cell is sufficient to function as a SRAM. Further implementation of accessing different register ports can be done by including additional transistors. Use of MOS technology in the SRAM implementation ensures low-power designs, which in turn reduce the heat dissipation of the circuit. Figure 6.3 shows an eight-transistor SRAM cell.

The 8T SRAM has two read-write (2WR) while 6T SRAM has only one read-write (1WR) That is, 6T SRAM performs either a 'read' operation or a 'write' operation at a given time, whereas the 8T SRAM can perform 'read' or 'write' independently [3, 4].

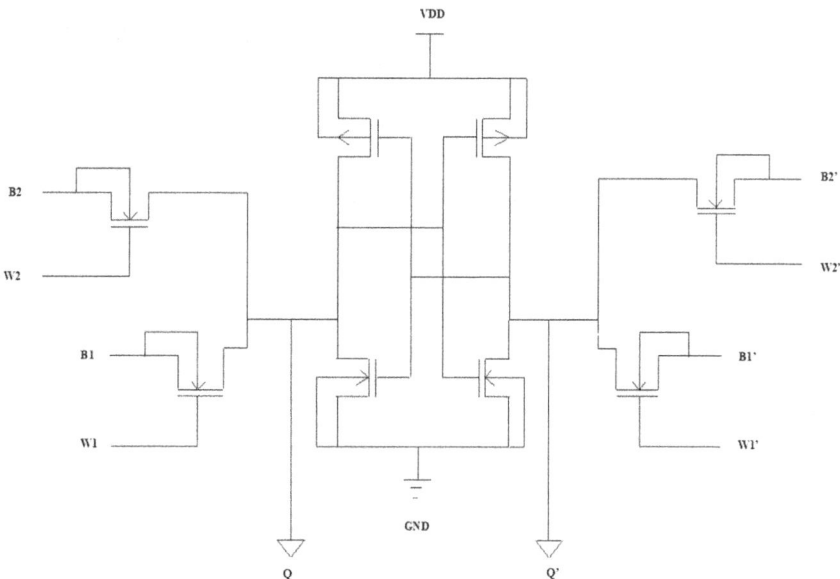

FIGURE 6.3 Cadence simulated 8t static random-access memory cell.

TABLE 6.1

Operation of an SRAM

W1	W2	Operation
0	0	Read
0	1	B2 selected for Write
1	0	B1 selected for Write
1	1	AND operation on B1 and B2

The 8T SRAM stores 2-bit information and is much more stable and flexible compared to a 6T SRAM. Figure 6.3 depicts a 'memory cell' in which the cells are combined and placed in 'rows' and 'columns' to form a memory bank. Each cell stores data and together they form a memory bank for bigger circuits, such as a processor. Each memory cell has a common 'row' and a common 'column' connection to its adjoining cells. The rows are known as 'word lines' and the columns are known as 'bit lines'. The output from the SRAM cell is obtained either as a true or its complementary form. Simulation of different SRAM structures can be seen in [5].

The different operations of the RAM can be seen in Table 6.1, where, depending on the value of the write lines, either B1 or B2 is used to perform the write operations. When both W1 and W2 are zero, it represents a memory read operation.

6.4.2 DRAM

The Dynamic RAM is also based on a MOS transistor which, with the addition of a capacitor (which acts as the memory cell), determines whether a logic '0' or a logic '1' is stored, depending upon the absence or presence of charge in the capacitor respectively as seen in Figure 6.4. The transistor count of the DRAM cell is less than that of the SRAM cell, and so it allows for higher packing density in memory modules. Due to the presence of the capacitor, which is prone to leakage, periodic refreshing of the circuit is necessary; this makes the DRAM slower compared to

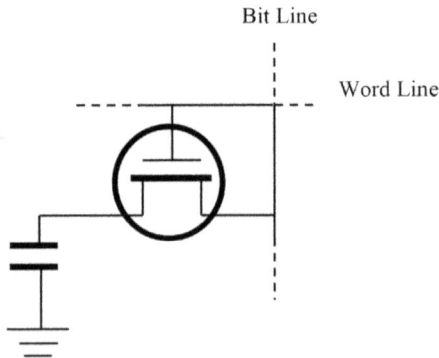

FIGURE 6.4   1T DRAM cell.

the SRAM cell. Sense amplifiers are also used to sense the value and write it into the memory cell. The word line and the bit line are used to read and write the data by accessing the row and column transistor, respectively. A DRAM array may be composed of millions of such cells, which increases the power density of the circuit.

SRAMs are faster than the fastest DRAMs, as SRAMs don't require 'refresh operations' as the DRAMs. Moreover, SRAMs are apt for low-power applications. Though SRAMs are costlier than DRAMs, they provide a better performance than DRAM. SRAM is not prone to noise and leakage current as DRAM and SRAMs are highly customizable, while DRAMs are not.

6.5 DESIGN EXAMPLE: LOW-POWER MEMORY COMPONENTS FOR A CUSTOM PIPELINED PROCESSOR

This section presents the design of a custom pipelined processor for real-time low-power applications based on complementary static CMOS design style due to its robustness, high immunity to noise, lower static power dissipation, high degree of automation, hassle-free design process and its amenability to voltage scaling [4].The pipelined processor has various components, such as the programme counter, register file, instruction memory, data memory, arithmetic logic unit (ALU), ALU control unit and control unit. The processor uses Harvard architecture that implements separate memory for data and instructions. This architecture is fast compared to others, and it has two separate memories for instructions and data. Therefore, there are two separate address/data buses between the control unit and the two memories. It has greater and more predictable bandwidth of memory and is effective due to its concurrency feature, i.e. its ability to read an instruction and perform data memory access concurrently with high speed.

6.5.1 PROGRAMME COUNTER (PC)

A PC is a register that stores the address of the memory location of the instruction to be fetched next. A PC automatically gets incremented so as to fetch the next instruction after executing the current instruction. PA C has no internal address, and it is the only register to be so. An n-bit PC can address up to 2^n bytes of memory.

6.5.2 REGISTER FILE (RF)

A RF is a bundle of registers and is a unit of array of memory cells that are used for the purpose of storage and computing buffers. The RF and the data memory (DM) together send the operands to the ALU for execution, and the resultants are then stored in the RF. The interface between the ALU and the RF comprises two terminals or ports, i.e., read port and write port, and a data bus. The RF implemented in the design has two read ports and one write port; therefore it allows the ALU to access two registers simultaneously.

6.5.3 INSTRUCTION MEMORY (IM)

The IM is an important part of the processor design as it contains all the op-codes (operation codes, or instruction). It is the memory that is called upon during the 'fetch' instruction. It stores the instruction that is currently being decoded or executed. The instruction to be executed is loaded into the IM, which stores it while it gets decoded, prepared and ultimately executed. It is a read-only at user level. This is where the instructions or applications are stored. The operation of the IM can simply be stated as each clock (CLK) cycle fetches the instruction from the address specified by the PC and increments PC by four at the same time.

6.5.4 DATA MEMORY (DM)

Another important memory component of the processor is the data memory (DM). The DM is the memory used to store variables temporarily. This consists of Rx space, I/O memory, extended I/O memory and internal SRAM. The Rx space comprises general purpose registers which have the fastest link to ALU operation. The I/O memory has addresses for the peripheral functions. Internal SRAM is for temporary storage and keeping intermediate results. DM has both read-write modes.

6.5.5 ALU AND CONTROL UNIT (CU)

The custom ALU design requires a control unit of its own that specifies what operation the ALU performs. It is a control unit associated with ALU, and it generates the appropriate control signals. The CU is the co-ordinator of the processor, and it handles the data transfer between the registers; the A-C handles the instructions for ALU. The CU controls the whole processor. The CU receives its instructions stored in memory. The CU performs fetching instructions, sending instructions to the ALU, receiving and transmitting results to memory. The data processing unit (DPU) is the unit that performs all the operations under the command of the CU. It implements the fetch-decode-execute cycle. The DPU performs the arithmetic operations and stores data. It processes data and addresses in the processor. The DPU is a combination of several functional units and, with the CU, forms the processor.

The design steps in implementing a pipelined processor can be seen in [5]. In order to implement the custom ALU (say, a 16-function ALU) we need to optimize the designs of the Adder, Subtractor, Multiplier, Divider, Incrementer, Decrementer, Universal Shift Register [6] and Magnitude Comparator circuits needed for it to complete the Arithmetic Block [7]. The implementation of the Logical Block of the ALU must be done by selecting suitable values of the transistor parameters so as to maintain the correct operation and minimizing the trade-off amidst timing, area and power [8].

A processor is not a stand-alone unit and needs to access external memory in some cases to enhance its capacity. A SRAM array based on the 8T SRAM or a DRAM-based memory module completes the design example. The implementation of each part of the processor can be carried out using various CAD tools from Cadence, such as Incisive, Encounter RTL Compiler, Encounter and Virtuoso.

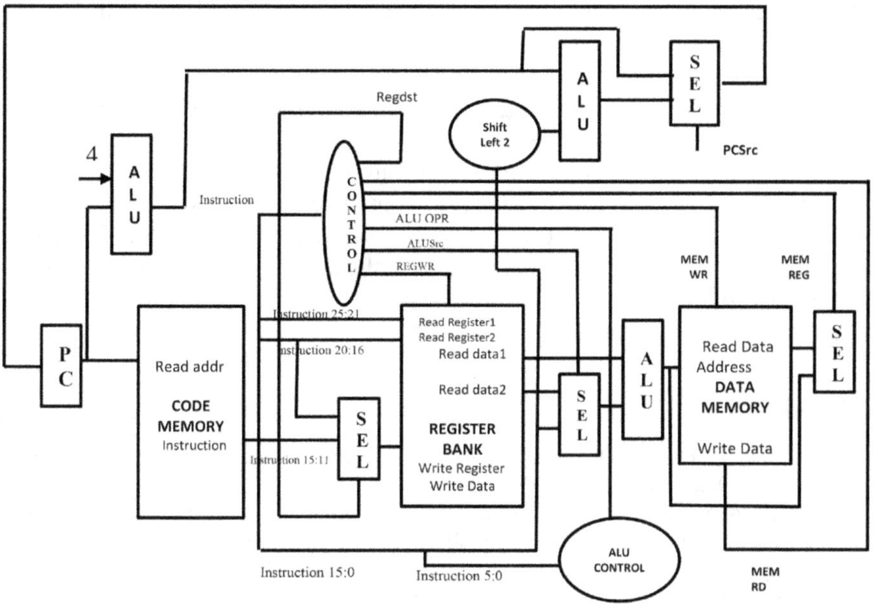

FIGURE 6.5   A pipelined processor plan.

Different configuration of the memory can be implemented with various low-power techniques that can be used for proper functioning of the pipelined processor.. Figure 6.5 shows the simple block schematic plan for a pipelined processor, with its various components, that allows parallelism within a processor, thereby facilitating faster throughput by breaking down the whole process into smaller modules. Here multiple tasks operate simultaneously by using different resources, which in turn helps throughput of the entire workload. The slowest stage decides the speed of operation of the processor, which in this case doesn't wait for the results from the previous stage. This design feature limits the number of interactions with the main memory and increases the overall speed of the circuit.

Autonomous and intelligent applications need to be independent from the human supervision; therefore, they need to be highly efficient and transparent in decision making. These systems need to be adaptive to all scenarios and learn to make a better decision that enhances the necessary trust factor from the users. Systems such as self-driving cars, UAVs, smart manufacturing robots and autonomous intelligent vehicles (AIVs) need to be designed accurately as part of Industry 4.0 [9].

6.6   SUMMARY

Use of MOS-based devices for integrated circuit design still rules the industry owing to their standardized process and design. Scaling of the device is one of the contributors in low-power design and, along with techniques such as LOCOS, U-groove, polysilicon gate, 3-D DRAM cells amongst others, design of low-power processors have gained momentum. Development of high-density memory units require highly efficient techniques and maintenance of the correct functionalities of SRAM and DRAM cells; implementing voltage scaling and threshold voltage scaling is an uphill task and might lead to soft errors. The designer would need to keep the architecture, external circuitry, signal-to-noise ratio and efficiency of the memory unit in mind while designing a memory module. In VLSI, the rate with which electrical energy is taken from the power supply and converted into heat is power dissipation and implementing a low-power technique to optimize the power dissipation leads to disturbances in the trade-off between other performance metrics. Any low-power techniques that were discussed earlier are not the end-all technique that a designer can blindly follow in order to optimize the design. Therefore, the designer's discretion is of utmost importance based on the low-power figure merits.

REFERENCES

1. Giraud B., Thomas O., Amara A. et al. 2009. SRAM Circuit Design. In Amara A., Rozeau O. (eds) *Planar Double-Gate Transistor.* Springer.
2. Sedra, A. S. and Smith, K. C. 2014. *Microelectronic Circuits.* Oxford University Press.
3. Rabaey, J.B., Chandrakasan, A., Nikolic, B. 2016. *Digital Integrated Circuits: A Design Perspective.* Pearson Education.
4. Dejan, M., Brodersen, R.W. 2012. *DSP Architecture Design Essentials.* Springer.
5. Sengupta, S. 2017a. Design and Implementation of Ultra-Low Power Custom 16-bit Multistage Pipeline Processor for High Speed Applications. M.Tech Thes, *Biju Patnaik University of Technology*, India.
6. Sengupta, S. and Dastidar, A. 2017b. Evaluation of the Performance Metrics of a 4-bit Universal Shift Register and its Implementation in an Arithmetic & Logic Unit. International Conference on Smart Technologies for Smart Nation (SmartTechCon 2017). IEEE. India.
7. Sengupta, S. and Dastidar, A. 2016. A Comparative Analysis of an UltraLow Voltage 1-bit Full Subtractor Designed in both Digital and Analog Environments. Proceedings of International Interdisciplinary Conference on Engineering, Science & Management. India.
8. Sengupta, S., Sarkar, P. and Dastidar, A. 2020. Design of a 4 bit Arithmetic & Logic Unit, Evaluation of its Performance Metrics & its Implementation in a Processor. International Conference for Emerging Technology (INCET 2020), IEEE, 2020. India.
9. Lynch, L., McGuinness, F., Clifford, J. et al. (2019). Integration of Autonomous Intelligent Vehicles into Manufacturing Environments: Challenges. International Conference on Flexible Automation and Intelligent Manufacturing (FAIM2019), Elsevier.

7 Performance Analysis on Low-Power, Low-Offset, High-Speed Comparator for High-Speed ADC
A Review

Krishan Mehra, Tripti Sharma and Simran Somal
Chandigarh University

CONTENTS

7.1 Introduction 99
    7.1.1 Comparator design and Operation 100
7.2 Classification of Comparator Architectures 102
    7.2.1 Static Comparator 102
    7.2.2 Dynamic Comparator 102
        7.2.2.1 Introduction to Latch Circuit 103
        7.2.2.2 Types of Dynamic Comparators 104
7.3 Design Metrics and Performance Analysis 105
    7.3.1 Offset Voltage 106
    7.3.2 Delay 107
    7.3.3 Area 107
    7.3.4 Power Consumption 107
Conclusion 109
References 109

7.1 INTRODUCTION

Nowadays low-power high-speed analog-to-digital converters (ADCs) are the most important part of the numerous handheld appliances. The comparator is a basic construction block for a different kind of ADC such as successive approximation register (SAR), pipelined, flash ADC, etc. The comparator is an electronic device that generates an output voltage or current whenever there is an alteration in magnitude between two signals. The comparator estimates the minute voltage difference at the input of an ADC. To change the continuous (analog) signal into its digital equivalent, it is essential to first sample the input of the analog-to-digital converter and then apply that to a group of converters. The comparators with optimum performance are widely preferred to various applications that require recovery of digital information from an analog signal, such as radio-frequency integrated circuit (RFIC), as a sense amplifier in memory circuits. The decision-making process of a comparator defines the speed of the comparator. In much of the literature, a comparator is defined as 1 bit ADC. Several years ago, an operational amplifier with very high gain was used as a comparator. But the use of this type of comparator lowers the speed of ADCs and also contributes to more power consumption [1].

7.1.1 COMPARATOR DESIGN AND OPERATION

The concept of the dynamic comparator was introduced to overcome the drawbacks of the static comparator, or simply operational amplifier-based comparators. The dynamic comparator operates at a lower supply voltage in comparison to a static comparator because of less stacking, and it reduces the overall power consumption. But the literature focuses on various parameters which limit the performance of dynamic comparators, such as large offset and delay, and high power consumption. This literature survey includes all these parameters and presents the various designs to overcome the disadvantage of all these parameters.

The input offset voltage is limiting the resolution of the comparator. The main source of the large input offset voltage is parameter variations. Using a linear amplifier, the offset cancellation with the medium resolution is achieved by applying any offset cancellation technique and offset cancellation for the latch stage of complementary metal oxide semiconductor (CMOS) comparator by relaxing the gain of the preamplifier stage. By improving the input-referred offset by an offset cancellation technique, a comparator with high resolution is achieved [2, 3]. A comparator circuit with low input offset voltage without any offset cancellation scheme is proposed with accurate clocking and also achieves low delay and low power consumption [4].

In a dynamic comparator there are two types of mismatch: static mismatch and dynamic mismatch. Static mismatch is due to misalliance in μnCox and the threshold voltage Vth. Dynamic mismatch is due to the inequality in load capacitances [4]. By increased conversion gain between the latch stage and amplification stage, and with reduced bias current, there is a reduction in offset voltage and overall power consumption [5]. In a fully differential dynamic comparator in which the preamplifier stage is separated from the latch stage and both stages use different clocks, this separation between these two stages helps in reducing the mismatch between the input transistor and hence reduce the offset voltage [6]. By tuning the body voltage, minimum noise and offset error is achieved [7]. The large common-mode input voltage improves the delay of the comparator, and it will lead to an increase in the speed of the comparator [8]. The large offset voltage of a comparator lowers the performance of the analog-to-digital converter. The dynamic latch comparator is more sensitive to load capacitance mismatch in comparison to static latch comparator [9].

The authors in Reference [10] propose a method to decrease power absorption by stopping the preamplifier stage. When there is sufficient differential voltage to activate the latch circuit, a two-stage dynamic comparator with XOR gate is used to stop the amplification phase and improve the performance of the comparator. The authors in Reference [11] achieve low-voltage operation for the hybrid comparator by using three stacked transistors; the dynamic amplification stage uses a p-type metal-oxide-semiconductor (PMOS) transistor at the input to enhance the positive feedback and to decrease the common-mode voltage; and the latch stage is quasi dynamic to reduce the latching time and decrease the overall power consumption. To decrease the power and improve the speed, the authors in Reference [12] propose a double-tail comparator with two extra transistors used in parallel to both of the tail transistors. The main motive to use these transistors is to decrease the time for which the circuit is inactive mode. To reduce the wastage of excess power and increase the resolution a dynamic comparator is proposed; when the comparator is near the finish of the comparison phase, there is no need for preamplification gain.

To stop the wastage of excess power, some extra circuitry is needed; a dynamic comparator with two PMOS transistors is used in series with V_{DD} and tail transistor at the end of the evaluation phase; one of these transistors turns on with V_{DD} and stops the current source of the preamplifier stage [13]. A comparator with 400µV resolution for high-speed SAR ADC is used in this configuration; two transistors are added, with two input terminals to reduce the kickback noise [14]. A two-stage dynamic comparator with forwarding body bias is designed for ultra-low power, in the preamplifier stage; the input PMOS transistor is body biased with the ground. And in the latch stage, the input negative-channel metal-oxide semiconductor (NMOS) transistors are body biased with supply voltage (V_{DD}). This forward body bias scheme also reduces the delay time [15]. To achieve low power and low delay, there is a shared charge technique in which a transistor is used to share the charge between the output nodes. Due to this charge sharing, output will not reach the threshold voltage, and this will lead to an increase in the speed of comparison [16].

The authors in Reference [17] propose a comparator with the bulk-driven load to increase the gain of amplification stage, which also causes an increase in speed, and an optimization technique for offset voltage and kickback noise with two extra circuits that are kickback noise control unit (KCU) and offset control unit (OCU). Two signals are generated during the simulation of the offset voltage: $V_{OS, R}$ $V_{OS, F}$. If these two signals are not equal, then the comparator experiences hysteresis. Due to inaccuracy in the SAR technique, a new resettable successive approximation register (RSAR) technique has been proposed [18]. In RSAR technique, the comparator is reset by applying predefined input before applying the next stage voltage and due to this correct value for the offset voltage, a rise and a fall is achieved. A new latch stage was proposed with enhanced total effective transconductance to accomplish high speed operation. The latch transistors are self-biased to improve the delay time of the comparator [19].

The authors in Reference [20] analyze offset voltage and noise for a regenerative dynamic comparator. To improve the headroom and accuracy with which a comparator makes decisions, a differential double-tail dynamic comparator has been proposed [21]. Fully differential double tail dynamic comparator (FDDTDC) improves the propagation delay in comparison to conventional double tail dynamic comparator (DTDC). The authors in Reference [22] propose removing large stacking of transistors to precharge the internal nodes of comparator with a single-tail transistor and due to removing the two clocked transistor power consumptions; delay is improved significantly.

7.2 CLASSIFICATION OF COMPARATOR ARCHITECTURES

The comparators can be classified into two categories depending on their architectures:

   i. Static comparator
  ii. Dynamic comparator

7.2.1 STATIC COMPARATOR

A static comparator is a very simple device which performs the comparison between two signals based on threshold detection of the input and reference signals. A static comparator is not time operated or by any clock signal. A static comparator is always enabled. It is constantly comparing two signals. A static comparator is similar to an operational amplifier with compensation [23]. The circuit of the static comparator is very simple, but it is always enabled because it has more power consumption and also suffers from inherent speed; for this reason static comparator is not used in the real world for high-speed ADCs and other comparator applications. Figure 7.1 shows a circuit diagram for the static comparator.

7.2.2 DYNAMIC COMPARATOR

A dynamic comparator enhances the speed and decreases the power utilization of the static comparator. To attain high-speed ADCs, high-resolution, low-offset

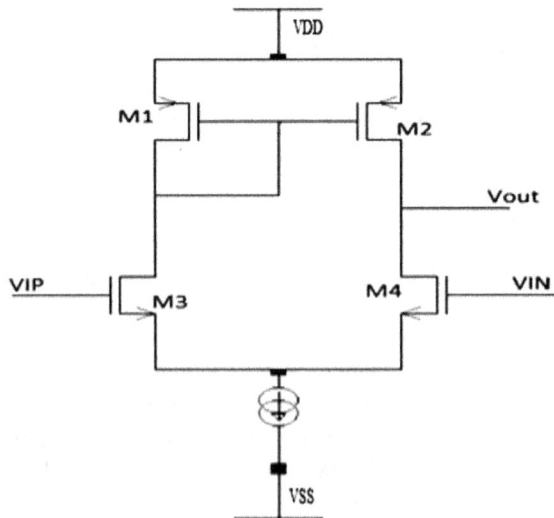

FIGURE 7.1   Static comparator [3].

dynamic comparators are preferred. The comparator is the heart of different types of ADCs. Several comparators are used in a single ADC to translate the analog signal into its digital equivalent. Therefore, a high-speed comparator for the ADC application is necessary. As stated in the previous section, static comparators are not used for high-speed ADCs due to their low operation speed and more power consumption, so to overcome the problems of the static comparator, dynamic comparators are introduced.

The clocked comparators are often called dynamic comparators. Generally, the clock system is used to perform the switching operation in the circuit. The speed of the clock defines the overall speed of the comparator. A dynamic comparator behaves as a non-linear device because the operating region for all transistors in dynamic comparator changes in each clock period. A dynamic comparator reduces the total power consumption in comparison to a static comparator because it contains a positive feedback circuit [24]. The literature presents various architectures for a dynamic comparator. In Reference [25], a one-stage dynamic comparator is designed in which the preamplification stage is cascaded with the latch stage, but this architecture is badly affected by kickback noise. Two-stage comparator circuits are efficient in comparison to single-stage comparators in terms of kickback noise. The impact of kickback noise has been shown to be reduced in two-stage comparators by declining the capacitive path between input and output [25, 26]. A high-speed, low-offset two-stage dynamic comparator with PMOS latch and PMOS preamplifier, by disabling the preamplifier stage after the best possible delay, reduces the total power expenditure of comparator circuit [24].

7.2.2.1 Introduction to Latch Circuit

In a dynamic comparator, there are two stages: preamplification stage and latching stage. The latch consists of two cross-coupled inverters, as shown in the Figure 7.2. The latching stage of the comparator must introduce offset and delay in the comparator. To reduce these effects, some literature introduces a new latching stage to enhance

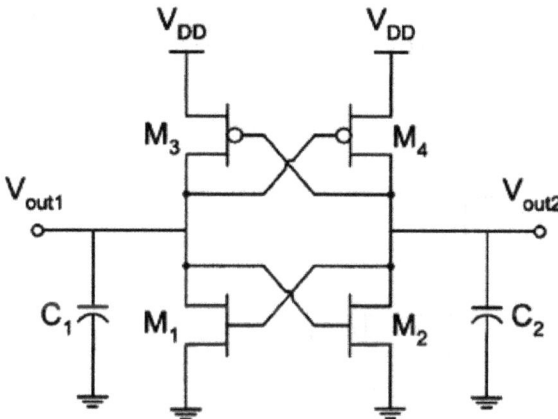

FIGURE 7.2 Schematic for latch circuit [1].

the transconductance by using separate gate biasing in the latch's transistors. The latch stage with improved transconductance in the comparator circuit achieves high speed and low power consumption. There are two modes of operation in a dynamic latch.

   i. Disable Latch: In the first mode, or in disabled mode, the positive feedback is disabled and input signals are applied in this mode.
   ii. Enable Latch: In second mode, the latch is enabled depending on the values of the initial voltage; after doing a comparison in this mode, one of the inputs goes high and the other will goes to low.

7.2.2.2 Types of Dynamic Comparators

   i. Single-tail dynamic comparator
   ii. Double-tail dynamic comparator

7.2.2.2.1 *Single-Tail Dynamic Comparator*

A single-tail dynamic comparator is demonstrated in Figure 7.3 and is mostly used for high-speed ADCs. As shown in the figure, the latch is made up of M (3–6) transistors to carry the regeneration process during the comparison phase. There are two phases of operation in a single-tail comparator: reset and evaluation. In the reset

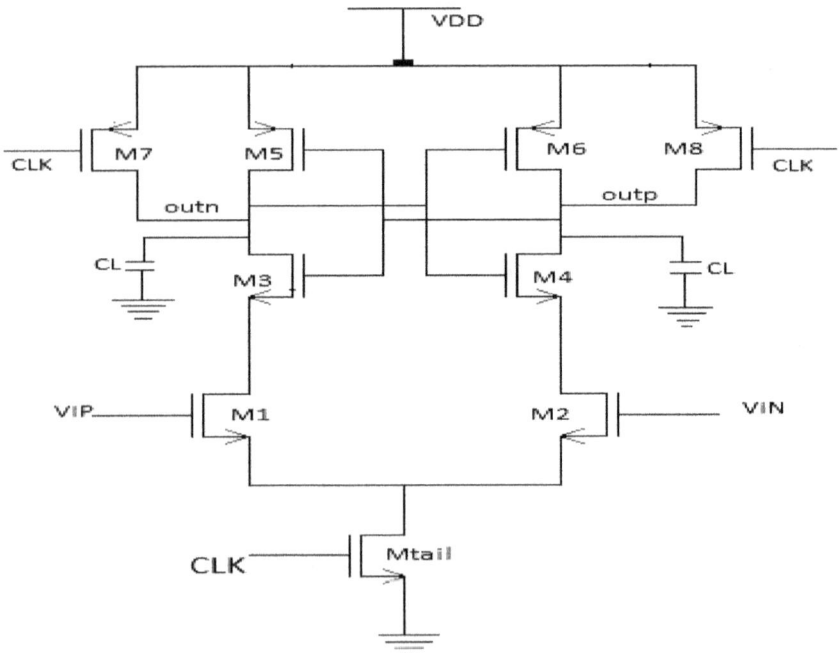

FIGURE 7.3 Conventional single-tail dynamic comparator [27].

phase, the clock (CLK) signal is at low level and the tail transistor is off. During this phase both the differential output node Outp and Outn are charged to V_{DD} through the transistor M7 and M8. And in evaluation (decision-making) phase, the CLK signal goes high. In this phase the tail transistor turns on and both the transistors M7 and M8 are turned off. If VIP is greater than VIN, the latch transistor is turned on and the node Outp discharges to V_{DD}-|V_{TH}| faster than Outn by transistor M1 and M2. The latch transistor M6 is turned ON before M5 to carry the regeneration process. In the last phase, the Outp pulls to Vss and Outn is pulled to V_{DD}.

One disadvantage of the single-tail comparator is that, due to a number of stacked transistors in the architecture, it needs a sufficiently high voltage supply to obtain optimum delay time. Another disadvantage of this comparator is that the current path for the latch and preamplifier stages is common, as the preamplifier requires small tail current to operate in a weak inversion region to achieve long integration interval and the latch stage requires large tail current for immediate regeneration. So, to reduce this problem, double-tail dynamic comparators are introduced [10].

7.2.2.2.2 Double-Tail Dynamic Comparator

The concept of the double-tail comparator is introduced to overcome the drawbacks of the single-tail or conventional dynamic comparators. The double-tail dynamic comparator operates at a lower voltage in comparison to a single-tail because of less stacking. Reference [28] proposed a double-tail dynamic comparator with low power consumption and small input-referred offset. To achieve parameters, two-control transistors are added in a cross-coupled manner in the preamplifier stage. The main idea is to increase voltage alteration at node Fn and Fp in order to increase the latch regeneration speed. This new latch stage improves the total transconductance in the reset phase due to the optimum delay time; also the overall speed of the comparator improves.

The typical double-tail dynamic comparator indicated in Figure 7.4 includes two levels: the preamplification level (M1, M2, M3, M4, Mtail1) and the latching level (M7, M8, M9, M10, S1, S2, Mtail2). In the reset segment, when (CLK =0) each tail transistor is in the off state. Clock-triggered PMOS transistors M3 and M4 are in on state and precharge the intermediate node to V_{DD}. Due to this the S1 and S2 turn on and discharge the output nodes to ground. And in the evaluation phase (CLK =VDD) each tail transistor turns on (M3 and M4 flip off), the voltage nodes at Fn and Fp begin to drop with the Mtail/Cf(n, p) charge. A voltage difference is constructed and MR1 and MR2 surpass the ΔVfn(p) to the latch degree. The positive feedback of the latch circuit begins the comparison phase. During the comparison among two inputs, one pulls out at V_{DD} and other to ground depending upon which one is greater in magnitude [27].

7.3 DESIGN METRICS AND PERFORMANCE ANALYSIS

1. Offset voltage
2. Delay
3. Area
4. Power consumption

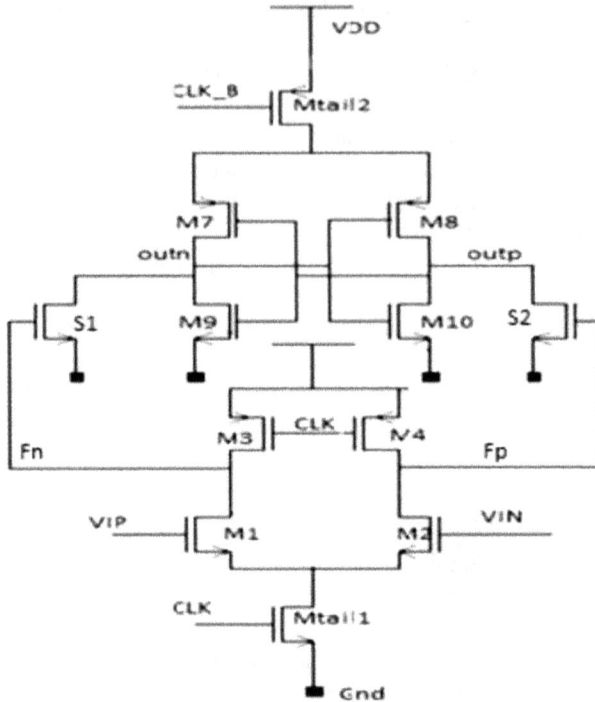

FIGURE 7.4   Conventional double-tail dynamic comparator [27].

7.3.1   OFFSET VOLTAGE

Offset voltage of the comparator is the most important parameter for analyzing the performance of any comparator. For any comparator the decision threshold is ideally at zero. But in practice the decision threshold shifts from zero due to mismatch and nonideal conditions. The offset voltage is the amount of voltage for which the decision threshold shifts from zero. In Reference [5], authors did an analysis for static and dynamic random offset voltage. There are two types of offset voltage: static offset and dynamic offset. The main source of static offset in a dynamic comparator is a misalliance in μCox and Vth. Dynamic misalliance occurs due to a mismatch in internal node capacitances in the dynamic comparator. In much of the literature, it is explained that the static mismatch in a preamplifier is reduced by increasing the transistor size. The gain of the preamplifier stage overcomes the offset at latch transistors [5]. The presence of a large offset in dynamic comparator limits the overall resolution of ADCs. So, it is necessary for a designer to attain a precise valuation of offset voltage during the simulation and to verify the offset voltage according to the specifications before the fabrication process. There are two methods to simulate the offset voltage; one uses staircase input and the other is a Monte Carlo simulation, but both methods require large simulation time. A fast and accurate technique has been proposed in Reference [19] to reduce the offset simulation time. Table 7.1 shows the offset voltages for various existing dynamic comparators.

TABLE 7.1

Comparison of Different Designs for Offset Voltage

Ref.	Technology Node (nm)	Supply Voltage (V)	Offset Voltage(mV)
[7]	180, 90	1	33
[27]	180	1.2	7.8
[11]	180	–	1.57
[14]	180	1.8	2.5
[16]	180	0.4	13.7
[17]	90	1.0	7.7
[18]	45	0.8	2.78
[20]	180	1.2	7.3
[32]	180	0.9	0.36
[24]	180	0.9	2
[29]	180	1.2	2.5

7.3.2 DELAY

The delay of a comparator circuit is described as the period between the starting of the comparison phase and half time of latch output [7]. The delay is one of the parameters on which the speed of the comparator depends. The total delay time is the sum of τ_0 and is τ_{latch}. Here τ_0 represents delay due to discharge of load capacitance (Cload). And τ_{latch} is due to the two pairs of inverters cross-coupled in the latch stage.

$$D_{Total} = \tau_0 + \tau_{latch}$$

The authors in Reference [27] have presented a delay analysis for a dynamic comparator. The reduction in common-mode voltage enhances the total delay time. A comparator with high common-mode voltage shows less delay and optimal speed. Both of the parameters – delay and offset – are related to each other. To reduce the offset voltage some literature recommends that the size of the tail transistor be reduced, but this will lead to an increase in delay, and if the size of the tail transistor is large, it results in low delay and small preamplifier gain but introduces large input offset voltage [29].

7.3.3 AREA

The area of the comparator is another important parameter for an efficient design. The area of the comparator depends on the number of transistors used in circuits. In some circuits to reduce the operating power, the use of stacking transistors is preferred, but it will increase the area. Table 7.2 shows the area consumption for various existing dynamic comparators.

7.3.4 POWER CONSUMPTION

The total power consumption in the chip is a major concern during design. In the comparator circuit, it is necessary to reduce the power consumption as much as

TABLE 7.2

Comparison of Different Designs for Area

Ref. No.	Technology Node (nm)	Supply Voltage V_{DD} (V)	Area
[7]	180,90	1	3.3
[31]	180	1.8	33×16
[14]	180	1.8	64.5
[16]	180	0.4	15×21
[17]	90	1.0	58.32
[18]	45	0.8	49.35
[20]	180	1.2	252
[32]	180	0.9	15.03×17.09
[24]	180	0.9	490
[29]	180	1.2	286
[30]	180	1.8	170

possible. In the comparator, offset voltage and power consumption are both important parameters, but there is a trade-off between these two parameters. There are two
main sources of total power consumption: static and dynamic power consumption.
The mathematical expression for total power consumption is given as follows

$$P_{Total} = f_{CLK}.C_L.V_{DD}^2 + V_{DD}.I_{leakage} \tag{7.1}$$

$$P_{Dynamic} = \frac{1}{2}\left(V_{DD}^2. \ f_{CLK}.C_L\right) \tag{7.2}$$

$$P_{Static} = V_{DD}.I_{leakage} \tag{7.3}$$

Numerous approaches are suggested in the literature to solve the problems
related to power dissipation. The authors in Reference [18] use a double-tail
dynamic comparator that uses bulk-driven PMOS transistors, which operates in
weak inversion and reduces the offset voltage as well as total power consumption.
A two-stage dynamic comparator with forwarding body bias is designed for ultra-
low power. In the preamplifier stage, the input PMOS transistor is body biased with
the ground [16]. The authors in Reference [30] proposed a dynamic comparator
with input PMOS for latch and preamplifier stage to achieve low power consumption. To reduce the total power consumption the authors in Reference [31] use two
transistors instead of one tail transistor in the latch stage. The benefit of using such
a configuration is that drive current is decreased to half and is divided into the two
transistors. The authors in Reference [12] use a hybrid comparator that includes two
stages with three stacked transistors, which sufficiently reduce the power consumption. In Reference [27] a double-tail comparator with minimum size transistor also
achieves minimum power. Table 7.3 shows the power consumption range for various existing dynamic comparators.

TABLE 7.3

Comparison of Different Designs for Power Consumption

Paper No.	Technology Node (nm)	Supply Voltage V_{DD} (V)	Power Consumed (W)
[5]	0.25μ,40n	1	–
[7]	180,90	1	51
[31]	180	1.8	242.6μ
[11]	180	–	0.26m
[12]	40	1.1	345.9
[14]	180	1.8	–
[16]	180	0.4	4.48n
[17]	90	1.0	32.62μ
[18]	45	0.8	24.92μ
[20]	180	1.2	268.6μ
[22]	180	0.9	309.73μ
[32]	180	0.9	265.25μ
[24]	180	0.9	0.20m
[29]	180	1.2	420μ
[30]	180	1.8	1.7m

CONCLUSION

This chapter presents a review of various architectures of dynamic comparator to compare the various performance parameters such as delay, power, offset and area. The dynamic comparator is mostly preferred for high-speed ADCs where the speed of the comparator shows the significant effect on operation of analog-to-digital converter.

REFERENCES

1. P. E. Allen and D. R. Holberg, *CMOS Analog Circuit Design*. Elsevier, 2011
2. R. S. Soin, F. Maloberti, and J. Franca, *Analogue-Digital ASICS: Circuit Techniques, Design Tools and Applications* (no. 3). IET, 1991.
3. M. Bruccoleri and P. Cusinato, "Offset reduction technique for use with high speed CMOS comparators," *Electronics Letters*, vol. 32, no. 13, pp. 1193–1194, 1996.
4. P. Cusinato, M. Bruccoleri, D. Caviglia, and M. Valle, "Analysis of the behavior of a dynamic latch comparator," *IEEE Transactions on Circuits and Systems I: Fundamental Theory and Applications*, vol. 45, no. 3, pp. 294–298, 1998.
5. J. He, S. Zhan, D. Chen, and R. L. Geiger, "Analyses of static and dynamic random offset voltages in dynamic comparators," *IEEE Transactions on Circuits and Systems I: Regular Papers*, vol. 56, no. 5, pp. 911–919, 2009.
6. Y. Jung, S. Lee, J. Chae, and G. Temes, "Low-power and low-offset comparator using latch load," *Electronics Letters*, vol. 47, no. 3, pp. 167–168, 2011.
7. M. Hassanpourghadi, M. Zamani, and M. Sharifkhani, "A low-power low-offset dynamic comparator for analog to digital converters," *Microelectronics Journal*, vol. 45, no. 2, pp. 256–262, 2014.

8. D. Xu, S. Xu, and G. Chen, "High-speed low-power and low-power supply voltage dynamic comparator," *Electronics Letters*, vol. 51, no. 23, pp. 1914–1916, 2015.

9. J. Gao, G. Li, and Q. Li, "High-speed low-power common-mode insensitive dynamic comparator," *Electronics Letters*, vol. 51, no. 2, pp. 134–136, 2015.

10. Y. Tao, A. Hierlemann, and Y. Lian, "A frequency-domain analysis of latch comparator offset due to load capacitor mismatch," *IEEE Transactions on Circuits and Systems II: Express Briefs*, vol. 62, no. 6, pp. 527–532, 2015.

11. A. Khorami and M. Sharifkhani, "Low-power technique for dynamic comparators," *Electronics Letters*, vol. 52, no. 7, pp. 509–511, 2016.

12. S. Huang, S. Diao, and F. Lin, "An energy-efficient high-speed CMOS hybrid comparator with reduced delay time in 40-nm CMOS process," *Analog Integrated Circuits and Signal Processing*, vol. 89, no. 1, pp. 231–238, 2016.

13. S. Hussain, R. Kumar, and G. Trivedi, "Comparison and design of dynamic comparator in 180nm SCL technology for low power and high speed Flash ADC," in *2017 IEEE International Symposium on Nanoelectronic and Information Systems (iNIS)*, 2017: IEEE, pp. 139–144.

14. A. Khorami and M. Sharifkhani, "Excess power elimination in high-resolution dynamic comparators," *Microelectronics Journal*, vol. 64, pp. 45–52, 2017.

15. S. Mahdavi, M. Poreh, S. Ataei, M. Jafarzadeh, and F. Noruzpur, "A New 1 GS/s Sampling Rate and 400 μV Resolution with Reliable Power Consumption Dynamic Latched Type Comparator," in *Fundamental Research in Electrical Engineering*. Springer, 2019, pp. 281–290.

16. Y.-H. Hwang and D.-K. Jeong, "Ultra-low-voltage low-power dynamic comparator with forward body bias scheme for SAR ADC," *Electronics Letters*, vol. 54, no. 24, pp. 1370–1372, 2018.

17. V. Savani and N. Devashrayee, "Design and analysis of low-power high-speed shared charge reset technique based dynamic latch comparator," *Microelectronics Journal*, vol. 74, pp. 116–126, 2018.

18. A. K. Dubey and R. Nagaria, "Optimization for offset and kickback-noise in novel CMOS double-tail dynamic comparator: A low-power, high-speed design approach using bulk-driven load," *Microelectronics Journal*, vol. 78, pp. 1–10, 2018.

19. H. Omran, "Fast and accurate technique for comparator offset voltage simulation," *Microelectronics Journal*, vol. 89, pp. 91–97, 2019.

20. Y. Wang, M. Yao, B. Guo, Z. Wu, W. Fan, and J. J. Liou, "A low-power high-speed dynamic comparator with a transconductance-enhanced latching stage," *IEEE Access*, vol. 7, pp. 93396–93403, 2019.

21. H. Xu and A. A. Abidi, "Analysis and design of regenerative comparators for low offset and noise," *IEEE Transactions on Circuits and Systems I: Regular Papers*, 2019.

22. P. P. Gandhi and N. Devashrayee, "Differential double tail dynamic CMOS voltage comparator," in *2017 International Conference on Intelligent Communication and Computational Techniques (ICCT)*, 2017: IEEE, pp. 8–11.

23. M. G. Johnson, (ed) "Static high speed comparator," Google Patents, 1996.

24. A. Khorami and M. Sharifkhani, "A low-power high-speed comparator for precise applications," *IEEE Transactions on Very Large-Scale Integration (VLSI) Systems*, vol. 26, no. 10, pp. 2038–2049, 2018

25. P. M. Figueiredo and J. C. Vital, "Kickback noise reduction techniques for CMOS latched comparators," *IEEE Transactions on Circuits and Systems II: Express Briefs*, vol. 53, no. 7, pp. 541–545, 2006.

26. T. B. Cho and P. R. Gray, "A 10 b, 20 Msample/s, 35 mW pipeline A/D converter," *IEEE Journal of Solid-State Circuits*, vol. 30, no. 3, pp. 166–172, 1995.

27. S. Babayan-Mashhadi and R. Lotfi, "Analysis and design of a low-voltage low-power double-tail comparator," *IEEE Transactions on Very Large Scale Integration (VLSI) Systems*, vol. 22, no. 2, pp. 343–352, 2013.
28. J. Lu and J. Holleman, "A low-power high-precision comparator with time-domain bulk-tuned offset cancellation," *IEEE Transactions on Circuits and Systems I: Regular Papers*, vol. 60, no. 5, pp. 1158–1167, 2013.
29. A. Khorami and M. Sharifkhani, "High-speed low-power comparator for analog to digital converters," *AEU-International Journal of Electronics and Communications*, vol. 70, no. 7, pp. 886–894, 2016.
30. H. Molaei, K. Haj Sadeghi, and A. Khorami, "Design of low power comparator-reduced hybrid ADC," *Microelectronics Journal*, vol. 79, pp. 79–90, 2018.
31. V. Deepika and S. Singh, "Design and Implementation of a Low-Power, High-Speed Comparator," *Procedia Materials Science*, vol. 10, pp. 314–322, 2015.
32. P. P. Gandhi and N. Devashrayee, "A novel low offset low power CMOS dynamic comparator," *Analog Integrated Circuits and Signal Processing*, vol. 96, no. 1, pp. 147–158, 2018.

8 Historical Review of DRA to MIMO-DRA
Designs and Advances

Madhusmita Mishra
NIT Rourkela

Gaurav Varshney
NIT Patna

CONTENTS

8.1 Introduction 114
8.2 Fundamentals of DRA.. 115
    8.2.1 Parameters of DRA... 116
        8.2.1.1 Dielectric Constant 116
        8.2.1.2 Quality Factor 117
        8.2.1.3 Resonant Modes and Radiation Behaviour....................... 117
        8.2.1.4 Methods of Excitation.. 118
        8.2.1.5 Methods of Bandwidth, Gain and Radiation
                Characteristics Enhancement.. 120
8.3 DRA Design Algorithms and Simulation Steps 121
    8.3.1 DRA Design Algorithm Steps in HFSS 123
    8.3.2 DRA Design Algorithm Steps in CST.. 123
    8.3.3 DRA Design Method in MATLAB 124
8.4 DRA Applications and Future Scope 124
    8.4.1 Dielectric Resonator Antenna for Ultra-Wide band Applications.... 124
    8.4.2 Hybrid Cylindrical Glass Dielectric Resonator Antenna for
        Indoor Communication... 124
    8.4.3 DRA for DVB-H Application 124
    8.4.4 Nano DRA for Photonics Applications 125
    8.4.5 Wide Dual-Band Rectangular DRA 125
    8.4.6 Dual-Band Stacked RDRA for 5G Applications 125
8.5 MIMO DRAs... 125
    8.5.1 Design Concepts and Advancements... 125
    8.5.2 Challenges in MIMO Antenna Design and Future Scope............... 126
8.6 Conclusion 127
References... 127

8.1   INTRODUCTION

Communication devices have an antenna as a critical element. Hence, there is always a requirement to reduce the size of the antenna to scale down the size of these devices. Current communication demands enhanced data rate and gain. Increase in data transfer rate is directly related to the channel capacity. Capacity is increased by increasing bandwidth. Working at higher frequencies, like THz, is a solution to this. Antenna design plays a vital role in wireless communication. This chapter represents an intensive review of the designs implemented on dielectric resonator antennas (DRAs) along with the evolution towards multiple-input multiple-output (MIMO) DRAs.

DRAs have various applications because of their prominent radiation efficiency, light weight, small size, low conductor loss, low profile, surface-wave loss and improved radiation performance. Hence, antenna engineers have viewed them as an alternative to conventional low-gain designs. Here we have given clear vision on the types of DRAs, different DR shapes, mode of operation, radiation behaviour of DRAs, wideband and ultra-wideband responses, methods of excitation, ways to enhance the radiation characteristics, polarization methods, gain and bandwidth enhancement techniques and the major advances in DRA technology [1–4]. The DRAs discussed here have various applications. We have focused on the underlying algorithms and simulation steps for simulation of DRA while providing a brief idea on the available design software [5–25].

MIMO systems enhance the channel capacity and, consequently, the data rate. Recently, several MIMO antennas have been designed using the features of DRA. MIMO and good-diversity performance are attainable if and only if there is prominent isolation amidst the multiple ports of an antenna and small envelope correlation coefficient (ECC). There can be single- and multi-radiator MIMO DRAs. The use of more input and output ports with the single-radiator MIMO DRAs is very difficult because obtaining the isolation between ports is difficult. An antenna with multi-radiators has the problem of complexity of size because of the increase in the number of separated radiators. Researchers have explored several geometries for DR to develop a wideband MIMO DRA structure. Still, there is a research direction towards advancing the ultra-wideband MIMO DRA. Antenna researchers are developing various devices for the THz band applications. In general, metals are used at the microwave ranges and lower frequencies. Although metallic antennas are preferred at the microwave frequencies, they are not useful at higher frequencies because of low conductivity. So it is necessary to replace metals for devices operating in the THz frequency range. Hence huge research has been done on the 2-dimensional (2D) materials. Recently, graphene was investigated for its electrical properties. The optical and electrical properties of graphene have enabled new device designs for applications at THz and optical frequencies. Here, we have given insight into the design of MIMO-DRA, taking into consideration the performance parameters: diversity gain (DG), isolation between multiple ports, the total active reflection coefficient, mean effective gain (MEG), envelop correlation coefficient (ECC) and channel capacity loss (CCL). All these parameters must maintain their acceptable values for the MIMO performance. Here we have investigated the designs

of frequency reconfigurable MIMO-DRA with an emphasis towards tunable radiation characteristics while providing an insight into the theoretical boundaries of the
radiation efficiency of graphene antennas. Currently, the situation of communication between nearly five billion wireless devices for voice and data transmission has
resulted in increased data utilization. This leads researchers towards fifth-generation
(5G) technology with more prominent features, providing high data rate, low latency
rate, effective EM spectrum utilization, vast machine-to-machine communication,
etc. For practical design of 5G technologies, efficient and compact antennas are
required. Here we have also discussed the effective design of DRAs and MIMO
DRAs for 5G applications [26–32]. The following sections are organized as follows.
In Section II, fundamentals of DRA are examined concisely in conjunction with
the design parameters. Section III discusses DRA design algorithms and simulation
steps. Section IV gives details on DRA applications and future scope. Section V
highlights the design concepts of MIMO DRAs along with design advancements,
challenges and future scope. Finally, concluding remarks are given.

8.2 FUNDAMENTALS OF DRA

A dielectric resonator (DR) acts as an energy storage element in microwave circuits.
But in the open condition it behaves like a radiator. DRA is the antenna which uses
the radiating mode of a DR. Single DRA is a three-dimensional device having any
of these shapes – triangular, rectangular, cylindrical, spherical, hemispherical, conical etc. to accommodate various design requirements. These shapes are shown in
Figure 8.1. The resonance frequency of any shape of DRA is decided by the dimensions and dielectric constant (ε_r). Microstrip patch antennas have limitations like
narrow bandwidth, smaller gain, less power handling capacity, weak polarization
purity, etc. This motivated antenna researchers to design DRAs.

A DRA is made from a dielectric resonator, ground and substrate with various
excited feeding methods. Microwave dielectric material with a low-loss profile is
used for implementing DRA. The size of a DRA is controlled by the dielectric constant of the materials. Using a wide range of dielectric permittivity values (2 to 100)
helps the antenna researcher to get reasonable control of size and bandwidth. This
means that low permittivity allows for wide bandwidth and high permittivity allows
for the compact size. The lower-frequency design parameters are the antenna size
and weight. Similarly, mechanical tolerances and electrical losses are design parameters at higher frequencies. Maximum dimensions (D) of dielectric resonator antennas are interrelated to the free-space resonant wavelength (λ_0) through the estimate
relation $D \propto \lambda_0 \varepsilon_r^{-0.5}$, where ε_r is the relative permittivity of the dielectric resonator
antenna. As the radiation efficiency of a DRA is not significantly affected by its
dielectric constant, a broad range of values can be exploited. On the other hand,
because the bandwidth of the DRA and the dielectric constant are inversely related,
this limits the choice of values for a given application. By using a high dielectric
constant material, the DRA size can be significantly lowered, thus making it feasible
for low-frequency operations.

Many designs have been published for DRAs, which operate at frequencies from
1 to 40GHz, with dimensions in the range from a few centimetres to few millimetres

FIGURE 8.1   Different shapes of DRA.

and dielectric constants approximately ranging from $8 \leq \varepsilon_r \leq 100$. The design parameters (input impedance, resonant frequency, permittivity, radiation pattern and coupling mechanisms) change for different shapes. So there exists a different analytical model for analyzing each geometrical configuration [1]. Although there are several shapes, the most popular are cylindrical and rectangular. The rectangular provides one extra degree of freedom for the purpose of designing [2]. Different shapes of the DR structure will generate different modes, resulting in different field patterns. The most prominent features which make the DR suitable for the antenna element are its simple structure and reduction in the Q-factor. The aspect ratio and shape of the DRA are the critical parameters, which determine the field pattern and modes generated within the DR structure. Each mode inside the DR has different field pattern and radiation efficiency. The important parameters of antenna like operating bandwidth and gain can be improved by adjusting these parameters. The DR structures in the antenna have no conductor losses, which improves efficiency. The coupling among the DR structure and transmission line can be checked by altering the DR locations. These characteristics have made DRA useful at microwave and millimetre-wave frequency applications [3].

When DRA is excited, the radio waves feed inside the DR and bounce back and forth within the resonating walls, resulting in standing waves. These standing waves store the energy in the form of electric and magnetic field components E & H, respectively. Due to accelerating current, the time-varying field diversifies off the DRA in the free space. Due to the fringing effect, the magnetic field leaks out into the environment through the walls of DRA into free space. Thus, DRA radiates.

8.2.1   PARAMETERS OF DRA

8.2.1.1   Dielectric Constant

The dielectric constant is defined as 'the ratio of the amount of electrical energy stored inside the material when an external voltage is applied to that of the energy stored in a vacuum'. It is denoted as ε_r. Dielectric polarization occurs when the electromagnetic field passes through the dielectric material. It defines the dielectric property of the DR.

8.2.1.2 Quality Factor

DRA Q-factor can be calculated as

$$Q = \frac{\omega W_S}{P_{rad}} \tag{8.1}$$

The Q-factor can compute DRA's impedance bandwidth as

$$BW = \frac{S-1}{Q\sqrt{S}} \tag{8.2}$$

where, S is the VSWR.

The normalized Q-factor can be expressed mathematically as follows:

$$Q_e = \frac{Q}{\varepsilon_r^{3/2}} \tag{8.3}$$

8.2.1.3 Resonant Modes and Radiation Behaviour

The radiation characteristics of DRA can be analyzed by using the resonant modes. The radiation phenomenon inside the DR is represented by using E- and H-field patterns, which determine the configuration of resonant modes. The field perturbation on the surface of DRA produces the E-field distribution pattern, which defines the resonant modes. These modes can be TE, TM and hybrid modes. The resonant modes can be calculated using the orthogonal Fourier basis functions. The resonant modes are responsible for determining the distribution of total current throughout the DRA to form the set of orthogonal solutions. The E_z and H_z fields are based on the concept of orthonormality. Based on boundary conditions, the modes are categorized as the confined and non-confined modes. The two conditions that define confined and non-confined modes are as follows:

i. $\hat{n}.E = 0$ and
ii. $\hat{n} \times H = 0$

Here, $\hat{n}$ is the unit vector perpendicular to the surface of DR. The modes which follow both of these conditions are the confined modes, and those which follow the first condition only are called non-confined modes. The surface of evolution like cylindrical and spherical can have the confined modes only. The mode $TM_{01\delta}$ is the confined mode. The fundamental confined and non-confined modes act as the electric and magnetic dipoles. The most common geometries of the DR are cylindrical and rectangular. In the case of cylindrical DR, transverse electric (TE), transverse magnetic (TM) and hybrid (HEM) modes exist. In case of the rectangular DRA, the existence of lower order *TM* modes was not verified experimentally. Hence, the resonant frequency of the lower order *TE* modes is calculated using transcendental equations. Hence, we can calculate the resonant frequency by using the wave number of free space, i.e., k_0.

$$k_0^2 = \omega_0^2 \mu_0 \epsilon_0 \tag{8.4}$$

The frequency of resonance of an isolated rectangular DR in free space can be obtained as follows:

$$(f)_r \ m,n, \ p = \frac{c}{2\pi \sqrt{\mu \epsilon}} \sqrt{\left[\left(\frac{m\pi}{a}\right)^2 + \left(\frac{n\pi}{b}\right)^2 + \left(\frac{p\pi}{d}\right)^2\right]} \tag{8.5}$$

The height of the DR is considered double in the z-direction of the equation 8.5 if the DR is fixed over the ground plane. The concept of the short magnetic and electric dipole is necessary for understanding the resonant modes in DRA. A simple electric current carrying wire can be considered as the short electric dipole. In a practical scenario, the electric loop antenna is considered as the magnetic dipole [2–4].

8.2.1.4 Methods of Excitation

The resonant frequency of DRA depends upon the physical dimensions and the material used in making the DR. Also, the order of the excited mode depends upon the method of excitation and position of the feeding technique. The generation of particular mode decides the operating characteristics of the antenna-like far-field radiation and the direction of radiation [5, 6]. There are a number of feeding techniques (Figure 8.2) which can be used for the excitation of DR [7, 8]. They are listed below.

8.2.1.4.1 Slot/Aperture-Coupling Methods

In the aperture-coupling method, a substrate is used with the feed-line at the bottom and a conducting ground plane at the top. The ground plane contains a slot that serves as the short magnetic dipole horizontal to the direction of radiation when the fundamental mode is excited. The current flowing through the feed-line is coupled to the slot, and the created short magnetic dipole couples the energy to radiating DR. The DR serves as the cavity, and the field is radiated into the free space. In the aperture-coupled feeding, the portion of the feed-line after the slot width is called the stub. The length of the stub is decided for providing better impedance matching. For impedance matching, this is the simplest feeding technique.

8.2.1.4.2 Coplanar Waveguide Lines

In this feeding technique, a feed-line is applied in the same plane as the ground plane, and the energy is coupled to DR direction from the feed-line to the DR. This is another simple method for the coupling of energy from the feed-line to the DR.

8.2.1.4.3 Dielectric Image Guide Excitation

This is the method free from the conductor loss because of the absence of the conducting feed-line. A dielectric-made feed-line is used for coupling energy from the

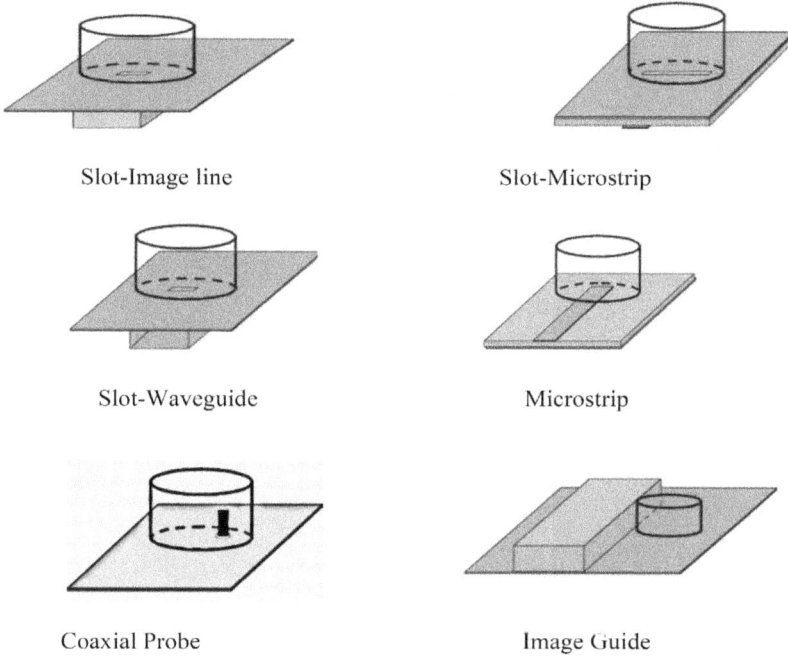

Slot-Image line   Slot-Microstrip

Slot-Waveguide   Microstrip

Coaxial Probe   Image Guide

FIGURE 8.2   Various feeding techniques of DRA.

feed to DR. This feed-line can be used at the higher frequencies in the range of milli-metre waves and above. Though the coupling between the guide and DR is generally small, it can be raised by running the guide nearer to the cut off frequency. Being similar to the waveguide slot method, it differs in that here the image line replaces the waveguide.

8.2.1.4.4   Coaxial Probe Feed Excitation

This is another feeding technique which generally excites the fundamental mode via excitation of vertical electric or magnetic dipole if placed in the centre of the DR. It usually excites the hybrid mode in the DR if placed at its edge. The length of the metallic pin used in the coaxial probe helps in the impedance matching. This feed-ing technique is a little complex in comparison to the aperture-coupled feeding tech-nique due to the requirement of the mechanical hole in the substrate. However, the compact antenna structures can be designed only by using this feeding technique.

8.2.1.4.5   Direct Microstrip Line Mechanism

This is another simple method after aperture coupling where a microstrip feed-line is direction connected to DR for its excitation. The impedance matching is a little poor in this feeding technique. However, a stub of the metallic line can be joined to the feed-line for impedance matching. The stub can also be lengthened along with the height of the DR. However, it may include the structural complexity.

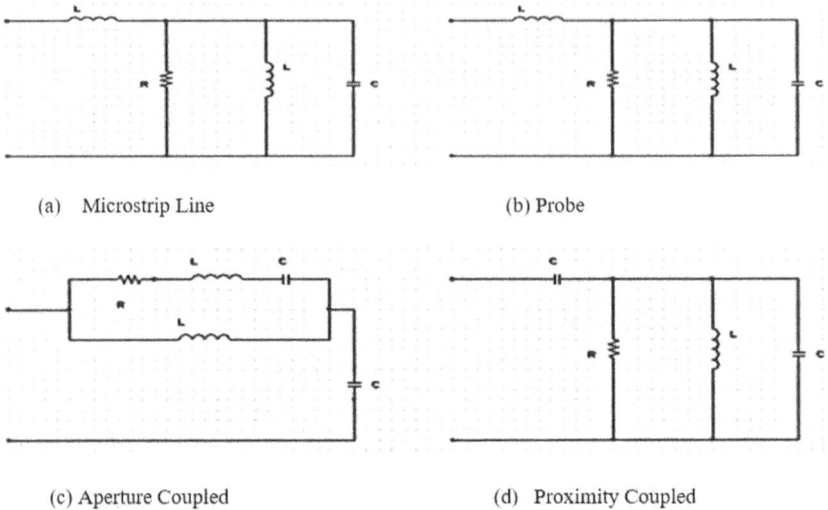

(a)   Microstrip Line   (b) Probe

(c) Aperture Coupled   (d) Proximity Coupled

FIGURE 8.3   Equivalent circuits of some feeding techniques for a DRA.

8.2.1.4.6   Proximity Coupled Microstrip Feed

This is a feeding technique which may provide the wide impedance bandwidth. In this feeding technique, a feed-line is sandwiched between two substrates, and the DR is placed at the top of the top substrate. The equivalent circuit diagram for this feeding technique is shown in Figure 8.3.

8.2.1.5   Methods of Bandwidth, Gain and Radiation Characteristics Enhancement

Bandwidth enhancement depends on modification in feed geometries and changes in the shape of the DRA. Use of unlike bandwidth enhancement techniques leads to the design of the different shape of dielectric resonator antennas. Bandwidth enhancement gives a big bandwidth for particular resonant frequencies. Designing DRAs for wireless applications depends mainly on bandwidth enhancement. When selecting a low dielectric constant and appropriate dimensions, a very low Q-factor can be obtained that indicates the theoretical possibility of designing an isolated cylindrical or rectangular DRA with very broad bandwidth. Increase in the impedance bandwidth of DRA is obtained by altering the shape of the DRA, using modified feed geometries, using multiple DRAs (stacked or array or embedded) and introducing air gap between the dielectric resonator and ground plane. Consideration of the high aspect ratio of DR leads to placing a significant number of modes nearer to each other. It thus enhances bandwidth due to merging of the adjacent modes. Embedding of DRAs within one another can improve the bandwidth of DRA. Enhancement in impedance bandwidth of DRA can be achieved by using matching networks (flat matching strips, loaded notched DRAs and Multi-segment DRAs). A single-feed

DRA with the regular shape like circular and rectangular provides a narrow bandwidth of circular polarization. Use of stair-shaped DR, rotated stair-shaped DR, and trapezoidal-shaped DR enhances the AR bandwidth.

The limitation of these structures is that the fabrication of such non-regular shapes of DR becomes difficult. To find the wide axial ratio (AR) bandwidth, multi-feed structures have been reported [9, 10]. Recently, a circularly polarised DRA (CPDRA) has been proposed, which consists of a rectangular DR and two half-split cylindrical DRs, while obtaining the excitation via a stair-shaped slot. Here the excitation of multiple orthogonal mode pairs was done for proper tuning, and merging of the orthogonal mode pairs. As a result, wider AR (41.01%) and impedance bandwidths (49.67%) are achieved [11]. Another proposed multiband DRA gives the 3 dB AR bandwidth in the lower and upper band as 19.98% and 3.07% respectively. To have triple-band CP characteristics, the antenna response is tuned accordingly [12]. The more the higher-order modes, the more will be the gain. A number of higher-order modes have been introduced by increasing the height of DR at the resonant frequency corresponding to the fundamental mode. Stacking of the DR elements also enhances the gain [13].

Use of hybrid patch and DRA is another technique to enhance the gain in addition to bandwidth [14]. Array antennas are used to enhance the radiation characteristics, which are not easily achievable by using a single antenna element. It consists of the combination of identical elements within the same antenna design for the improvement of the radiation properties. Antenna array design depends on the consideration of construction and destruction of the electric field in some specified directions. If there is constructive interference in the desired direction, there will be field addition, and if there is destructive interference, then there will be field cancellation. Since full cancellation is not possible practically, minor (side) lobes and back lobes are created. Use of optimization techniques and change in distance, amplitude and phase between elements can reduce the side lobes. Side lobe diminution of the pattern is also done by checking the amplitude of each element of the array. Satellite, RADAR and other communication systems prefer this method for improving the signal to noise ratio and consequently the system performance.

Another useful technique suitable for satellite and microwave communication systems is the sole control of the excitation phase of each element of the array. Control of displacement of an individual element of antenna is another way to reduce side lobe pattern since this makes a highly directive antenna. This method is meant for the specific purpose only. This is because it is practically impossible for the antenna element to be shifted from one place to the other every time [15].

8.3   DRA DESIGN ALGORITHMS AND SIMULATION STEPS

The main information required for the design of DRA is the radiation field, resonant frequency, bandwidth, resonant modes inside the DR structure, radiation Q-factor and field distribution [3, 16]. The design and simulations of the DRA antenna can be carried out using software packages like FEKO, Empire XCcel, HFSS and CST Microwave Studio suite. The simulated antennas then can be fabricated using a photolithographic method. Real-time measure of S-parameter/VSWR characteristics is done by a network analyzer. The real-time antenna gain and radiation patterns can be computed in anechoic chamber. Full wave solvers use either analytical or numerical methods for getting the solutions. The analytical methods give the exact solution, and they include variable separation, expansion in series, conformal mapping, integral solution (e.g., Laplace and Fourier transformation) and perturbation methods. Numerical methods give approximate solutions, and they include finite difference method [time domain approach], method of moments (MoM) [frequency domain approach], finite element method (FEM) [frequency domain approach], finite integration technique (FIT) [time domain approach] and transmission-line modelling [time domain approach].

HFSS uses FEM and thus gives more accurate results while designing antennas. CST uses FIT, and it provides the designer ease in simulations. Results of both the simulators are not identical due to the use of the different computational technique. On the one hand, HFSS results are nearer to experimental results, with more emphasis on the structure available. On the other hand, in CST, perfect boundary approximation (PBA) can be done and it is suitable for 3D antenna simulation. The FEM methods provide solutions in the frequency domain, while the FIT methods give solutions in the time domain. FEKO uses the MoM method. The MoM can be used in combination with the geometric optics approach (GO), the unified theory of diffraction (UTD) and the multilevel fast multimode method (MLFMM). It can genuinely model arbitrary 3D structures. With the FE method modelling of dielectric volumes can be done with an SIE approach or VIE approach or with a hybrid approach.

Empire XCcel uses the finite-difference time-domain (FDTD) technique. Due to adaptive on-the-fly code generation, it comes with a highly accelerated kernel, providing very fast simulations. It gives more accurate results for curved structures by the perfect geometry approximation (PGA) algorithm. It also provides frequency-dependent loss calculation and special algorithms for modelling thin conducting sheets. Selection of antenna design software depends on the structure's geometry and the intended accuracy of the solution. For example, ZELAND IE3D uses MoM solution, which provides brilliant analysis accuracy for the frequency domain, but it cannot admit very fine details on the geometry of the structure. Hence for simple structures like rectangular or circular, IE3D would be the best. ZELAND Fidelity uses FDTD analysis. It uses a combination of specific geometries. It is mainly suitable for regular shapes like cylindrical DRA for an example. MoM and FDTD methods are not desirable for big structures like large antenna arrays or reflector antenna. Ansoft HFSS and CST provide a better interface to allow very fine inside information regarding the geometry of the simulated structure. HFSS uses FEM, and CST uses FIT, which is similar to FDTD. Both methods are preferable for small or moderate objects in comparison with the working wavelength.

CST provides the advantage of getting the results on a wideband as it begins in the time domain and changes the results to FD through Fourier transform. Dissimilar operations are found in HFSS, though they are FD solution. Accuracy of MoM is slightly more than the accuracy of finite element. Thus results in HFSS and Zeland IE3D differ slightly for regular shapes like rectangular patch antenna. In this case the result of HFSS is less accurate than Zeland IE3D. But in the case of complicated geometry, the accuracy of HFSS and CST are much better than IE3D due to the geometrical approximations taken for the structure.

FEKO has two main solvers, based on MoM and geometrical theory of diffraction (GTD). The portion of FEKO related to GTD cannot be substituted with Zeland, HFSS or CST because it is basically meant for large structures similar to reflector antennas. To summarize the debate over antenna algorithms, a good designer must be able to use different computer-aided design (CAD) tools with clear understanding of the bounds of their numerical techniques and modelling interface. Recently CST has gained popularity. But with inappropriate tool settings, the problem of ripple in the frequency response may occur. Using HFSS and CST requires a good knowledge of the numerical techniques they use; that is why most users get incorrect results, especially when the error is concerned with feed modelling and source-port definition. This software can simulate very large structures depending on the available hardware resources. For some structures, HFSS is faster and leads to better results, while CST may be better for some other structures [17, 18].

8.3.1   DRA Design Algorithm Steps in HFSS

- Start.
- Open HFSS software.
- Insert geometrical project design.
- Form basic fields. (Assign material for each element, radiation boundary for the structure and then give the required excitation.)
- Discretize the finite element and form the surface integral for fields. (In the analysis setup assign frequency sweep range and solution frequency, then run and analyzed the results.)
- Stop.

8.3.2   DRA Design Algorithm Steps in CST

- Start.
- Open CST Studio Suite software and select 'New Project'; then click 'Create Project Template'; then select 'Antennas' under the option 'MW & RF & Optical'. Then click on 'Dielectric Resonator' from workflow.
- Then select 'Solver' in time domain to enter fields in time and spatial domain. (Select dimensions, frequency, time, temperature, voltage, current, resistance, conductance, inductance, capacitance.)
- Implement Yee's algorithm by selecting field settings (Freq Min, Freq Max, E & H Field, Far Field, Power flow and loss).
- Estimate cell size and time step.
- Analyze the dielectric resonator antenna by providing source signal and feed modelling.
- Cut short the computational area by a virtual limit, and then run and analyze the results.
- Stop.

8.3.3 DRA Design Method in MATLAB

In MATLAB analysis, the resonant frequency and the unloaded Q-factor are calculated. Then estimation of the bandwidth of various resonant modes of a dielectric resonator placed on an infinite ground plane is done. The resonator dimensions are set according to the input specifications on aspect ratios, resonant frequency, resonant mode, minimum bandwidth, dielectric constant, etc. The user can quickly find the resonator dimensions in the design mode. MATLAB software provides supports for analyzing and designing various resonator shapes like hemispherical, cylindrical, cylindrical ring and rectangular cross-section. Optimization of antenna design can also be done with MATLAB. All the simulations of E-field, H-field, far-field can also be done in MATLAB, as can be done in HFSS and CST. The difference lies in writing your own MATLAB programmes using the MATLAB toolbox [19].

8.4 DRA APPLICATIONS AND FUTURE SCOPE

A substantial number of recent publications centre on designing DRA for particular applications.

8.4.1 Dielectric Resonator Antenna for Ultra-Wide Band Applications

An antenna having a conical shape dielectric resonator fed through microstrip line can give the results for ultra wide band application, meeting the bounds decided by FCC for UWB application, i.e., 3.110.6 GHz. It is implemented without considering higher-order modes to get the desired characteristics [20].

8.4.2 Hybrid Cylindrical Glass Dielectric Resonator Antenna for Indoor Communication

DRA radiation is mainly of two types: directional and omnidirectional. Because of larger coverage area, the omnidirectional DRA is preferred over the directional one. Bandwidth of the omnidirectional DRA can be widened either by employing higher-order modes or by applying the hybrid DRA. Using the recently proposed hybrid-glass DRA, getting excitation from a probe platform achieves performance nearly the same with and without patterns. The impedance bandwidth of this hybrid DRA can be checked by the height of the glass DRA in a wide range of frequencies, without altering the feeding size [21].

8.4.3 DRA for DVB-H Application

Recently, a stacked dielectric resonator antenna, designed with switchable frequency and radiation pattern, has been proposed for DBV-H application. The parasitic metallic plate is used for ground elongation and it causes a shift in the resonant frequency from 0.47 GHz to 0.89 GHz [22].

8.4.4   NANO DRA FOR PHOTONICS APPLICATIONS

In nano DRA, a nano-strip feed-line excites a cylindrical DR. An input Gaussian pulse is applied at the corner of feed-line, and thus the field components are split into two parts. Edges of the feed-line carry the split field components, which have a quarter wavelength path difference between them. Thus their coupling to DR gives the circularly polarized response. The designed antenna operates in the optical C-band. Different physical parameters and their effects are investigated by the authors for nano-scale practical implementation. The antenna gives 3dB AR of 5.72% and 10dB impedance of 11.58% [23].

8.4.5   WIDE DUAL-BAND RECTANGULAR DRA

Recently authors have proposed a wide dual-band rectangular souvenir DRA. The lower band is designed by employing $TE_{111}{}^y$ mode of rectangular DRA and the U-slot resonator. The upper band is designed by employing higher-order $TE_{113}{}^y$ and $TE_{311}{}^y$ modes. The wide dual-band rectangular DRA is designed for WLAN applications. The word "LOVE" has been cut up on the sidewall to build it as a souvenir DRA [24].

8.4.6   DUAL-BAND STACKED RDRA FOR 5G APPLICATIONS

For 5G applications, researchers recently have designed three types of stacked rectangular dielectric resonator antennas (RDRAs). First is the basic stacked RDRA, and the second is the same with a single notch. The third one is combination of first and double notches. Authors have authenticated the comparison of all the proposed designs using 3D simulation tools by monitoring the performance of bandwidth, reflection coefficient and gain [25]. The achievement of wideband and multiband CP response with the facility of tuning the CP response shows development in the field of CPDRA. The CPDRA is the advanced version of DRA. Nowadays, MIMO DRAs are the research topic to enhance the data transfer rate in wireless communication systems. The introduction of circular polarization in MIMO DRAs provides polarization diversity, which improves the signal strength.

8.5   MIMO DRAs

8.5.1   DESIGN CONCEPTS AND ADVANCEMENTS

MIMO technology is the ultimate way to improve wireless throughput, i.e., the total number of data sent per second. Increased throughput is the reason behind having multiple antennas on the router at home. It is also reason that smartphones are now using MIMO for both WIFI and LTE. The design of MIMO-DRA takes into consideration the following performance parameters.

The ECC is a main metric for designing MIMO antennas for mobile communications. It characterizes the correlation between every two ports. MIMO antennas should have low ECC to indicate that channels are independent, and hence the system provides a high diversity gain (DG) and high capacity [26–29]. To identify that the design of antenna is 'good' when there are multiple antennas on a single product, the antennas must be independent in transmitting simultaneous and independent data streams. Here *independent* pertains to good isolation between antennas and dissimilar radiation patterns or at least not very 'correlated'. For this we need to know a 'radiation pattern correlation metric', which is referred to in the industry as ECC. ECC speaks to the independence between radiation patterns of two antennas. ECC considers radiation pattern shapes, polarizations and relative phases of the fields between the two antennas. Typically, correlation can be decreased by providing distinct polarization, by making more separation distance and by ascertaining that peak radiation direction is distinguishable. For highly efficient antennas, antenna isolation (by measuring s_{12} only) is sufficient for the determination of ECC. Thus ECC can be determined without taking into account the radiation patterns of the antenna. So without any math, we can determine ECC. This is because for same (or highly correlated) radiation pattern produced by the antenna, there will be tight coupling (or low isolation). Diversity gain (DG) is the measure of advances obtained from multiple to single antenna systems. Diversity gain is calculated from DG equation [29]. The MEG measures the received power from the diversity antenna relative to the received power from an isotropic antenna. In MIMO systems channel capacity is linearly proportional to the number of antenna elements. Channel capacity loss (CCL) is used to measure system degradation. The total active reflection coefficient (TARC) is the proportion of the square root of full reflected power to the square root of full incident power. TARC is computed from the scattering matrix.

8.5.2 Challenges in MIMO Antenna Design and Future Scope

The challenges in designing a MIMO antenna system are as follows: system integration, enhancement of port-isolation and field-isolation, and proper performance metric characterization of MIMO antenna in a laboratory environment. So designing the MIMO antenna system is not straight forward, and it involves evaluation of new performance metrics [30, 31]. The importance of MIMO antenna systems lies in the increase of their application in current wireless devices and appliances, and this drive will extend because 5G wireless standards rely on MIMO technology. But still, there are some major misconceptions and inadequate understanding of the fundamental aspects while designing, characterizing, and assessing MIMO antenna systems. Though there are a lot of recently published works on MIMO antennas, various common misconceptions and misuses of metrics still exist, and this nullifies the conclusions made in many studies [31]. Some of the designs suffer from wrong assumptions in MIMO antenna geometries, thus leading to limitations in the practicability of the proposed models. Such common misuses and misinterpretations with relation to printed and non-printed MIMO antenna systems have been highlighted [31] to give some correct guidelines for researchers. Some aspects of the application of MIMO-based antenna systems in future wireless standards with particular emphasis towards the millimetre-frequency range and massive MIMO architectures are as follows. Massive MIMO technology is under substantial investigation for application in 5G wireless systems. So future designs demand the computation of the correlation coefficient among MIMO antenna elements using the field equation. The use of millimetre-wave is a great feature for 5G. The integration of MIMO solutions within this band is now a field of potential research. MIMO antennas have a large number of wireless applications, such as base stations, cell phones, tablets, laptop PCs, wireless routers and access points. The MIMO antennas have huge application in cognitive radio platforms. An international mobile telephony standards organization has declared OFDM as the main waveform for the 5G mobile service radio access [32]. The combination of MIMO and OFDM techniques results in a system with enhanced performance and more resilience to communication errors. The main drawback of a MIMO-OFDM implementation is the cost, in terms of size, power consumption and hardware complexity. To overcome this challenge, strategies that can lower the energy consumption and the cost of implementation and operation are required.

8.6 CONCLUSION

In this chapter, we have presented an intensive review of the designs implemented on DRAs along with the evolution towards multiple-input multiple-output (MIMO) DRAs. The chapter gives a clear vision of the design parameters of DRA, challenges and advancements in the respective designs. A clear comparison of software required for designing DRA is also given. Addressing the need to implement MIMO DRA, the chapter has also highlighted common misuses and misinterpretations with relation to printed-MIMO antenna systems. Finally, the chapter ends with the realization of the critical aspects while designing, characterizing and evaluating MIMO antenna systems.

REFERENCES

1. D. Soren et al. 2014. Dielectric Resonator Antennas: Designs and Advances. *Progress In Electromagnetics Research B*, Vol. 60, pp: 195–213.
2. A. Petosa. 2007. *Dielectric Resonator Antenna Handbook*. Artech House Antennas and Propagation Library.
3. R. S. Yaduvanshi. 2014. Design and Analysis of Superstrate Embedded Dielectric Resonator Antenna. *Int. J. Ultra Wideband Communications and Systems*, Vol. 3, No. 1, pp: 31–37.
4. J. R. James. 2003. *Dielectric Resonator Antennas*. Research Studies Press Ltd. Baldock, Hertfordshire, England.
5. F. Wang et al. 2019. Ultra-Wideband Dielectric Resonator Antenna Design Based on Multilayer Form. *International Journal of Antennas and Propagation*. pp: 1–10.
6. F. Elmegri et al. 2013. Dielectric Resonator Antenna Design for UWB Applications. Loughborough Antennas & Propagation Conference (LAPC), Loughborough. pp: 539–542.
7. C. Sahoo. 2013. Design and Analysis of Dielectric Resonator Antennas for WLAN Applications. M. Tech Thesis, NIT Rourkela.
8. R. K. Khan, and S. K. Behera. 2013. A Four Element Rectangular Dielectric Resonator Antenna Array for Wireless Applications. IEEE International Conference ON Emerging Trends in Computing, Communication and Nanotechnology (ICECCN), Tirunelveli. pp: 670–673.
9. C. Han and L. Wang. 2016. Array Pattern Synthesis Using Particle Swarm Optimization with Dynamic Inertia Weight. *International Journal of Antennas and Propagation*. pp: 1–7.

10. A. Iqbal and O. A. Saraereh 2016. Design and Analysis of Flexible Cylindrical Dielectric Resonator Antenna for Body Centric WiMAX and WLAN Applications. Loughborough Antennas & Propagation Conference (LAPC), Loughborough. pp: 1–4.
11. G. Varshney et al. 2017. Wide Band Circularly Polarized Dielectric Resonator Antenna With Stair-Shaped Slot Excitation. *IEEE Transactions on Antennas and Propagation.* Volume: 65, Issue: 3. pp: 1380–1383.
12. G. Varshney et al. 2018. Inverted-Sigmoid Shaped Multiband Dielectric Resonator Antenna With Dual-Band Circular Polarization. *IEEE Transactions on Antennas and Propagation*, Vol. 66, No. 4, pp: 2067–2072.
13. G. Varshney et al. 2015. Gain and Bandwidth Controlling of Dielectric Slab Rectangular Dielectric Resonator Antenna. Annual IEEE India Conference (INDICON), New Delhi. pp: 1–4.
14. G. Varshney et al. 2016. Enhanced Bandwidth High Gain Micro-Strip Patch Feed Dielectric Resonator Antenna. International Conference on Computing, Communication and Automation (ICCCA), Noida. pp: 1479–1483.
15. M. S. A. Abousheishaa. 2016. PhD Thesis. Design and Optimization of Dielectric Resonator Antenna Arrays Based on Substrate Integrated Waveguide Technology for Millimeter-Wave Applications. University of Ontario Institute of Technology.
16. G. A. E. Vandenbosch 2009. A Practical Guide to 3D Electromagnetic Software Tools. *IEEE Antennas Propagat. Magazine.* Vol. 51, No. 1.pp: 23–38.
17. M. Hadjloum et al. 2015. An Ultra-Wideband Dielectric Material Characterization Method using Grounded Coplanar Waveguide and Genetic Algorithm Optimization. *Applied Physics Letters. Vol.* 107. pp: 142908–142908-4.
18. S. K. K. Dash. 2018. Modeling of Dielectric Resonator Antennas using Numerical Methods Applied to EPR. Chapter 5, Topics From EPR Research, Ahmed M. Maghraby, Intech Open.
19. A. Perron. 2020. Dielectric resonator antenna (DRA) design and analysis utility (https://www.mathworks.com/matlabcentral/fileexchange/22480-dielectric-resonator-antenna-dra-design-and-analysis-utility), MATLAB Central File Exchange. Retrieved May 17, 2020.
20. G. Varshney et al. 2015. Conical Shape Dielectric Resonator Antenna for Ultra Wide Band Applications. International Conference on Computing, Communication and Automation. IEEE. pp:1304–1307.
21. X. S. Fang et al. 2017. Wideband and Bandwidth-Controllable Hybrid Cylindrical Glass Dielectric Resonator Antenna for Indoor Communication. *IEEE Access.* pp:1–9.
22. R. Khan et al. 2016. A Reconfigurable Dielectric Resonator Antenna with Pattern Diversity for DVB-H Application. International Conference on Intelligent Systems Engineering (ICISE), Islamabad, IEEE. pp: 222–225.
23. G. Varshney et al. 2019. Obtaining the Circular Polarization in a Nanodielectric Resonator Antenna for Photonics Applications. *Semiconductor Science and Technology Letter.* IOP Publishing. pp: 1–8.
24. X. S. Fang1 et al. 2019. Design of the wide dual-band rectangular souvenir dielectric resonator antenna. *IEEE Access.* Vol. 7, pp: 161621–161629.
25. A. I. Bugaje et al. 2018. Design of Dual Band Stacked RDRA for 5G Applications. *TELKOMNIKA Telecommunication, Computing, Electronics and Control.* Vol. 16, No. 3. pp: 31–37.
26. A. M. Elshirkasi et al. 2019. Envelope Correlation Coefficient of a Two-Port MIMO Terminal Antenna under Uniform and Gaussian Angular Power Spectrum with User's Hand Effect. *Progress In Electromagnetics Research C*, Vol. 92. pp: 123–136.
27. Y. A. S. Dama et al. 2011. *International Journal of Antennas and Propagation.* Hindawi Publishing Corporation. pp: 1–7.

28. A. A. Khan et al. 2016. Design of a Dual-Band MIMO Dielectric Resonator Antenna with Pattern Diversity for WiMAX and WLAN Applications. *Progress in Electromagnetics Research M*, Vol. 50. pp: 65–73.
29. B. Collians. 2005. Antenna for MIMO systems. Next Generation Wireless Networks Workshop. U of Edinburg.
30. M. S. Sharawi. 2017. Advancements in MIMO Antenna Systems. *Chapter in the Developments in Antenna Analysis and Synthesis.* IET.
31. H. S. Singh 2019. Investigations of MIMO Antenna for Smart Mobile Handsets and Their User Proximity. *Medical Internet of Things (m-IoT): Enabling Technologies and Emerging Applications, Intechopen.* pp: 23–38.
32. M. S. Sharawi 2017. Current Misuses and Future Prospects for Printed Multiple-Input, Multiple-Output Antenna Systems. *IEEE Antennas and Propagation Magazine.* pp: 162–170.

9 Frequency Synthesizers and Their Applications in Signal Processing

Govind Singh Patel
IIMT College of Engineering Greater Noida

CONTENTS

9.1 Introduction 131
9.2 Phase-Locked Loop 132
    9.2.1 Phase Detector 132
    9.2.2 Loop Filter 132
    9.2.3 Voltage-Controlled Oscillator 133
    9.2.4 Divider 133
    9.2.5 Application of PLL 133
9.3 Direct Digital Synthesis 134
    9.3.1 Phase Accumulator 134
    9.3.2 Phase-to-Amplitude Converter (ROM/LUT) 135
    9.3.3 Digital-to-Analog Converter and Filter 135
    9.3.4 Application of DDS 136
9.4 Direct Analog Synthesis 136
9.5 Summary 137
References 138

9.1 INTRODUCTION

This work presents modern techniques used to modulate and generate signals for many applications, such as communication, electronic imaging, radar, etc. These techniques generate sine waveform in terms of phase, frequency and amplitude. In the next subsection, the types of frequency synthesis techniques are presented.

There are three types of frequency synthesizers.

1. Phase-locked loop
2. Direct digital synthesis
3. Direct analog synthesis

9.2   PHASE-LOCKED LOOP

Phase-locked loop is a device that is used to provide digital and modulate sine waves to many applications. It consists of a phase detector, loop filter, voltage-controlled oscillator and divider. Phase means that it compares phases between reference input and loop feedback input [1–3]. Locked means it compares frequencies, whether in range or not. If there is no difference in frequency, then no action is required; otherwise it needs to match this frequency difference through the divider. Third is loop. The second input of the phase detector is in the form of a feedback loop. So it is called phase-locked loop.

9.2.1   PHASE DETECTOR

A phase detector is a device which is used to detect inputs in the form of frequency or phase. It produces corresponding output to filter for further processing. There are many types of phased detectors, such as logic gate–based phase detector, charge pump–based phase detector, gate and flip-flop-based phase detector. A logic gate-based phase detector is used to compare digital inputs and give corresponding output. The next detector is a charge pump; it is a combination of D flip-flop, capacitor and gates. And the last detector is the flip-flop-based detector; it is used in both digital and analog types of PLL.

9.2.2   LOOP FILTER

The architecture of the PLL is shown in Figure 9.1. It is second part of the PLL. It filters phase or frequency according to the requirement of the output. There are many types of filters, such as low-pass filter, high-pass filter and band-pass filter.

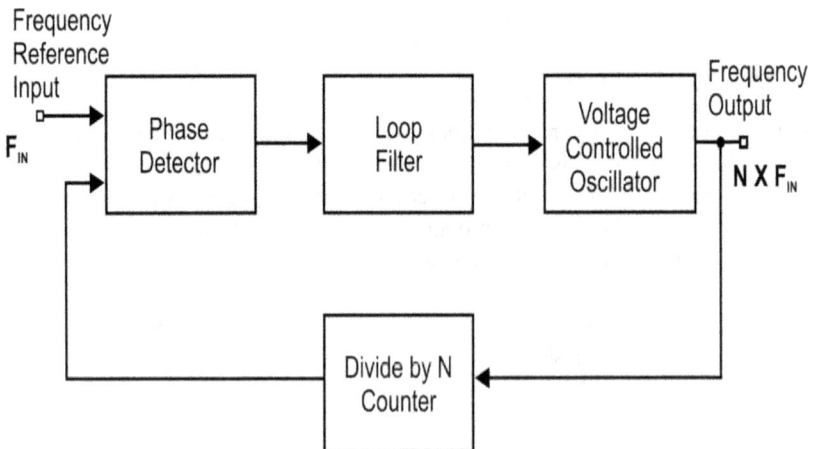

FIGURE 9.1   Architecture of phase-locked loop.

Sometimes loop filter works in conjunction with a phase detector. That type of filter is called a charge pump phase detector. It is generally used whenever active and passive poles can be analyzed to determine the stability of the system. Location of these poles may be determined according to the stability of the circuit/system.

9.2.3 VOLTAGE-CONTROLLED OSCILLATOR

The voltage-controlled oscillator (VCO) is the third part of the PLL. It is a very important device which is used to generate frequency corresponding to voltage input that is applied to it. Output of the VCO is used in many applications, such as receivers, wireless communication systems, encoding techniques, radar, etc. Many typed of VCO are available in the market, such as crystal oscillator, cross-coupled oscillator, RC phase-shift oscillator, Wein bridge oscillator, Colpitts oscillator, etc. The RC phase-shift oscillator is used to design this proposed architecture.

9.2.4 DIVIDER

A divider is a device which is used to average high frequencies and produce feedback-loop output to a phase detector. A fractional frequency synthesizer uses a delta modulator for higher-frequency range synthesis.

9.2.5 APPLICATION OF PLL

Figure 9.2 shows an application of PLL in a communication system. It synchronises input data with the transmitter for information transmission through the channel.

A PLL has a reference and transmission clock to synchronize the transmission. Let us consider that Tx data is the input data and Tx clk is the clock apply to it. Rx clk is the clock that is used at receiver. Input data may be 8 bit or 16 bit for the transmission. This data may be transmitted through the channel and received by the receiver. In both sides, the clock for transmitter and receiver should be synchronized to achieve synchronized output. Figure 9.2 shows the waveform of Tx data with their respective Tx clk and Rx clk. How data is transmitted from transmitter to receiver

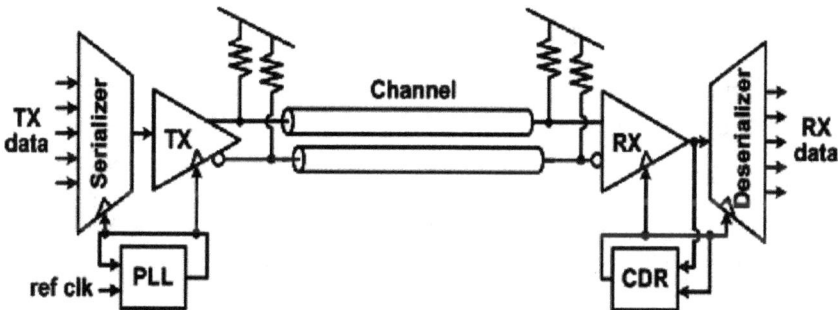

FIGURE 9.2   Application of PLL in communication.

through channel is shown in the waveform. How Tx clk and Rx clk clocks are synchronised with channel is presented in the waveform. This is the most important application of signal processing using frequency synthesizers [4, 5].

Other applications PLL are as follows:

- PLL can be used as frequency multiplication.
- It can also be used as frequency translation.
- It is used in amplitude detection.
- It is used in frequency detection.
- PLL generally is used in frequency shift keying demodulation.
- It is used in in control system to control DC motors.
- It is used as a frequency synthesizer.
- It is also used to generate different clocks.
- It is used as a spread spectrum frequency synthesizer.
- Many demodulators use PLL in demodulations such as SSB, QAM, QPSK, FSK, FM.
- Reduced jitter of the circuit also can be measured using PLL.

9.3   DIRECT DIGITAL SYNTHESIS

Direct digital synthesis (DDS) consists of a phase accumulator, look up table (LUT) and digital to analog converter (DAC). It is used to generate and modulate frequency for digital communication applications. In the next subsection, phase accumulators, memory and digital-to-analog converters have been analyzed with their applications [6–9].

9.3.1   PHASE ACCUMULATOR

A phase accumulator is an important part of the DDS shown in Figure 9.3. It accumulates frequency control word inputs on each clock. The accumulated outputs can be calculated using the following mathematical equation as follows:

$$f_o = \frac{FIW}{2^M} f_c$$

where

f_o = Frequency output
FIW = Frequency of input word
f_c = Clock frequency
M = Accumulator length
T = Clk period

Output of the phase accumulator depends upon frequency of input words, clock frequency and accumulator length.

The output of the accumulator is given as follows:

$$S(n) = S(n-1) + FIW$$

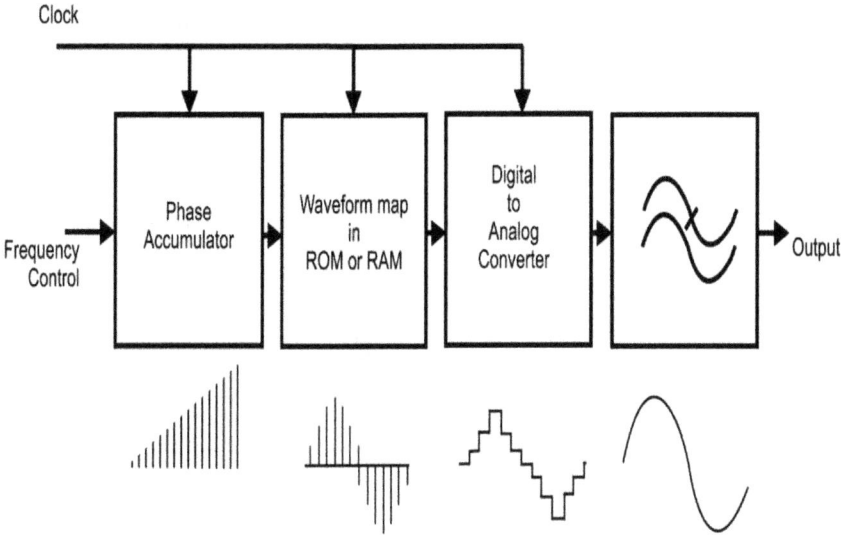

FIGURE 9.3 Block diagram of DDFS.

where
  $S(n)$ = Acc. O/P
   n = Clk tick

$S(n)$ is the accumulative output; it depends upon clock tick and frequency input words. It is accumulated at each clock tick to a specified time period.

9.3.2 PHASE-TO-AMPLITUDE CONVERTER (ROM/LUT)

It is a LUT. This memory is used to store the accumulated output of the phase accumulator. The storage capacity can be improved by designing novel architecture of the ROM. That design may consist of multiplexers, adders, subtractors and registers. A multiplexer takes multiple inputs and gives a single output, but output depends upon multiple factors corresponding to inputs. These multiplying factors are control by the clock control unit. Many registers, adders and subtractors are used to store more data in less space properly. New techniques have been developed using these components.

9.3.3 DIGITAL-TO-ANALOG CONVERTER AND FILTER

A digital-to-analog converter and filter converts digital signals into analog signals for further processing. That analog output is used to control other applications. Sometimes direct digital signals can be used in many applications. Sample and hold circuits are used to convert an analog signal. There are three techniques to complete conversion process. First is sampling, a process to sample digital signal using approximate technique. Second is quantization, a process to quantize sampled signals for the further processing. Finally is encoding, a process to encode digital signals into

FCLOCK (fc) DDS DAC OUT RECONSTRUCTION FILTER FILTER OUT LIMITER CLOCK OUT

IDEAL TIME DOMAIN RESPONSE

IDEAL FREQUENCY DOMAIN RESPONSE

ODD HARMONIC SERIES

"REAL WORLD" FREQUENCY RESPONSE

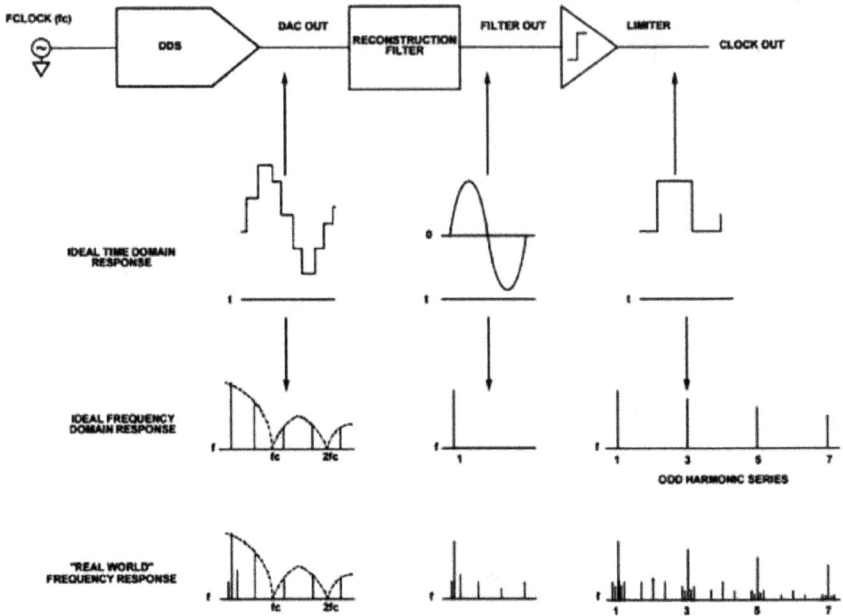

FIGURE 9.4 Application based on DDS.

analog signals. The combination of these three techniques is called digital-to-analog conversion. The converted output is given directly to the corresponding application for which it is being designed.

9.3.4 APPLICATION OF DDS

Figure 9.4 shows an application of DDS in digital communication. In this application, the frequency control word plays important role to divide higher frequencies for averaging.

It is also called a fractional-N PLL frequency synthesizer. Its application is shown in Figure 9.4. Output frequency is divided by a digital frequency selection logic circuit, which averages high-frequency output in terms of feedback-loop input. All digital architecture is also available for particular applications [10–13]. In that case all parts are digital components which may improve performance in terms of power, area and noise. In a few of these architectures, DDS is used as feedback-loop component. That type of circuit is also called as hybrid frequency synthesizer.

9.4 DIRECT ANALOG SYNTHESIS

Direct analog synthesis (DAS) uses arithmetic operations to convert reference input to required output in the frequency domain. But this technique does not allow closed-loop feedback. It generates different frequencies with multiplexers, mixer, filtering and division.

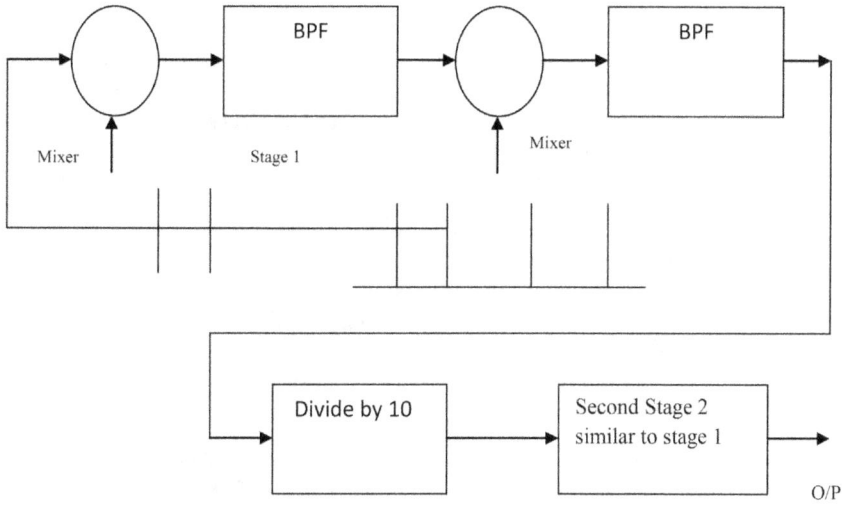

FIGURE 9.5 Direct digital design using multiple references.

This design covers 16.0 to 16.99 MHz output frequency range, 10 MHz reference frequency with step size 10-kHz(0.01 MHz). This design requires following references frequencies such as 14, 16, 18, 130, 131MHz. Figure 9.5 shows DDS using multiple references.

The spectral purity depends on references input and the speed of the synthesizer. There are two stages; in the first stage 1, a mixer is used to mix feedback input and references input. After mixing, a band-pass filter is used to filter the required frequency [14, 15].

This operation is complex; therefore it can be divided in many stages. In the first stage, the reference input and feedback-loop input are mixed by a mixer and filter out required signal using a band-pass filter. In the second stage, frequency is divided by a divider for second-stage processing at a different frequency range.

This work analyzed three main frequency synthesis techniques, in which the parameters of PLL, DDS and DAS have been presented and analyzed. The work shows interfacing of different architectures with their applications. It found better results in terms of analysis of parameters and interfacing with the applications corresponding to the respective architecture of frequency synthesizer.

9.5 SUMMARY

This chapter presented various types of frequency synthesizers and their applications in the fields of communication, digital encoding techniques, transceivers, radar, etc. The parameters of this technique have been analyzed using ADS and MATLAB tools. Mathematical equations of these techniques have also been derived for respective applications. The parameters in terms of frequency, phase, noise, area and power have been analyzed to correspond to the applications.

REFERENCES

1. Goldberg B. G. (1999). *Digital Frequency Synthesis Demystified*. 3rd ed., USA, LLH Technology Publishing.
2. Hsich G. C. & Hung J. C. (1996). Phase locked loop techniques: A survey, *IEEE Trans. Ind. Electron.*, vol. 43, no. 6, pp. 609–616.
3. Prasad V. & Sharma C. (2012). A review of phase locked loop, *Int. J. Emerging Tech. & Advanced Eng.*, Bangalore, vol. 2, no. 6, pp. 2250–2459.
4. Lata K. & Kumar M. (2013). All digital phase-locked loop: A survey, *Int. J. Future Computer & Commun.*, vol. 2, no. 6, pp. 551–555.
5. Smedt B. D. & Gielent G. (1998). Nonlinear behavioural modeling and phase noise evaluation in phase locked loops, in Proceedings of IEEE Custom Integrated Circuits Conf., ESAT-MICAS., Santa Clara, CA, pp. 53–56.
6. Larson D. C. (1998). High speed direct digital synthesis techniques and applications, in Proceedings of 20th Annu. IEEE Gallium Arsenide Integrated Circuits Symp., Atlanta, pp. 209–212.
7. Caro D. D. & Strollo A. M. (2005). High-performance direct digital frequency synthesizers using piecewise-polynomial approximation, *IEEE Trans. Circuits Syst. I, Reg. Papers*, vol. 52, no. 2, pp. 324–337.
8. Langlois J.M.P. & Khalili D. A. (2003). Piecewise continuous linear interpolation of the sine function for direct digital frequency synthesis, in Proceedings of IEEE Microw. Symp. Digest, Philadelphia, vol. 1, pp. 65–68.
9. Mohieldin A. N., Emira A. A. & Sinencio E. S. (2202). A 100-MHz 8-mW ROM-less quadrature direct digital frequency synthesizer, *IEEE J. Solid State Circuits*, vol. 37, no. 10, pp. 1235–1243.
10. Mortezapour S. & Lee E. K. F. (1998). Design of low-power ROM-less direct digital frequency synthesizer using nonlinear digital-to-analog converter, *IEEE J. Solid State Circuits*, vol. 34, no. 10, pp. 1352–1359.
11. Yang B. D., Choi J. H., Han S. H., Kim L. S. & Yu H. K. (2004). An 800-.hz low-power direct digital frequency synthesizer with an on-chip d/a converter, *IEEE J. Solid State Circuits*, vol. 39, no. 5, pp. 761–767.
12. Ashrafi A. & Adhami R. (2005). A direct digital frequency synthesizer utilizing quasi-linear interpolation method, in Proceedings of 37th IEEE South Eastern Symp., Alabama Univ., USA, pp. 144–148.
13. Eltawil A. M. & Daneshrad B. (2002). Piece-wise parabolic interpolation for direct digital frequency synthesis, in Proceedings of IEEE Custom Integrated Circuits Conf., Los Angeles, pp. 401–404.
14. Saber M. S., Elmasry M., & Abo-Elsoud M. E. (2006). Quadrature direct digital frequency synthesizer using FPGA, in Proceedings of IEEE Int. Conf. Comp. Eng. Syst., Cairo, pp. 14–18.
15. Hsu L. W. & Chang D. C. (2005). Design of direct digital frequency synthesizer with high ROM compression ratio, in Proceedings of 12th IEEE Int. Conf. Electron., Circuits Syst., Gammarth, pp. 1–4.

10 Design of Ultra-Low Power OTA Based on Subthreshold Operation with High Gain, Large Transconductance and Small Area

Simran Somal, Tripti Sharma and Krishan Mehra
Chandigarh University

CONTENTS

10.1 Introduction 139
10.2 Conventional Operational Transconductance Amplifier 140
10.3 Proposed Operational Transconductance Amplifier 141
10.4 Experimental Setup 142
10.5 Layout of Proposed OTA Circuit 142
10.6 Calculations of Parameters of the Circuit.. 143
    10.6.1 Gain and Gain Bandwidth 143
    10.6.2 Slew Rate 143
    10.6.3 Power Efficiency 144
10.7 Implementation and Results 144
10.8 Conclusion 146
References 146

10.1 INTRODUCTION

If we look towards the latest reports of semiconductor industry, the voltage supplied to drive circuits are in the range of 0.5 V. For portable/battery-operated devices, bio-medical devices, mixed signal circuits and sensor applications, voltage supplies are targeted to be 0.3 V by 2022. The basic usable circuit in all analog circuits and mixed signal circuits, is an operational amplifier. Amplifiers are circuits which require high power. Various kinds of operational amplifiers such as cascode, folded cascode and telescopic, etc., are used. From all these, operational transconductance amplifiers are a popular choice for designers. For making ultra-low power operational amplifiers, a weak inversion or subthreshold region is an efficient approach.

Folded cascode circuits, which used p-type metal-oxide-semiconductors (PMOS) as driver circuits, were considered previously. They have advantages such low flicker noise and high non-dominant pole. But due to low gain these circuits are not used currently [1].

Low-power OTAs are also present, using bulk-driven input stages. In this type of circuit, bulk terminals are used for input rather than applying input on a gate terminal. But due to lack in transconductance (as g_{mb} is 5 times smaller than g_m), this technique also has less gain and unity gain bandwidth [2].

For enhancing the phase margin, a recycling folded cascode (RFC) technique with current mirrors is used. Current mirrors are circuits which require high power. So they are not suitable for low-power applications. Current mirrors also require a larger area. Tank and colleagues [3] present an RFC op-amp which consumes only 724 nW power. But this circuit lacks phase margin and transconductance. For making very high-speed circuits, these types of amplifiers are not useful [4–12].

As we decrease the supply voltage level, the feature size is also going to decrease. By this $g_m r_o$ tends to decrease rapidly. If we are going to reduce the technology, the gain of the circuit is reduced by 3–6 dBs. This paper proposes the OTA circuit which develops in 45 nm technology under Cadence Virtuoso environment [3].

The proposed circuit works on 0.5 V power supply that significantly reduces power consumption without decreasing gain and unity gain bandwidth. In this circuit, the biasing current is recycled again and again, and all transistors work under a weak inversion region. The paper first presents the conventional circuit and then the enhanced OTA circuit. The layout of the proposed OTA is also presented, followed by results that are concluded after simulation.

10.2   CONVENTIONAL OPERATIONAL TRANSCONDUCTANCE AMPLIFIER

The circuit reported in Reference [13] is shown in Figure 10.1. The circuit is a type of OTA which uses current split technique for enhancing the gain and transconductance. The split current is provided to the two pairs of differential amplifiers: M1a, M1b, M2a and M2b. The transistors which are used for input stage are selected in W/L ratio as (1-x)/x for M1a and M1b. In similar way, for differential pair M2a and M2b the W/L ratios are selected as (1-y)/y, where x and y are selected as x+y=1.

Transconductance of this circuit shows 450% improvement compared to various folded cascode (FC) and RFC OTAs. By taking the smaller value of x, transconductance shows further improvement. But the choice of a small value will show decay in the phase margin. The value of transconductance should be regulated, because a lower value of g_m will make the amplifier unstable. The values of parameter x and y selected for this circuit are x=1/3 and y=2/7. VC and VB are selected for providing current to the M0 and MB transistors.

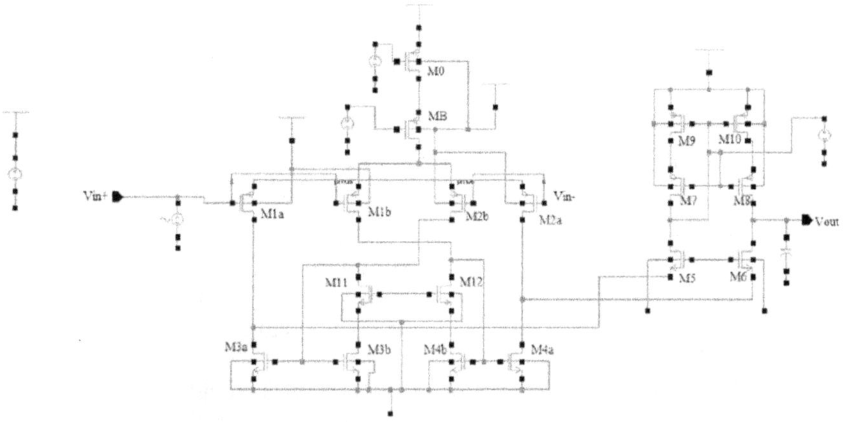

FIGURE 10.1   Conventional RFC OTA. (From [13].)

10.3 PROPOSED OPERATIONAL TRANSCONDUCTANCE AMPLIFIER

The circuit shown in Figure 10.2 is an enhancement in the circuit discussed in Reference [14]. This circuit is based on the same split-current technique. VB is applied to provide current to M1a and M2a. Similarly, VC is applied to provide current to M1b and M2b.

In this circuit, all transistors are working in a subthreshold region. A subthreshold region is an ultra-low power region, so if we need to design a circuit that has low

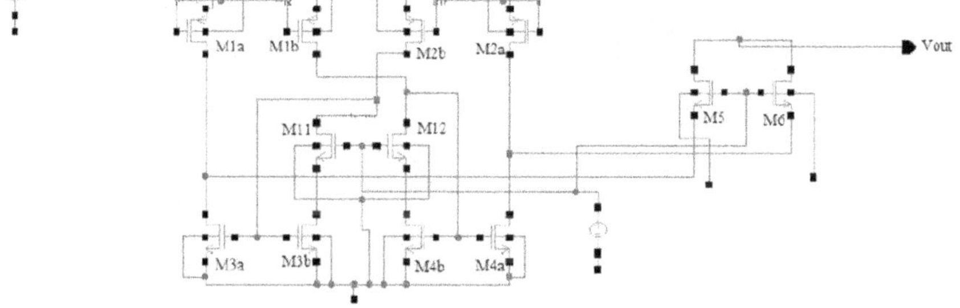

FIGURE 10.2   Proposed operational transconductance amplifier.

power consumption, we need to design the circuit in a subthreshold region, in which all the conduction is going through leakage current only.

10.4 EXPERIMENTAL SETUP

First, the circuit works on a voltage of 0.5 V. A DC voltage source VB is used to provide biasing to M0; VC is used to provide biasing to MB. Another DC voltage source Vs is used between gates M11, M12, M5 and M6.

When VB and VC are biased properly to an operating point, they start driving a current which is provided to the M0 and MB. The current drawn by VB is provided to the differential pair M1a and M2a; similarly, current drawn by VC is provided to differential pair M1b and M2b.

An AC signal input voltage of amplitude 250 mVpp is provided with a frequency of 1 GHz to the input pair M1a and M1b through Vin+ port. Similarly, a DC voltage source with AC magnitude of 220 V and DC magnitude of 300 mV–700 mV is provided through Vin-port to the M2a and M2b input-stage transistors.

The current from these differential pairs is collected at the drains of M3a and M4a. M11 and M12 are cross coupled with M1b and M2b. The cross coupling is used for increment in gain of the circuit. The current collected at M3a and M4b is provided to M5 and M6, respectively. That current drives the both transistors and we get output voltage through Vout port.

10.5 LAYOUT OF PROPOSED OTA CIRCUIT

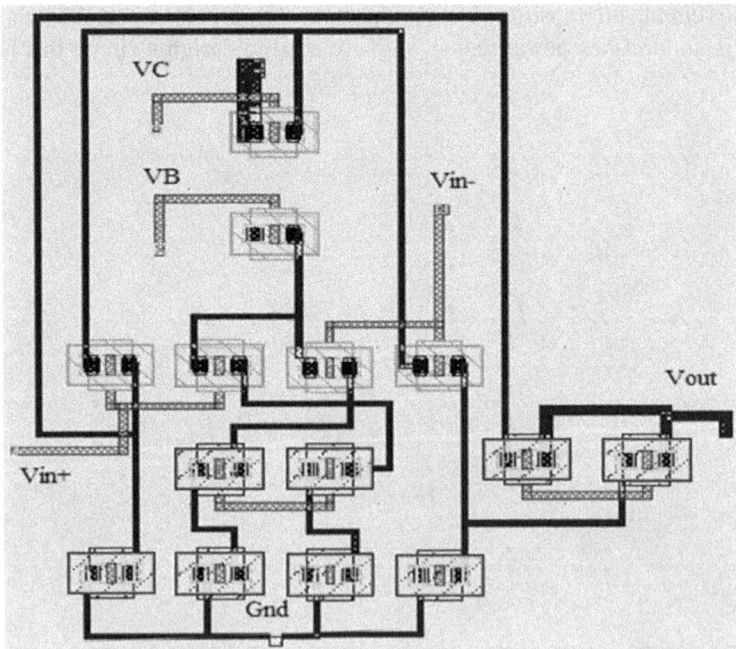

FIGURE 10.3   Layout of proposed OTA circuit.

10.6   CALCULATIONS OF PARAMETERS OF THE CIRCUIT

10.6.1   GAIN AND GAIN BANDWIDTH

For any amplifier circuit which is built using MOS transistors, gain of the circuit is always product of transconductance and the output resistance.

$$a_v = g_m r_o \tag{10.1}$$

where g_m represents small signal transconductance, r_o is small output resistance and av is small signal gain. This is the small signal transconductance and can be calculated only for an individual transistor.

$$Av = GmRout \tag{10.2) [13]}$$

From here transconductance can be calculated as,

$$Gm = \frac{Av}{Rout} \tag{10.3) [13]}$$

where Av is the gain of an amplifier, Gm is transconductance which is calculated for whole amplifier circuit and Rout is the output resistance which can be calculated by DC analysis while simulating the circuit.

Also the gain bandwidth (GBW) product can be calculated from here,

$$GBW = \frac{Gm}{2\pi C} \tag{10.4) [31]}$$

As there is no load capacitance in the circuit, C is the value of parasitic capacitances. As parasitic capacitances are very small in value, the value of C can also be neglected. GBW is equally important for an amplifier as from here we can calculate after which frequency level gain is going to be decreasing.

10.6.2   SLEW RATE

When there is any high frequency of input signals, transistors M1a and M1b go in the cut-off region. Therefore all the transistors that are taking current from these two transistors, such as M4a, M4b and M5, will move in the triode region. So the entire DC current, which is double that of the biasing current, will pass through M2b, M3a, M3b and M6 and give the output voltage. So, the slew rate for this condition can be calculated as,

$$SR = 2\frac{(2-x)Ib}{(1-y)C} \tag{10.5) [13]}$$

where *Ib* is the biasing current and *C* is the parasitic capacitances, x is taken as 1/3 and y is taken as 2/7.

10.6.3 POWER EFFICIENCY

Power efficiency is an important parameter for an ultra-low power amplifier. Power efficiency of the amplifier can be calculated as the ratio of transconductance to the biasing current.

$$PE = \frac{Gm}{2Ib}$$ (10.6) [13]

As biasing current is only leakage current, which is provided to the circuit, power efficiency of the circuit shows better results than the previously reported amplifier circuit.

10.7 IMPLEMENTATION AND RESULTS

The simulations of present OTA are done on Cadence software and a Spectre simulator. The technology node used for circuit is 45 nm. The circuit is designed in such a way that only M1a, M1b, M2a and M2b are working in the linear region. Remaining all transistors are working in the subthreshold region. The biasing current is only the leakage current, which is calculated by DC analysis of the circuit. Biasing current is the bulk current of an input-stage transistor. The sizing used for transistors is reported in Table 10.1.

The proposed circuit provides 21.83 dB, a larger gain than the TRFC (transconductance recycling folded cascode) OTA circuit. It shows approximately 700% improvement in GBW of the circuit. The phase margin shows not much improvement because of working in subthreshold region. Transconductance is calculated by

TABLE 10.1
Aspect Ratio for TRFC and OTA Circuit

Transistor	TRFC (μm) [13]	OTA (μm)
M0	128/0.5	70/0.12
M1a, M2a	64/0.36	45/0.12
M1b, M2b	64/0.36	45/0.12
M3a, M4a	24/0.5	45/100
M3b, M4b	8/0.5	45/100
M5, M6	16/0.18	120/100
M7, M8	64/0.18	-
M9, M10	64/0.5	-
M11, M12	8/0.18	45/100
MB	64/0.18	120/100

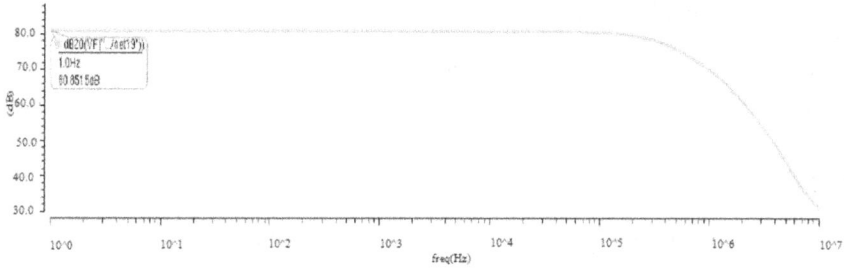

FIGURE 10.4 Gain response of OTA circuit.

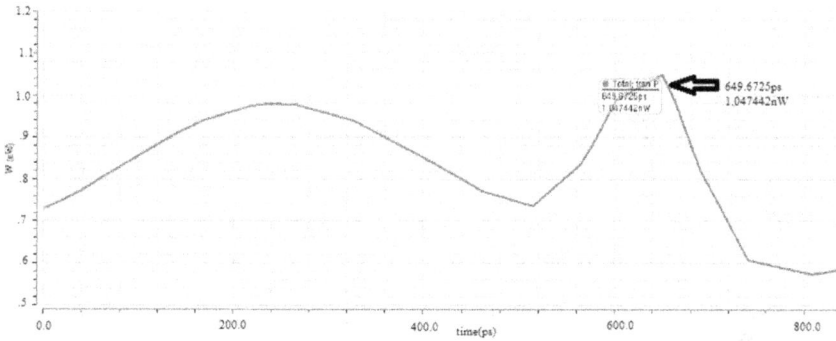

FIGURE 10.5 Power response of OTA circuit.

TABLE 10.2
Comparison of OTA with Published OTA Circuits

Parameter	[15]	[16]	[1]	[13]	OTA
Technology (nm)	180	180	50	180	45
Supply Voltage (V)	0.6	0.5	0.5	0.5	0.5
Gain (dB)	82	106.3	74	59.02	80.851
GBW (MHz)	0.0191	0.033	4.8	0.131	71.08
Slew Rate (V/μs)	.012	.028	4.4	0.000072	0.024
Phase Margin (°)	60	102.4	49	72.56	78.96
Power (W)	401n	369n	100μ	23.7μ	1.047n
Transconductance (μS)	-	-	-	6.03 to 70	80 to 452.96
Area (μm^2)	-	-	-	315.02	19.2
Power Efficiency	-	-	-	7MW	1411.74MW
FoM1 (KHz.pF/μW)	716.25	1365	960	166	113.98
FoM2 (V/ms.pF/nW)	450	1135	680	162	102.3n

variations in the VB voltage source. It is reported as by variating the voltage from −100 mV to 300 mV. Transconductance is calculated as 80 μS for lower levels and 452.96 μS for maximum voltage level. Thus, we can say that the circuit is completely tunable and works for any analog application.

10.8 CONCLUSION

The proposed OTA circuit gives a greater enhancement in performance than the conventional transconductance amplifiers. Power efficiency shows that this circuit is better than all other circuits for using in ultra-low power regions. Gain shows very much improvement over others. As the circuit is built in a lower-technology node, the area of the circuit is very small, which is required nowadays. Transconductance is tunable, so any type of analog circuit is built using this amplifier. Further enhancement can be achieved by varying the phase margin.

REFERENCES

1. G. Di Cataldo, A. D. Grasso and S. Pennisi, "Two CMOS current feedback operational amplifiers," *IEEE Transactions on Circuits and Systems II: Express Briefs*, vol. 54, pp. 944–948, 2007.
2. M. Yavari, "Single-stage class AB operational amplifier for SC circuits," *Electronics letters*, vol. 46, pp. 977–979, 2010.
3. N. Tang, W. Hong, J.-H. Kim, Y. Yang and D. Heo, "A sub-1-V bulk-driven opamp with an effective transconductance-stabilizing technique," *IEEE Transactions on Circuits and Systems II: Express Briefs*, vol. 62, pp. 1018–1022, 2015.
4 F. Rezaei and S. J. Azhari, "Ultra low voltage, high performance operational transconductance amplifier and its application in a tunable Gm-C filter," *Microelectronics Journal*, vol. 42, pp. 827–836, 2011.
5. R. Singh, Y. Audet, Y. Gagnon, Y. Savaria, É. Boulais and M. Meunier, "A laser-trimmed rail-to-rail precision CMOS operational amplifier," *IEEE Transactions on Circuits and Systems II: Express Briefs*, vol. 58, pp. 75–79, 2011.
6. M. R. V. Bernal, S. Celma, N. Medrano and B. Calvo, "An ultralow-power low-voltage class-AB fully differential opamp for long-life autonomous portable equipment," *IEEE Transactions on Circuits and Systems II: Express Briefs*, vol. 59, pp. 643–647, 2012.
7. T. Kulej, "0.5-V bulk-driven CMOS operational amplifier," *IET Circuits, Devices & Systems*, vol. 7, pp. 352–360, 2013.
8. F. Esparza-Alfaro, S. Pennisi, G. Palumbo and A. J. Lopez-Martin, "Low-power class-AB CMOS voltage feedback current operational amplifier with tunable gain and bandwidth," *IEEE Transactions on Circuits and Systems II: Express Briefs*, vol. 61, pp. 574–578, 2014.
9. M. Akbari and O. Hashemipour, "Enhancing transconductance of ultra-low-power two-stage folded cascode OTA," *Electronics Letters*, vol. 50, pp. 1514–1516, 2014.
10. Z. Alsibai and S. Bay Abo Dabbous, "Ultra-low-voltage low-power bulk-driven quasi-floating-gate operational transconductance amplifier," *Advances in Electronics*, vol. 2014, 2014.
11. M. Akbari, S. Biabanifard, S. Asadi and M. C. Yagoub, "Design and analysis of DC gain and transconductance boosted recycling folded cascode OTA," *AEU-International Journal of Electronics and Communications*, vol. 68, pp. 1047–1052, 2014.

12. R. Y. Chen and Z.-Y. Yang, "CMOS transimpedance amplifier for visible light communications," *IEEE Transactions on very large scale integration (VLSI) systems*, vol. 23, pp. 2738–2742, 2014.

13. A. S. Khade, V. Vyas, M. Sutaone and S. Musale, "Transconductance enhancement of a low voltage low power recycling folded cascode OTA using an asymmetrical current split input stage," *Microelectronics Journal*, vol. 91, pp. 53–60, 2019.

14. A. Ragheb and H. Kim, "Ultra-low power OTA based on bias recycling and subthreshold operation with phase margin enhancement," *Microelectronics Journal*, vol. 60, pp. 94–101, 2017.

15. M. Akbari and O. Hashemipour, "A 0.6-V, 0.4-µW bulk-driven operational amplifier with rail-to-rail input/output swing," *Analog Integrated Circuits and Signal Processing*, vol. 86, pp. 341–351, 2016.

16. T. Sharan and V. Bhadauria, "Sub-threshold, cascode compensated, bulk-driven OTAs with enhanced gain and phase-margin," *Microelectronics Journal*, vol. 54, pp. 150–165, 2016.

11 Artificial Intelligence for Precision Medicine

R.A. Rayan
University of Alexandria

C. Tsagkaris
University of Crete

CONTENTS

11.1   Introduction 149
11.2   Precision Medicine 150
11.3   Big Data and PM 151
11.4   AI and Advancing PM 152
11.5   Overlap between AI, ML and DL 153
11.6   Application of AI in Genetic Interpretation 153
11.7   Application of AI in Image-Based Diagnosis 153
11.8   Application of AI in Data-Driven Medicine 154
11.9   Limitations 154
11.10 Conclusions 155
References 155

11.1   INTRODUCTION

In 2015, precision medicine (PM), also known as precision health, came to light when President Obama announced the Precision Medicine Initiative (PMI). PM is an innovative technique to prevent and treat illnesses and promote health according to subjective variables such as genes, lifestyle, socioeconomic criteria and ecosystems. It is a substitute for conventional one-way medicine and facilitates customizing personal health care. Technology and data could add innovative perspectives in health and illness, hence artificial intelligence (AI) and big data would play significant roles in PM [1].

Data is the basic element of PM, and nowadays technology enables us to create and assemble more data. All principal fields have endorsed big data, for instance, genetic data, for gaining more perspectives. Given the progress in high-throughput sequencing techniques and human genomics, it is feasible to produce vast amounts of sequencing, such as whole-genome sequencing (WGS) data, with relatively less expense. PM initiatives worldwide attempt to sequence millions of volunteers. Broadly implementing electronic health records (EHRs) has driven a fast buildup of data that created a wealthy repository of genetic studies and other biomedical research fields. Integrating data from the EHRs and from genomics has led to a novel surge of innovations. Modern techniques allow collection of immense volumes of data from mobile wearable devices, the internet and social media or other tools that contain personal living data, socioeconomic factors or ecological exposures. Interpreting big data for application needs big data analytics, leading to potential advances from such interpretations [2].

AI, a significant technology in the big data context, is broadly applied to several disciplines. Despite the lack of knowledge about the operating algorithms, AI is becoming a key element of everyday living. Many applications of AI involve voice, face or script recognition; natural language processing; robotics; and auto-driven cars. AI is steadily transforming medicine and influencing the decisions of healthcare stakeholders as well, being deployed in image-centred diagnosis in several clinical specialties, including radiology, pathology, ophthalmology and dermatology. At Aravind Eye Hospital in India, Google applied AI technology to detect retinal disorders from retina photos of diabetics. AI has likewise excelled in the genetic variants for sequencing the next-generation data [3].

11.2 PRECISION MEDICINE

The concept of PM is essentially Hippocratic. Unravelling the aetiological factors behind any disease and offering evidence-based treatment to reverse those factors summarizes the manifesto of Hippocratic medicine. To our knowledge, it also represents the first traces of the aspiration that treatment could be provided on a precise and personalized basis. It was also when Protagoras raised the claim that "of all things, the measure is a man" consolidating the notion of personalization. Throughout the next centuries, medicine was dominated by an interplay and sometimes even a conflict between human intelligence and supernatural beliefs. William Harvey's work in circulation and the anatomical dissections of Andreas Vesalius highlighted the potential of human intelligence in biomedical science and healthcare provision. Throughout the years, sound scientific methods can promote innovation better than eminence or expert status [4].

PM, also called precision health or personalized medicine, is an innovative technique to treat and prevent illnesses and promote health by considering disparities among subjects – for example, behaviours, genes, lifestyle, ecosystem and socioeconomics – for fostering customized personal health care rather than applying a similar care approach for everyone. Technology and data can provide modern perspectives on health and illness; hence, AI and big data will contribute significantly to PM, leveraging the evolving data and technologies to better control patients' health and illnesses through more accurate diagnosis and prioritized treatments. This will enable health care to achieve optimum results with the delivery of better services and minimize expenses of repeated examinations and useless therapies. Several nations, supported by large funds, have invested in big data collections and genomics to develop PM initiatives [1].

For instance, the United States (US) committed a large sum of money to start the PMI in 2015, where the "All of US Research Program" targeted recruiting millions of participants to create a vast research population nationwide and gather various types of data on health, including genetics, urinalysis or data from EHRs, questionnaires and monitoring recreational activities, for applying PM for all illnesses. The United Kingdom (UK) started the 100 000 Genomes Project for sequencing the genes of all the patients in the National Health Service (NHS) and introduced genomic medicine service in 2012. Patients' EHR data has been assembled and integrated into the genomics data where this de-identified data is accessible for exploring in a controlled and protected setting. The United Kingdom invested in PM using the AI approach to enhance data for early diagnosis and better patient outcomes; whereas the government began financing the gene sequencing of a huge number of volunteers for better insights into the mechanisms of illnesses in 2018. The Chinese started a huge and expensive 15-year programme in 2016 for gene sequencing of 100 million humans by 2030 and built a centrally coordinated data hub for PM. China endorses PM initiatives that use big data analytics, involving AI and machine learning (ML) to enhance diagnosing and managing illnesses [5].

11.3   BIG DATA AND PM

In the past few decades, the osmosis of medicine with a matrix of biomedical scientific disciplines from biology and chemistry to computer science has revealed a new variable in the equation: data. The ongoing digitalization of medicine is adding an enormous amount of data to the existing knowledge. Collecting and analyzing data through proper studies comprise the conceptual fundamentals of PM. However, there is still a deficit pertaining to human intelligence. Human intelligence is inherently incapable of processing such datasets promptly and in due time before new data emerge. AI comprises the mental workforce that is expected to cope with this flow of big data [1].

Big data refers to huge datasets that are not simply analyzed by traditional techniques. Several examples of big data in our daily life include social media, global positioning systems, purchase transaction records and others. And several examples of big data exist in medicine such as EHRs, imaging data and gene sequencing. Big data needs both the data and efficient data analytics to recognize and extract the obscure information and trends and leverage the power of big data for effective decision-making [1].

Genomics is big data. The genome of a human involves 3 billion base pairs of nucleotides. The Human Genome Project (HGP), a 13-year worldwide project, costs billions of dollars to map and sequence the entire human genome by the year 2003. The achievements of the HGP generated a good deal of genome-wide association studies (GWASs) and led to several advances in human genomics. Results from such research are assembled in the GWAS Catalogue database that is maintained by the European Bioinformatics Institute (EMBL-EBI) and the National Human Genome Research Institute (NHGRI). These GWASs have dramatically enhanced insights into the genetics of human disorders and contributed to developing novel drug targets. Over the past 15 years, the cost of WGS has decreased significantly. Such declines in expenses promote several kinds of research to adopt sequencing techniques. Lately, the UK Biobank (UKB), a longitudinal study of 500 000 UK resident, has published the data on whole-exome sequencing (WES) for the initial sample of 50 000 participants [6].

The EHR is big data and a prosperous field for medicinal studies. Health-care providers used to record patients' medical data on paper. In the United States, efforts to shift to automated recording systems and the 2009 Health Information Technology for Economic and Clinical Health (HITECH) Act achieved broad worldwide deployment of accredited EHRs. EHRs offer a cost-effective and wide repository of patient clinical data for biomedical studies besides allowing the adoption of AI into the clinical process. The EHR data is ongoing, so researchers can execute prospective studies including the crucial aspects of diagnosis and prognosis. The EHR data hold unique kinds of information, such as demographics, diagnosis, imaging, laboratory findings, medical notes, health behaviour and treatments. Yet, EHR data might be limited for research because of random followups, lack of high-quality data, incomplete patient records and the enormous volumes of unstructured data. However, EHRs offer a wide data warehouse of phenotypes for genetic analysis. Several pieces of research have adequately applied phenotypes obtained from the EHR to perform genetic association studies [1].

EHR-related biobanks are triggering a fresh tide of genomics innovations. GWASs have grown from analyzing single phenotypes to testing a range of phonemes, such as those in phenome-wide association studies (PheWAS). Combining phenotype-genotype studies is critical to identify the genomic structure and attributes of human disorders and the pleiotropy data. There are several exceptional models on EHR-related biobanks, for example, the UKB, Kaiser Permanente's Research Program on Genes and Environment, and Health (RPGEH), Genomics (eMERGE) Network and the NIH's Electronic Medical Records. The RPGEH of Kaiser Permanente allows wide investigations of genomic and ecological elements that impact human health. The eMERGE Network, sponsored by the NHGRI, integrates data from EHRs and DNA bio-repositories for genomic research. An additional noteworthy model is the UKB database, among the biggest existing biobank datasets, which researchers can access worldwide. From 2006 to 2010, fundamental data was gathered about lifestyle, dietary habits, medical history and different bodily readings, along with saliva, blood and urine samples [7].

11.4   AI AND ADVANCING PM

The history of AI can be traced back to the 1960s when comprehensive research led to the construction of Dendral, the premiere problem-solving application, that was invented by a multidisciplinary team at Stanford University and was capable of evaluating hypotheses formation. It first saw action in organic chemistry, where it was used to recognize unknown samples, taking into account their mass spectra and characteristics. The first AI system with health-oriented applications was MYCIN. This system was also developed at Stanford and it was primarily used to recognize bacteria and the dosology of needed antibiotics for treating infections caused by them [8]. These past events promoted the growth of AI in health care, using algorithms to adapt scattered data and conquer additional stages of assessing approaches needed in health.

Alan Turing defined AI concepts while John McCarthy first used the 'artificial intelligence' terminology in 1956 at a Dartmouth conference. In AI, machines could execute jobs that require human intelligence, for example, learning, reasoning and diagnosing diseases. In 1959, while working at IBM, Arthur Samuel first introduced the 'machine learning' (ML) terminology. ML and AI are often used alternatively; however, ML is a means of fulfilling AI. With conventional programming, humans compose a particular algorithm to allow machines to carry out a certain job. However, for complicated jobs, such as image recognition, particular transcribed rules might be neither pragmatic nor inclusive enough. ML enables computers to learn such rules instantly with no obvious programming, which renders ML the desirable framework for AI applications [9].

11.5 OVERLAP BETWEEN AI, ML AND DL

Nowadays, medically oriented AI is a vibrant field of development. In conjunction with ML, it can revolutionize medical imaging. Deep learning (DL), the greatest boom in ML techniques, includes training artificial deep neural networks via several layers of linked neurons, which enables machines to discover instantly representations and pick up models from unprocessed data. The neural network techniques outpace several traditional DL techniques in domains, such as voice recognition, image analysis and natural language processing. DL algorithms are promising in finding complex structures and models from high-dimensional and huge amounts of data. There have been astounding breakthroughs in AI/ML applications, for example, in genetic interpretation, image-based diagnosis and data-driven medicine [1].

Existing evidence has highlighted its impact on skin lesions and treatable retinal disease diagnosis. AI algorithms have already been used for detecting a heart attack with an electrocardiogram (ECG) record (sensitivity of 93.3% and specificity of 89.7%), outpacing cardiologists [10]. During the 4th Industrial Revolution, which is also expected to be the era of PM, AI and its infrastructure are expected to pave the way.

11.6 APPLICATION OF AI IN GENETIC INTERPRETATION

Misfolded proteins' data is vital to recognize their biologic roles and correlations to disorders. Google's DeepMind team first became involved through their AI application, AlphaFold, depending on only sequencing genes; it ranked high in the 13th Critical Assessment of Structure Prediction (CASP), a global contest to detect the structure of the three-dimensional (3D) protein in 2018 [11].

11.7 APPLICATION OF AI IN IMAGE-BASED DIAGNOSIS

Fundus photography, typically examined and assessed by ophthalmologists, is a non-invasive technique to capture images of the optic disc, the retina or the macula and discover or track eye diseases, for instance, age-related macular degeneration (AMD), glaucoma and diabetic retinopathy (DR). For instance, a DL algorithm was developed to find DR using 128 175 retinal images and reached analogous functioning to ophthalmologists in datasets of two independent examinations. Deep convolutional neural networks have been used to find glaucoma and instantly rank AMD using fundus photos, and the process reached a precision close to that of professional ophthalmologists [12].

Another non-invasive imaging technology, called optical coherence tomography (OCT), is usually applied by experts in glaucoma and retinal disorders. For example, a deep-learning technique applies transfer learning to detect choroidal neovascularization and diabetic macular oedema. With 37 206 OCT photos for choroidal neovascularization, 11 349 for diabetic macular oedema, 8617 for drusen and 51 140 normal photos, the algorithm reached outcomes analogous to those of trained professionals [13].

Another application, a novel DL framework used with 3D OCT photo data, achieved a 5.5% error rate and functioned equivalently to two major retina experts. Following training on fewer than 15 000 images, the algorithm segmented tissues and rendered coherent and competing diagnoses and consultation suggestions, particularly for sight-threatening retinal disorders without missing one critical case. Interestingly, the algorithm offered information on the components of the scans to reach diagnosis accompanied by its reliability and thereby enhanced the applied neural networks' transparency and mitigated the 'black-box' matter of artificial neural networks [14].

In 2018 the Food and Drug Administration (FDA) authorized the premiere licensed AI application, IDx-DR, for detecting DR. The software examines OCT-scanned photos to deliver medical decision support for both mild and severe DR. Although these retinal OCT image studies and fundus photography target primarily optic disorders, such images could also open a window for other medical disorders, for instance, the anticipation of cardiovascular risk factors and the early detection of dementia [15].

11.8 APPLICATION OF AI IN DATA-DRIVEN MEDICINE

In innovative AI and for the availability of such data, more interesting progress of AI in medicine is being witnessed in both industry and research. ML used big data repositories to draw practical observations, serving as a substitute for time-consuming and expensive research. For instance, ML algorithms were used to examine outcomes of neovascular AMD therapy, applying intravitreal vascular endothelial growth factor inhibitors (anti-VEGF treatments) through universal records and applying real-world findings instead of trial results. Additionally, SOPHiATM GENETICS, Inc. (Boston, MA) is using AI in genetics for improving data-driven medicine, and many pharmaceutical companies, for example, Novartis, Pfizer, GlaxoSmithKline and Merck, have been strengthening their AI expertise in data-based medicine [16].

11.9 LIMITATIONS

Despite all the promises of AI; it has some constraints where ML algorithms need enormous quantities of basic factual training data. Besides some black-box problems, particularly with deep neural networks, it is often tricky to realize the algorithm's method in deriving such outcomes. Data privacy and security, along with health inequities of representing minorities in datasets, are challenging for fully using AI potential. Many scholars have pointed out the need to integrate AI tools under the umbrella of human-controlled procedures. The development of such elements in parallel with adequate safety valves is essential. Till now, the validation of the output at an expert level of efficacy has not been possible. Evidence suggests that the data verification process is not inherent to the system's performance on benchmark datasets. There is a great likelihood that this applies to real-life conditions as well [1].

11.10 CONCLUSIONS

Medicine is the science and practice of detecting, managing and preventing illnesses to promote health. The development of novel technologies speeds up understanding genetics and medicine with enhanced precision. Adopting novel techniques, such as automatic learning from big data to investigate subtle models, would improve PM. In a few decades, PM and AI may be part of routine practice. There will always be room for improvement and contemplation on how human-centric science joined forces with non-human intelligence to make these advances happen. However, with all the promises of AI, it has some limitations, hence the need to control over the AI tools.

REFERENCES

1. Gao, X. R., Cebulla, C. M. and Ohr, M. P. (2020). Chapter 19: Advancing to precision medicine through big data and artificial intelligence. In *Genetics and Genomics of Eye Disease* (X. R. Gao, ed) pp 337–49. Academic Press.
2. Kohane, I. S., Drazen, J. M. and Campion, E. W. (2012). A glimpse of the next 100 years in medicine. *N. Engl. J. Med.* 367 2538–9.
3. Poplin, R., Chang, P.-C., Alexander, D., Schwartz, S., Colthurst, T., Ku, A., Newburger, D., Dijamco, J., Nguyen, N., Afshar, P. T., Gross, S. S., Dorfman, L., McLean, C. Y. and DePristo, M. A. (2018). A universal SNP and small-indel variant caller using deep neural networks. *Nat. Biotechnol.* 36 983–7.
4. Marsico, G. (2019). Artificial intelligence: A benefit for patients? *Soins Rev. Ref. Infirm.* 64 40–1.
5. Zhang, L., Wang, H., Li, Q., Zhao, M.-H. and Zhan, Q.-M. (2018). Big data and medical research in China. *BMJ* 360.
6. UK Biobank. (2019). Exome sequence data available on 50,000. Available at https://www.ukbiobank.ac.uk/2019/03/new-data-available-exome-sequence-data-on-50000-participants/.
7. Sudlow, C., Gallacher, J., Allen, N., Beral, V., Burton, P., Danesh, J., Downey, P., Elliott, P., Green, J., Landray, M., Liu, B., Matthews, P., Ong, G., Pell, J., Silman, A., Young, A., Sprosen, T., Peakman, T. and Collins, R. (2015). UK biobank: An open access resource for identifying the causes of a wide range of complex diseases of middle and old age. *PLOS Med.* 12 e1001779.
8. Mazhar, A. (2019). Evolution of Artificial Intelligence in Healthcare. Available at https://scientiamag.org/evolution-of-artificial-intelligence-in-healthcare/.
9. Alpaydin, E. (2009). *Introduction to Machine Learning.* MIT Press.
10. Uddin, M., Wang, Y. and Woodbury-Smith, M. (2019). Artificial intelligence for precision medicine in neurodevelopmental disorders. *Npj Digit. Med.* 2 1–10.

11. Sample, I. (2018). Google's DeepMind predicts 3D shapes of proteins. *The Guardian*.

12. Pead, E., Megaw, R., Cameron, J., Fleming, A., Dhillon, B., Trucco, E. and MacGillivray, T. (2019). Automated detection of age-related macular degeneration in color fundus photography: a systematic review. *Surv. Ophthalmol.* 64 498–511.

13. Kermany, D. S., Goldbaum, M., Cai, W., Valentim, C. C. S., Liang, H., Baxter, S. L., McKeown, A., Yang, G., Wu, X., Yan, F., Dong, J., Prasadha, M. K., Pei, J., Ting, M. Y. L., Zhu, J., Li, C., Hewett, S., Dong, J., Ziyar, I., Shi, A., Zhang, R., Zheng, L., Hou, R., Shi, W., Fu, X., Duan, Y., Huu, V. A. N., Wen, C., Zhang, E. D., Zhang, C. L., Li, O., Wang, X., Singer, M. A., Sun, X., Xu, J., Tafreshi, A., Lewis, M. A., Xia, H. and Zhang, K. (2018). Identifying medical diagnoses and treatable diseases by image-based deep learning. *Cell*. 172 1122–1131 e9.

14. Fauw, J. D., Ledsam, J. R., Romera-Paredes, B., Nikolov, S., Tomasev, N., Blackwell, S., Askham, H., Glorot, X., O'Donoghue, B., Visentin, D., Driessche, G. van den, Lakshminarayanan, B., Meyer, C., Mackinder, F., Bouton, S., Ayoub, K., Chopra, R., King, D., Karthikesalingam, A., Hughes, C. O., Raine, R., Hughes, J., Sim, D. A., Egan, C., Tufail, A., Montgomery, H., Hassabis, D., Rees, G., Back, T., Khaw, P. T., Suleyman, M., Cornebise, J., Keane, P. A. and Ronneberger, O. (2018). Clinically applicable deep learning for diagnosis and referral in retinal disease. *Nat. Med.* 24 1342–50.

15. Poplin, R., Varadarajan, A. V., Blumer, K., Liu, Y., McConnell, M. V., Corrado, G. S., Peng, L. and Webster, D. R. (2018). Prediction of cardiovascular risk factors from retinal fundus photographs via deep learning. *Nat. Biomed. Eng.* 2 158–64.

16. Adams, J. (2019). Artificial Intelligence: A New Era of Data-Driven Medicine. Available at https://blog.blackswan-analysis.co.uk/artificial-intelligence-a-new-era-of-data-driven-medicine.

12 Review on Pupil Segmentation Using CNN-Region of Interest

A. Swathi and Aarti
Lovely Professional University

Sandeep Kumar
Sreyas Institute of Engineering and Technology

CONTENTS

12.1 Introduction 157
12.2 Literature Survey 158
12.3 ROI Selection.. 160
    12.3.1 Face Detection 161
    12.3.2 Finding ROI Region.. 162
12.4 Available Datasets 164
12.5 Conclusion 165
References... 165

12.1 INTRODUCTION

Applications such as border control, identity-based systems, biometric systems, iris-recognition systems, computer graphics, security-based systems, etc. are highly dependent upon segmentation. With increasing challenges in security issues, the iris is going to be a lead character in many applications. The pupil, as part of iris, is the basic and most tangible part to treat and dependable criteria for many of the applications that were developed on iris [1, 2]. Applications range from gaze, movement and drowsiness detection to age prediction. Therefore segmentation of the pupil is the source of finding and extracting information about a person. This kind of information is also used in psychology and cognitive science applications. It has applications in virtual reality too based upon gaze [3]. Machine learning is a trending technology that has already come into our lives with many challenges. It's been already proven for its efficiency and learning rates with many stand-alone algorithms. Convolutional neural networks (CNNs) are sure to be popular for the same reason. Combining CNN efficiency with graphics processing unit (GPU) based systems, the training and testing rate of CNNs became reliable in developing the machine learning based applications. Fully connected CNNs started showing a lot of improvement in training with data loss and accuracy loss values. Another approach is semantic segmentation, which is capable of identifying 50 000 objects at a time – a real value of using CNN. These methods can show accuracy of 99% when used with a GPU in less than a second [4]. The region of interest (ROI) technique has also become very popular because of its simplicity. It is best suited for applications such as identifying whether a person is happy, sad, anxious, surprised, etc. [5]. Variations in facial landmarks such as eyes, pupils, eye lids, mouth and nose play a major role in identifying facial expressions. So localizing these facial land marks using ROI has given the best results in training the datasets. These methods usually depend upon CNN, support vector machines (SVMs), cascading algorithms, boosting classifier algorithms, etc. [6] and proposed an edge-based localization method using ROI selection.

When the vision technology first started, the Haar cascade classifier was commonly used to detect the eye area. Later, SVMs using traditional datasets were used to train the network to predict the emotions. Working environments such as the Rasberry Pi and python computer language were chosen as the best environments for these applications.

There is much less research on pupil segmentation. We tried our best to collect as much information as possible to implement an ROI model. The current paper discusses the literature survey in Section 12.2. The flow chart and usable CNN structures are discussed in Section 12.3.2.

12.2 LITERATURE SURVEY

Edge-based techniques [3, 7, 8] have been the most-used techniques to predict the pupil; on the other hand, threshold-based methods have come into the picture as edge-based methods show a saggy nature when the image is blurred. Threshold-based techniques show good accuracy compared to video streaming and blurred images with low resolution. There are techniques such as adaptive training, which combines the threshold-based training and segmentation to produce the results. These techniques show accuracy in various conditions. False prediction rate (FPR) and time consumption to predict the pupil is a troublesome task to predict. Due to this CNNs have evolved into the area of detecting objects. The advantage with CNN is efficiency, except in initial training. But the main challenge is to find a proper dataset to train images.

Some existing systems use a camera which continuously monitors the face with visible light to predict the eye movement. In Reference [9], authors implemented pupil detection using SVM. They first detected circles using the Haar cascade classifier in an input image. Later they used SVM to detect the pupil, which produced an accuracy of 90%. They used the face recognition grand challenge (FRGC) 1.0 database to train the data. It had 94% accuracy when the eye is close to the camera. Using BioID Face Database, the authors in Reference [10] implemented detection and localization with three algorithms. They used a web camera to process the images in real time from the trained dataset.

Reference [9] implemented segmentation using the random walker algorithm, an SVM algorithm that labels using background prediction at pixel-level intensities. They say that the prediction is accurate when each pixel is assigned a label. The authors in Reference [11] developed a CNN-based technique to detect pupil, pupil centre and eyelid landmark. They provided a dataset with five lakh (5 00 000) images also. He used L1 loss and logloss functions to perform segmentation. The learning rate of the algorithm is lower, and they used 2000 epochs to gain the accuracy.

The authors in Reference [12] performed feature extraction by using a reduced-pixel block algorithm on CASIA version 3, with 91% accuracy in matching. They used edge detection and Hough transform algorithms to draw the iris and pupil. Local binary patterns were used to represent texture patterns which in turn used the Log-Gabor filter. Further they extended it to other features by supplying this as input to SVM.

The authors in Reference [9] experimented on 500 images of makeup database to first detect the face and then locate the pupil on the face to predict a person's gaze by calculating pixel intensities. They claimed 99% accuracy. The authors in Reference [13] proposed a fully connected CNN to perform pupil segmentation. They developed VOG – an open-source eye-tracking method. They trained the CNN with a dataset of 3946 images.

The authors in Reference [14] developed a CNN to classify left and right eyes using ROI and extracting pupil data based on selected ROI. The algorithm works even in poor illumination and with facial occlusion. They used two CNN architectures. The first was to determine eye region and the second to locate centre of the eye. They used GI4E –a public database with 1236 images – and BioID database with 1521 images to train the net. They obtained the accuracy of 85.6%.

Segmentation based on deep learning is been proposed by Reference [15]. It first preprocesses the input image and finds the ROI by using rough boundary detection. On ROI CNN is been applied to approximate the pupil to calculate the ratio of contraction and dilation of pupil. The authors used NICE-II and MICHE datasets to perform segmentation.

First an image is downscaled to divide into overlapping subregions. Each subregion is evaluated by CNNs. The highest CNN response is known to be the pupil region, according to Reference [16]. The CNNs are trained using the gradient descent method with 10 epochs, with a batch size of 500. The authors used the MATLAB deep learning tool box for the implementation.

The authors in Reference [17] used a random sample consensus (RANSAC) algorithm to perform segmentation. They used three layers, namely, a preprocessing layer, a feature extraction layer and a clutter-removal layer, to fit the RANSAC on a CASIA version 3 dataset. They trained the dataset with 35 epochs, with a batch size of 30 images with precision of 0.34, recall of 0.94 and F1 score of 0.50.

Reference [18] took a different approach, using a fully automatic vestibular neuritis diagnostic system to detect the pupil. The authors used a nystagmus parameter, a parameter of benign positional vertigo estimation. They used a specific mask for the region of interest to propose a geodesic active contour method to detect the pupil; the proposed method has shown 95.5% accuracy.

In a health care application developed by Li et al. [19], using Chinese traditional beliefs, predicted the pupil and iris from the facial input image. They performed iris segmentation using machine-learning and edge-based detection methods. They used RCNN on a CASIA database to perform training and a Gaussian mixture model to find the pupil area, with 82.9% accuracy, 0.90 precision, 0.95 recall, and 0.92 F1 score.

In point-of-gaze estimation [20] determining pupil centre is mandatory. In the proposed system, the authors performed ROI detection and ran the edge-segmentation algorithm to perform comer detection for pupil boundary detection. The authors in Reference [21] performed ROI-based segmentation on CASIA database.

12.3 ROI SELECTION

Due to several factors, such as illumination conditions and quick, unexpected movements, eye detection of the pupil and corneal area is difficult in real time. They cause noise errors. So before we process any image of eye, it has to be filtered and de-noised. This process is called preprocessing.

The quick and easy method to analyze the pupil area is to find the eye area first, because it is difficult to locate the exact eye in the acquired image. In several traditional methods, finding the circle of the iris in the image is almost impossible, and the error rate is high. Hence, finding the eye region in an image will fixate the pupil position more quickly than the other techniques. Finding the ROI is not difficult as the grey area in the eye will be high at the pupil position. In traditional techniques, edge detection can be used to extract the pupil location. One can apply the Prewitt operator to find the edge of the image. It uses 3×3 kernels to calculate horizontal and vertical derivatives defined as follows:

$$G_x = \begin{matrix} -1 & -1 & -1 \\ 0 & 0 & 0 \\ 1 & 1 & 1 \end{matrix}, G_y = \begin{matrix} -1 & 0 & 1 \\ -1 & 0 & 1 \\ -1 & 0 & 1 \end{matrix}$$

In non-traditional techniques, such as using CNN to perform segmentation, detection of pupil is pipelined, as shown in Figure 12.1.

From the literature survey we made, the flowchart describes the process that best suited to detect the pupil. We will address each sub part in detail here. The basic flowchart explains the process of pupil detection from the input image. The image can be a real-time full image of a subject or just the eye part from the existing ROI of interest boundary box is drawn around the iris part, and the CNN is applied to perform the pupil feature extraction. The detected pupil region is classified by labelling or drawing an ellipse around it.

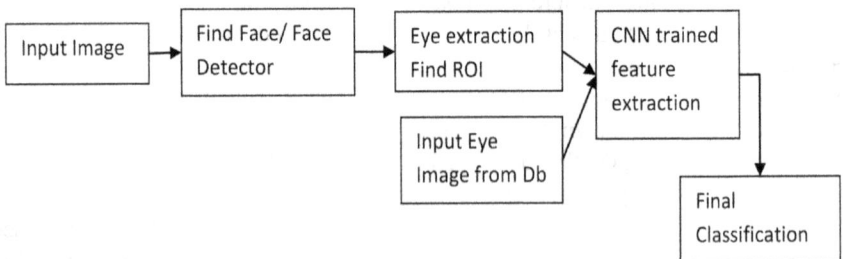

FIGURE 12.1   Flowchart of basic pupil detection.

12.3.1 FACE DETECTION

Several techniques have been built to detect the face region in an image [22]. The famous Viola-Jones algorithm, the histogram of oriented gradient (HOG) and principal component analysis (PCA) are famous traditional techniques to detect faces. The Eigenface vector, LCA (Latent Channel Activation), SIFT (scale invariant feature transform), SURF (speed up robust feature), BRIEF (robust independent elementary features), LPQ (Local Ohase Quantization) and LBP (Local Binary Pattern) are the widely used algorithms to extract facial features. Keypoint-based techniques such as SIFT, SURF, etc. detect the points of interest in the face. Several authors have used various techniques to perform facial recognition. These are listed in Table 12.1.

There are several classifiers built for facial recognition. These depend upon one of the techniques from Bayesian, K means, SVM, CNN, ANN (Artificial Neural Netwoks), etc. But this technology is best suited for facial recognition technology. As

TABLE 12.1
Discussion of Face Detection Techniques Used by Different Authors

Author	Method Used	Database	Result/ Accuracy	Comments
Khoi et al. [23]	Local Binary Pattern (LBP)	LFW	90%	Feature extraction in frontal face
Xi et al. [22]	Local Binary Pattern (LBP)	FERET	97%	CNN to find cosine similarity
Bonnen et al. [24]	MLBP	FERET	86%	Feature extraction failed in many cases as database taken was with wearing glasses
Karaaba et al. [25]	HOG	FERET (Face Recognition Technology)	68.5%	Less accuracy
Gowda et al. [26]	PCA	LFW (Labeled Faces in Wild)	78.9%	Checked with cosine matching
Vinay et al. [27]	Surfa and SIFT	LFW	78.86%	High processing time
Taigman et al. [28]	DeepFace	FB	97.35%	Loss function is softmax
Schroff et al. [29]	FaceNet	Google	99.53%	Tripplet loss using Google net -24 architecture
Liu et al. [30]	L-softmax	CASIA-WF	98.7	VGG Net-18 architecture with loss of softmax
Liu et al. [31]	Coco loss	Random Set	99.1	-
Liu et al. [32]	Sphere Face	CASIA-WF	99.4	Softmax layer with ResNet-64 architecture
Zheng et al. [33]	Ring loss	Random Set	99.5	Ring loss with ResNet-64

it's not the goal of our current discussion, we did not introduce these methodologies, as they need a lot of time for training.

12.3.2 FINDING ROI REGION

One of the best ways of accurately finding an object from an image is placing a bounding box around the region of interest. Unlike classification, which detects whether the object is present in the image or not, ROI finds multiple objects in a given image and gives the location. So, in image-processing applications, the current way of detecting the object is using ROI. ROI needs a trained network to recognize the object. From the literature survey we observed that Raspberry Pi is the most-used technology to design ROI algorithms. The trained CNN detects five different feature points of eye, including the iris. The performance of eye-tracking algorithms is constructed on 0.5 million frames of a database collected from a subject count of 38. The five feature points are inner and outer corners, centre of upper and lower lids, and centre of the iris. Augmented CNN is trained using four datasets, Swirski ExCuSe, LPW, pupil Net and ElSe, with a detection rate of 84.5%. There are various CNNs that were implemented in different years. The most-used ones are described in Figure 12.2.

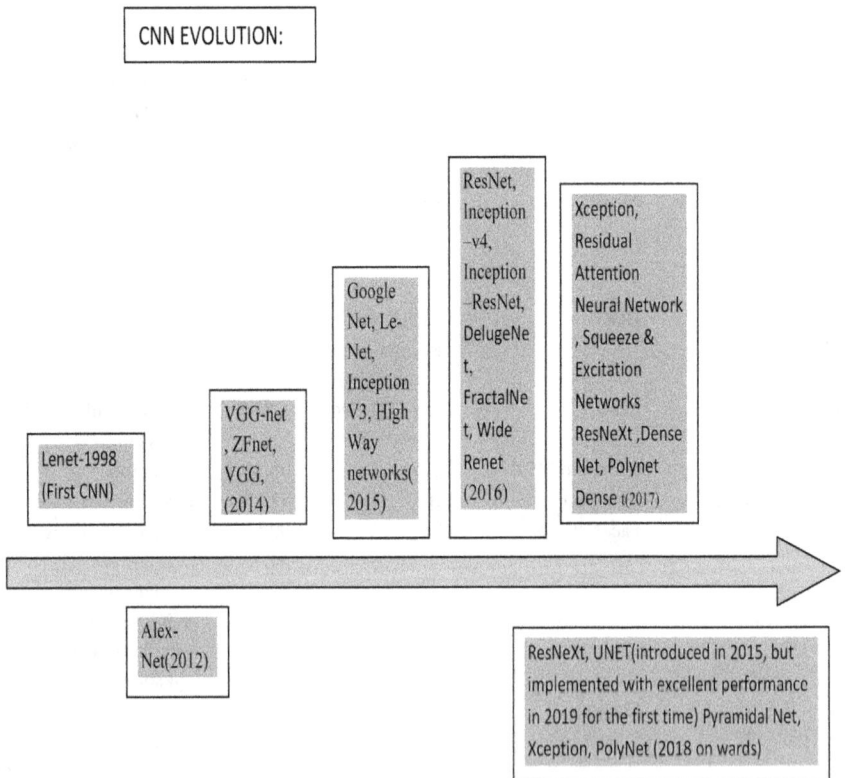

FIGURE 12.2   CNN evolution from the beginning.

Finding the region of interest fining is done by first creating a model to detect the pupil. Otherwise one can use an existing model created for the same purpose. Then import the model into the existing library.

The weights of the model can be checked by the command 'Model.py model.cfg model.weights' in python prompt. Save the model once the weights are ready. The ROI selection flowchart is described in Figure 12.3. The proposed system accuracy

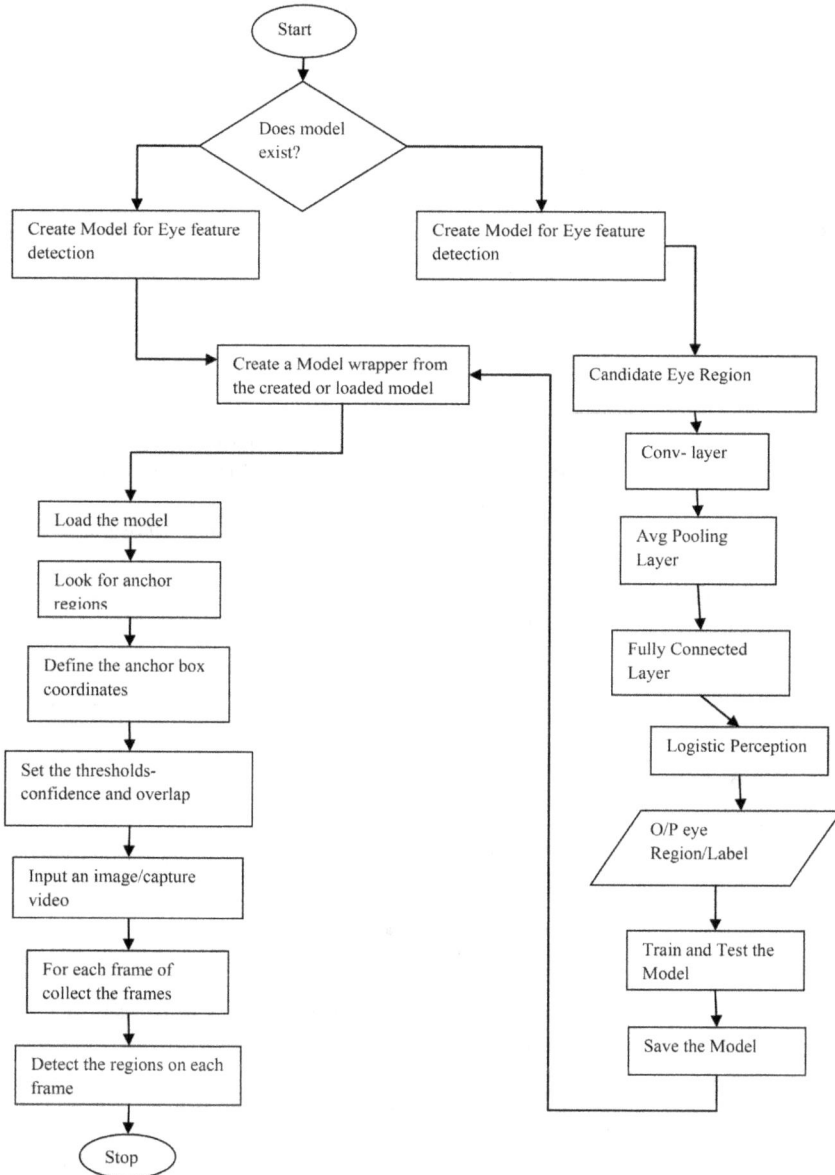

FIGURE 12.3 Flowchart of pupil segmentation.

is 83.6%. Once the network is built, it will be ready to get trained. Researchers used benchmark datasets as well as their own databases to perform the training process.

12.4   AVAILABLE DATASETS

There are limited datasets available for pupil segmentation according to age groups as described in Table 12.2.

The other common datasets used for pupil can be taken from the iris datasets. Table 12.3 discusses the available datasets.

TABLE 12.2
Datasets Available for Pupil with Age Group Information

Dataset Name	Comments
CASIA-IRIS thousands	Includes irises of different age groups
CASIA-Iris version 4	-
CASIA-IRIS twins	Twins irises of different age groups
VISAGE dataset	200 images per gender with the age group of 0–19, includes meta data of landmarks
LPW	Dataset of 22 participants of different age at different illuminations
VOG dataset	4000 images of 20–40 year age group of males and females
DHS dataset	From demographic health survey dataset, collected from 22 different countries of 4 different age groups.
Notre Dame db	Collected from Notre Dame University; consists of 50 subjects of different age groups

TABLE 12.3
Datasets Available for Iris for Pupil Detection

Dataset Name	Comments
CASIA-IRIS V1	756 iris images developed by Chinese Academy of Science
CASIA-IRIS V2	2400 NIR images
CASIA-IRIS V3	22 034 iris images of different resolutions. It's called a Lamp dataset as images were taken under dark and light background conditions and with different dilation ranges in pupil
CASIA-IRIS V4	2576 iris images. Contains three different sets. One of them is Iris thousands
BATH IRIS DATABASE	32 000 NIR iris images of left and right eye
MMU1 dataset	450 images of different angles of eyes
MMU 2 dataset	995 images taken from different distance locations
ICE iris dataset	2954 NIR images of irises with poor lighting and with high occlusion
WVU dataset	1852 with different-angled NIR images and with poor lighting images
UPOL dataset	With visible light 384 images

12.5 CONCLUSION

Compared to the traditional methods explored in the literature survey, pupil detection using CNN models was quick. CNN models are also good at accuracy and efficiency. This survey is made for the purpose of different methods and different architectures present in CNN as well as different datasets that exist for experimentation. Going forward, we would like to perform segmentation using Unet, which includes five layers of CNN and has proven to perform tumour detection on a medical image database in less than one second. We would also like to experiment with pupil characteristics based on dilation.

REFERENCES

1. Liu, X., Xu, F., Fujimura, K.: Real-time eye detection and tracking for driver observation under various light conditions. *In Intelligent Vehicle Symposium* 2, pp. 344–351. IEEE (2002).
2. Braunagel, C., Rosenstiel, W., Kasneci, E.: Ready for take-over? A new driver assistance system for an automated classification of driver take-over readiness. *IEEE Intelligent Transportation Systems Magazine* 9(4), 10–22 (2017).
3. Swirski, L., Bulling, A., Dodgson, N.: Robust real-time pupil tracking in highly off-axis images. In *Proceedings of the Symposium on Eye Tracking Research and Applications*. pp. 173–176. ACM (2012).
4. Arashloo, S. R., Kittler, J.: Efficient processing of MRFs for unconstrained-pose face recognition. In *2013 IEEE Sixth International Conference on Biometrics: Theory, Applications and Systems (BTAS)*. pp. 1–8. IEEE (2013).
5. Saha, P., Bhattacharjee, D., De, B. K., Nasipuri, M.: An Approach to Detect the Region of Interest of Expressive Face Images. *Procedia Computer Science* 46.pp. 1739–1746. (2015).
6. Erbilek, M., Fairhurst, M., Abreu, M.C.D.C.: Age prediction from iris biometrics. In *5th International Conference on Imaging for Crime Detection and Prevention (ICDP 2013)*. pp. 1–5. IET(2013).
7. Fuhl, W., Kübler, T., Sippel, K., Rosenstiel, W., Kasneci, E.: Excuse: Robust pupil detection in real-world scenarios. In *International Conference on Computer Analysis of Images and Patterns*. pp. 39–51. Springer (2015)
8. Fuhl, W., Santini, T.C., Kübler, T., Kasneci, E.: Else: Ellipse selection for robust pupil detection in real-world environments. In *Proceedings of the Ninth Biennial ACM Symposium on Eye Tracking Research & Applications*. pp. 123–130. ACM (2016)
9. Mohsin, H., Abdullah, S.H.: Pupil detection algorithm based on feature extraction for eye gaze. In *2017 6th International Conference on Information and Communication Technology and Accessibility (ICTA)*, pp. 1–4. IEEE(2017).
10. Peng, W., Green, M.B., Ji, Q., Wayman, J.: Automatic eye detection and its validation. In *Proceedings of the 2005 IEEE Computer Society Conference on Computer Vision and Pattern Recognition (CVPR'05)* (2005).
11. Fuhl, W., Rosenstiel, W., Kasneci, E.: 500,000 images closer to eyelid and pupil segmentation. In *International Conference on Computer Analysis of Images and Patterns*, pp. 336–347. Springer, Cham (2019).
12. Kulkarni, S.R.B., Hegadi, R.S., Kulkarni, U.P.: ROI based Iris segmentation and block reduction based pixel match for improved biometric applications. In *International Conference on Advances in Computing, Communication and Control*, pp. 548–557. Springer, Berlin, Heidelberg(2013).

13. Yiu, Y.-H., Aboulatta, M., Raiser, T., Ophey, L., Flanagin, V.L., zu Eulenburg, P., Ahmadi, S.-A.: DeepVOG: Open-source pupil segmentation and gaze estimation in neuroscience using deep learning. *Journal of Neuroscience Methods* 324, pp. 108–307 (2019).

14. Li, B., Fu, H.: Real time eye detector with cascaded Convolutional Neural Networks. *Applied Computational Intelligence and Soft Computing* (2018). https://doi.org/10.1155/2018/1439312

15. Arsalan, M., Hong, H.G., Naqvi, R.A., Lee, M.B., Kim, M.C., Kim, D.S., Kim, C.S., Park, K.R.: Deep learning-based iris segmentation for iris recognition in visible light environment. *Symmetry* 9, no. 11, pp. 263, (2017).

16. Fuhl, W., Santini, T., Kasneci, G., Kasneci, E.: Pupilnet: Convolutional neural networks for robust pupil detection. *arXiv preprint arXiv:1601.04902* (2016).

17. Morley, D., Foroosh, H.: Improving ransac-based segmentation through CNN encapsulation. In *Proceedings of the IEEE Conference on Computer Vision and Pattern Recognition*, pp. 6338–6347 (2017).

18. Slama, A. B., Mouelhi, A., Sahli, H., Manoubi, S., Mbarek, C., Trabelsi, H., Fnaiech, F., Sayadi, M.: A new preprocessing parameter estimation based on geodesic active contour model for automatic vestibular neuritis diagnosis. *Artificial Intelligence in Medicine* 80, pp. 48–62 (2017).

19. Li, Y.-H., Aslam, M.S., Yang, K.-L., Kao, C.-A., Teng, S.-U.: Classification of body constitution based on TCM philosophy and deep learning. *Symmetry* 12, no. 5, pp. 803 (2020).

20. Topala, C., Akinlara, C.: An adaptive algorithm for precise pupil boundary detection using the entropy of contour gradients. (2013). Elsevier preprint.

21. Kortli, Y., Jridi, M., Al Falou, A., Atri, M.: Face recognition systems: A survey. *Sensors* 20, no. 2, pp. 342 (2020).

22. Xi, M., Chen, L., Polajnar, D., Tong, W.: Local binary pattern network: A deep learning approach for face recognition. In *2016 IEEE International Conference on Image Processing (ICIP)*, pp. 3224–3228. IEEE (2016).

23. Khoi, P., Thien, L.H., Viet, V.H.: Face retrieval based on local binary pattern and its variants: A comprehensive study. *International Journal of Advanced Computer Science and Applications (IJACSA)* 7, no. 6, pp. 249–258 (2016).

24. Bonnen, K., Klare, B.F., Jain, A.K.: Component-based representation in automated face recognition. *IEEE Transactions on Information Forensics and Security* 8, no. 1, pp. 239–253 (2012).

25. Karaaba, M., Surinta, O., Schomaker, L., Wiering, M.A.: Robust face recognition by computing distances from multiple histograms of oriented gradients. In *2015 IEEE Symposium Series on Computational Intelligence*, pp. 203–209. IEEE (2015).

26. Gowda, H.D.S., Kumar, G.H., Imran, M.: Multimodal biometric recognition system based on nonparametric classifiers. In *Data Analytics and Learning*, pp. 269–278. Springer, Singapore (2019).

27. Vinay, A., Hebbar, D., Shekhar, V.S., Murthy, K.N.B., Natarajan, S.: Two novel detector-descriptor based approaches for face recognition using sift and surf. *Procedia Computer Science* 70, pp. 185–197 (2015).

28. Taigman, Y., Yang, M., Ranzato, M.'A., Wolf, L.: Deepface: Closing the gap to human-level performance in face verification. In *Proceedings of the IEEE Conference on Computer Vision and Pattern Recognition*, pp. 1701–1708 (2014).

29. Schroff, F., Kalenichenko, D., Philbin, J.: Facenet: A unified embedding for face recognition and clustering. In *Proceedings of the IEEE Conference on Computer Vision and Pattern Recognition*, pp. 815–823 (2015).

30. Liu, W., Wen, Y., Yu, Z., Yang, M.: Large-margin softmax loss for convolutional neural networks. In *ICML*, vol.2, no.3, p. 7 (2016).

31. Liu, W., Wen, Y., Yu, Z., Li, M., Raj, B., Song, L.: Sphereface: Deep hypersphere embedding for face recognition. In *Proceedings of the IEEE Conference on Computer Vision and Pattern Recognition*, pp. 212–220(2017).
32. Liu, Y., Li, H., Wang, X.: Rethinking feature discrimination and polymerization for large-scale recognition. *arXiv preprint arXiv:1710.00870* (2017).
33. Zheng, Y., Pal, D. K., Savvides, M.: Ring loss: Convex feature normalization for face recognition. In *Proceedings of the IEEE Conference on Computer Vision and Pattern Recognition*, pp. 5089–5097 (2018).

13 An Ensemble Classification-Based Model for Automatic Lung Cancer Detection Using CT Images

Shivam Modgil, Bobbinpreet Kaur and Nitin Sharma
Chandigarh University

CONTENTS

13.1 Introduction 169
13.2 Literature Review 171
13.3 Research Methodology 172
    13.3.1 Ensemble Classifier... 175
13.4 Pseudo Code 176
13.5 Results and Discussion 176
13.6 Conclusion 181
References... 181

13.1 INTRODUCTION

Image processing is a technique used for the enhancement of unprocessed images captured from various cameras from various origins. With the help of image processing, the significant data can be retrieved efficiently. In recent years, various methods have been developed to effectively extract complex information in image-processing techniques. Several methods are used by image processing technology in the last few years. This approach is widely used in army, clinical and investigational areas. Some associations also use image processing for simplifying the manual workload and execution of positive actions [1].

Image processing is applied inside numerous applications in order to improve the optical description of pictures. For the preparation of pictures, various calculations are implemented. Digital image processing is another name for this. Processing a digital image includes both visual and analogy images. There are different methods of digital image processing. Another name for this is image acquisition [2].

In lung cancer, anomalous cells multiply and grow in the form of a tumour. The lymph fluid which surrounds lung tissue carries the cancerous cells from lungs to blood. Lymph fluid flows via lymphatic vessels. The lymph fluid drains into the lymph nodes in the lungs and in the centre of the chest. The increase of lung tumours always takes place in the centre of the chest because of the regular lymph fluid flow to the centre of the chest. When a cancer cell leaves its origin, it is called metastasis. This cancerous cells now go towards a lymph nodule or to different body part with the help of blood flow. The prime lung tumour is a kind of cancer which originates from the lung [3].

Lung cancer is a leading cause of cancer death. It is tough to diagnose it since the signs appear at the end stage. However, early diagnosis and treatment of the disease may minimize mortality. CT imaging is effective for diagnosing lung cancer as suspected tumours can be identified as unimagined nodules in lung cancer [4]. Nonetheless, by reviewing a CT scan image and anatomical differences in intensity, the doctor and radiologist can reduce structural misjudgements in identifying cancer cells [5]. Recently, machine-aided diagnostics have been a backup and effective tool to help radiologists and doctors reliably diagnose the cancer [6]. Several programmes and studies on lung cancer have been developed. However, some devices are not sufficiently accurate to detect, and others must be modified to attain 100% accuracy. Identification and classification of lung cancer has been carried out using imaging and machine-learning methods. Recent systems designed to identify cancer using CT scan images of the lungs have been examined. Figure 13.1 shows the current best solutions with some limitations.

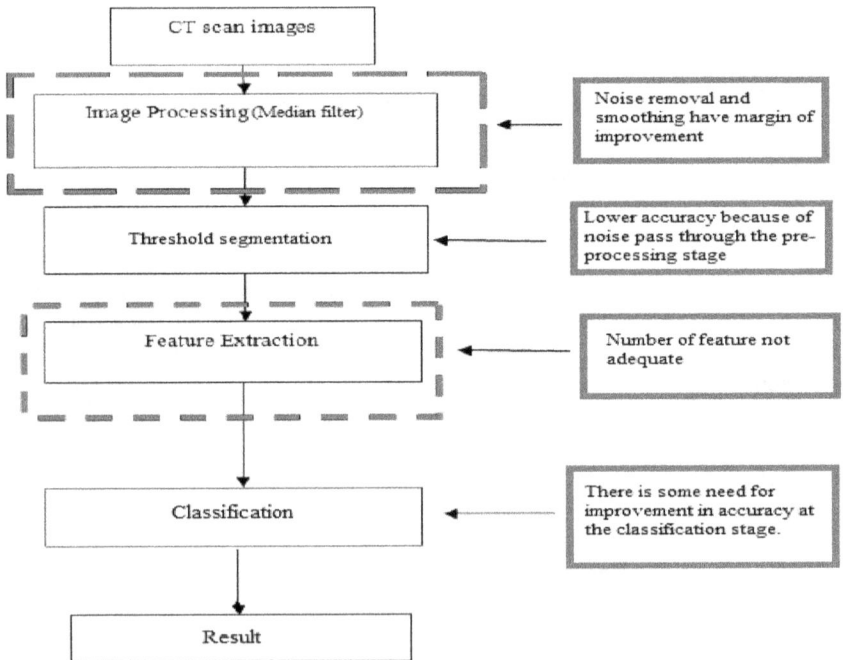

FIGURE 13.1 Broad process for lung cancer detection with some limitation.

13.2   LITERATURE REVIEW

Reference No.	Publication/ Year	Author Name	Finding
[7]	Hindawi/2017	Qing Zeng Song, Lei Zhao et al.	Three kinds of deep networks – CNN (convolutional neural network), DNN (deep neural network) and SAE (stack autoencoder)– are included in this article on calcification of lung cancer. Experimental findings show that the highest performance has been archived on the CNN network, with an exactness of 84.15%.
[8]	IEEE/2015	Aggarwal, Furquan et al.	Suggests a model that classifies nodules with normal structure of lung anatomy. The method extracts characteristics of geometry, statistics and grey levels. LDA (linear discriminant analysis) is used for classification and maximum segmentation threshold. The accuracy of the system is 84%.
[9]	IEEE/2016	Gonzalez et al.	To measure the region of interest (ROI), the system uses the prior knowledge and the Hounsfield unit (HU). The qualified and graded support vector machine determines whether the nodule is benign or malignant, with formal features such as area, circularity and textural aspect such as mean, variance, energy, entropy, skewness, contrast and smoothness.
[10]	IEEE/2015	Ignatious, Joseph et al.	A watershed segmentation system was developed. It uses Gabor filters to improve image quality in pre-processing. It is 90.1% more accurate than a segmentation model using a neural fuzzy model and a region-growing method. The limitation of this model is that cancer is not classified as benign or malignant. It is accurate, but is still not satisfactory.
[11]	IEEE/2015	Roy, Sirohi et al.	This model is an active contour system to detect the nodule of lung cancer by means of the fuzzy interference system. The grey transformation system uses a fuzzy inference method for the enhancement of image contrast. Features such as area, mean, entropy, correlation, major axis and a minor axis length are extracted for the classification training.
[12]	IEEE/2016	Sangamithraa et al.	This uses K means unsupervised clustering or segmentation learning algorithm. It groups the pixel dataset by specific characteristics. This model implements a backpropagation network for classification. Features such as entropy, correlation, homogeneity, PSNR (peak signal to noise ratio) and SSIM (structural similarity method) are extracted using the GLCM method.

[13]	IEEE/2016	Jin, Zhang et al.	To diagnose lung cancer, a neural convolution network was used as a classificatory with this CAD scheme. The device has 84.6% accuracy, 82.5% sensitivity and 86.7% specificity. The advantage of this model is that it uses a circular filter during the extraction process of the region of interest (ROI), minimizing training costs and identification.
[14]	IEEE/2015	K. Punithavathy, M. M. Ramya et al.	These authors used PET/CT images to design a novel technique for detecting lung cancer in an automatic manner. The artefacts within the image had been removed using various image pre-processing techniques. Morphological operators were used in this work for extracting the region of interest (ROI) from lung images. The regions were classified as normal or abnormal using Fuzzy C means (FCM) clustering algorithm.
[15]	Springer/2019	Chethan Dev, Kripa Kumar et al.	The proposed method helps physicians to discover cancer in its early stage to increase the survival rate among patients. DICOM (digital imaging and communications in medicine) format images used in the proposed system prove to provide more efficient results. And SVM classifiers are used for classification of cancerous or non-cancerous images
[16]	IEEE/2017	Lilik Anifah, Haryanto et al.	The proposed work used artificial neural network backpropagation–based grey level co-occurrence matrices (GLCM) features to detect lung cancer.
[17]	Elsevier/2018	Jing Songa, Menghan Hua et al.	Authors implemented a microscopic hyper spectral imaging method in this work. Identifying Anaplastic lymphoma kinase (ALK) positive lung cancer cells and evaluating the remedial impact was the main aim of this technique. A household microscopic hyperspectral imaging system was use in this work to capture the hyperspectral images of five sets of lung tissues

13.3   RESEARCH METHODOLOGY

This research work is related to lung cancer detection from the CT scan images using image-processing techniques. The proposed methodology has four phases for lung cancer localization and characterization (Figure 13.2).

1. **Pre-processing:** The pre-processing is the first phase in which a CT scan image is taken as input. The technique of image denoising will be applied, which will remove noise from the input image. For the removal of noise and contrast enhancement, we use median filters and histogram equalizers.
2. **Segmentation:** In the second phase, the approach of region-based segmentation will be applied, which will segment the similar and dissimilar regions from the CT scan image. The procedure of picture segmentation allocates a

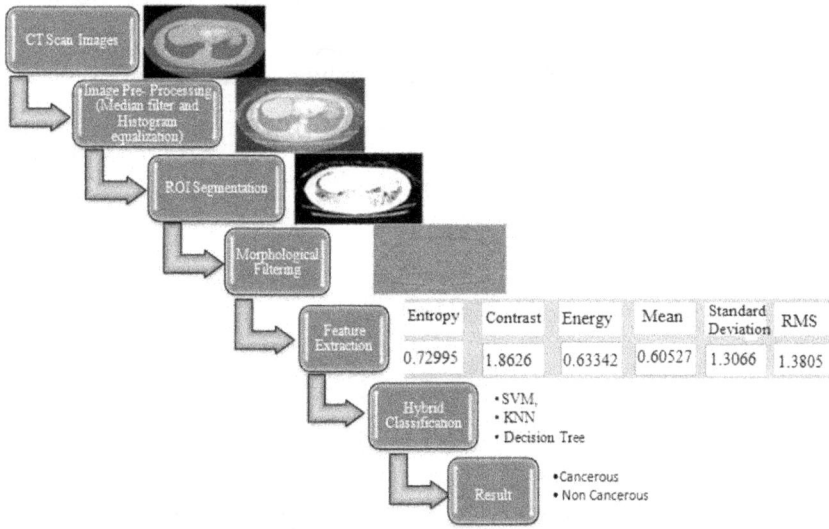

	Entropy	Contrast	Energy	Mean	Standard Deviation	RMS
	0.72995	1.8626	0.63342	0.60527	1.3066	1.3805

FIGURE 13.2   Proposed model.

digital picture into different sections like sets of pixels, also recognized as super-pixels. The chief objective of this process is the alteration of a picture demonstration in an easier investigative manner. Picture sectioning is used for identifying the location of objects, limits and borders in pictures. In this process, a label is assigned to each pixel in a picture and pixels with identical labels share definite features. Several techniques are available for image segmentation: thresholding, K means clustering, watershed segmentation, and region-based segmentation.

In our proposed model we have applied threshold-based segmentation. The most basic methods for segmenting images are thresholding methods. Such methods segment the pixels in terms of intensity. These techniques are used in light object images rather than background images. Such methods can be chosen manually or automatically, i.e., based on previous expertise or information in the image.

3. **Feature Extraction:** The feature extraction is the third phase, in which a GLCM algorithm will be applied for the feature extraction of the CT scan image. In the region of image processing, features play an extremely significant role. Different image pre-processing approaches, such as binarization, thresholding, normalization, masking approach, etc., are implemented on the sampled picture before the computation of features. Then for the attainment of features, feature-withdrawal approaches are executed for the classification and detection of descriptions. Following are the techniques involved in feature extraction. The GLCM algorithms extract textual features of the image. Six features are extracted for lung cancer detection. The formulas for calculation of these features [18] are given in Table 13.1.

TABLE 13.1

Mathematical Formula to Calculate Texture Feature from GLCM

Sr. No.	GLCM Feature	Formula
1	Entropy	$\sum_{i,j=0}^{N-1} P_{i,j}(-\ln P_{i,j})$
2	Contrast	$\sum_{i,j=0}^{N-1} P_{i,j}(i-j)^2$
3	Energy	$\sum_{i,j=0}^{N-1} P_{i,j}^2$
4	Mean	$\mu_i = \sum_{i,j=0}^{N-1} i(P_{i,j})\,,\ \mu_j = \sum_{i,j=0}^{N-1} j(P_{i,j})$
5	Standard Deviation	$\sigma_i = \sqrt{\sigma_i^2}\,,\ \sigma_j = \sqrt{\sigma_j^2}$
6	RMS	$\dfrac{\sqrt{\sum_{i=1}^{M}\lvert\mu_{ij}\rvert^2}}{M}$

4. **Classification:** In the last phase, the approach of hybrid classification will be applied, which can categorize and localize the cancer. The hybrid classification algorithm is the combination of SVM, k-nearest neighbours (KNNs) and decision tree. SVM stands for support vector machine and it is a classification algorithm based on optimization theory. As it maximizes the margin it is also known as a binary classifier. The best hyperplane will classify all data points of a single class and the classification provided by the SVM. In the SVM, the larger hyperplane of the two groups represents the highest margin. There are no interior data points when there is maximum width between the slabs parallel to the hyperplane, which are also known as margin. The SVM algorithm separates the maximum margin in hyperplane. This classifier depends on learning by similarity. The n-dimensional arithmetic qualities are used for the description of training sets. Each training sample in the n-dimensional region represents a point. The superior element of the training samples is amassed in an n-dimensional sample space besides these lines. The k-nearest neighbour classifier is used the k training samples that is near to the unidentified model in the case of an unrecognized training sample. Euclidean distance is the term used for description of 'closeness'. These classifiers are considered as nonparametric supervised learning techniques and are used for categorization and deterioration. The main aim of this approach is the development of a model for the accurate prediction of an intended variable in accordance

with several key variables. In this approach, each core nodule commu-nicates with one of the key variables. Each side demonstrates the value of the target variable. In this classifier, every interior (non-leaf) nodule is labelled with the help of a key trait. The rounded sections created from a nodule named with a trait are labelled with every probable characteristic value. The output of SVM, KNN and decision tree is combined through the bagging process. The bagging process selects output of the best classi-fier for lung cancer detection. Hybrid classification is based on the concept of ensemble classifiers.

13.3.1 ENSEMBLE CLASSIFIER

The creation of a fair model from a dataset is a major task of machine-learning algo-rithms. Learning, or training, is the process of creating models from data, and the learned model is called hypothesis, or learning system. A group of classificators is called the learning algorithm, which identify new data points with a choice of their predictions. This is known as the ensemble method.

Ensembles have also been found to be much more reliable than individual classi-fications. The ensemble method, also known as committee-based learning or multi-classifier learning systems, include a variety of assumptions for solving the same problem. The random forest trees are among the most popular examples of ensemble modelling, with many decision trees for predicting outcomes [19, 20]. Figure 13.3 shows a general ensemble architecture.

There are various algorithms for an ensemble classifier. Some of the commonly used ensemble algorithms are bagging, boosting and stacking. Here we used the stacking ensemble technique in our proposed work. We discuss this technique below:

Bagging Algorithm: The combination of bagging, or bootstrap aggregation, is secure, effective and quick. The methodology employs multiple iterations for a train-ing set – that is, samples of substitution, which can be used for classification or regression for any type of model. The process of bootstrap bagging is useful only in non-linear unstable models (i.e., a slight adjustment in the training set will lead to significant model change) [21, 22]. Figure 13.4 illustrates the different stages of a bagging algorithm.

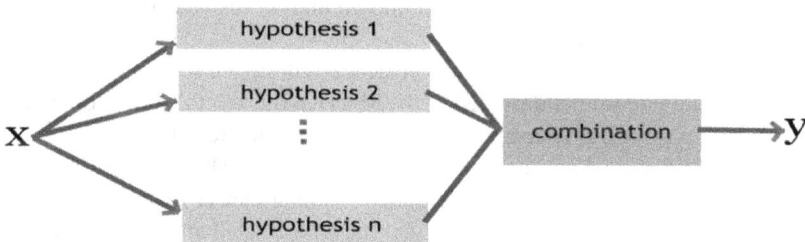

FIGURE 13.3 A general ensemble architecture.

13.4　PSEUDO CODE

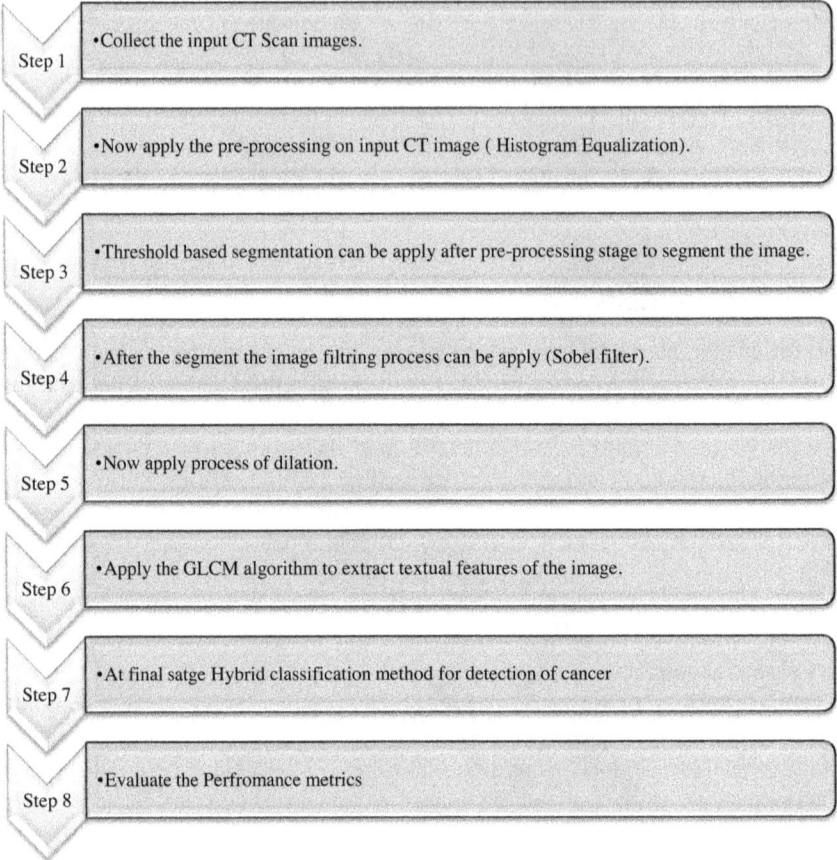

Step 1 • Collect the input CT Scan images.

Step 2 • Now apply the pre-processing on input CT image (Histogram Equalization).

Step 3 • Threshold based segmentation can be apply after pre-processing stage to segment the image.

Step 4 • After the segment the image filtring process can be apply (Sobel filter).

Step 5 • Now apply process of dilation.

Step 6 • Apply the GLCM algorithm to extract textual features of the image.

Step 7 • At final satge Hybrid classification method for detection of cancer

Step 8 • Evaluate the Perfromance metrics

13.5　RESULTS AND DISCUSSION

For evaluation of the proposed model, the implementation is done using MATLAB, and a graphical user interface (GUI) is developed. This research is based on lung cancer detection. The datasets are collected from the various internet sources (LIDC, LIDC-IDRI) [23]. The performance of proposed algorithms is analyzed in terms of accuracy, precision and recall.

As shown in Figure 13.5, the default interface is designed for lung cancer detection. The default interface is designed in the MATLAB using the guide toolbox. The guide toolbox will help produce good representation of the results.

As shown in Figure 13.6, the interface is processed in which various operations used in hybrid classification are also applied for lung cancer detection. The technique of histogram equalizer, segmentation, filtering, dilated and filtering is applied for the detection of lung cancer from the CT scan images.

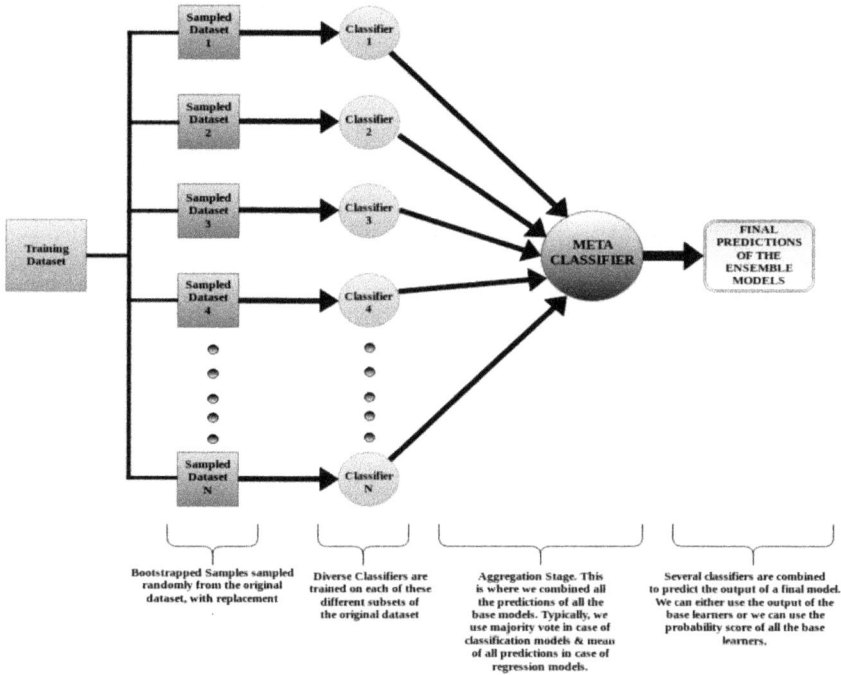

Bootstrapped Samples sampled randomly from the original dataset, with replacement

Diverse Classifiers are trained on each of these different subsets of the original dataset

Aggregation Stage. This is where we combined all the predictions of all the base models. Typically, we use majority vote in case of classification models & mean of all predictions in case of regression models.

Several classifiers are combined to predict the output of a final model. We can either use the output of the base learners or we can use the probability score of all the base learners.

FIGURE 13.4 Different stages of a bagging algorithm.

We have implemented the method from Reference [7] on the dataset we are using, so there is some improvement in the parameter. But in our proposed model there is more improvement than existing method as shown in Tables 13.2 and 13.3.

In Figure 13.7, the accuracy of proposed and existing methods are compared for lung cancer detection. The hybrid classification method has high accuracy compared to existing methods.

As shown in Figure 13.8, precision and recall parameters are used for performance analysis. It is analyzed that precision and recall values of the proposed method are high compared to existing methods. The precision and recall factors define the ratability of the proposed model. The reliability of proposed model refers to how much accuracy is correct for the lung cancer detection.

TABLE 13.2
Performance Analysis [7]

Parameter	Existing Method [7]	Proposed Method
Accuracy (%)	89	95
Precision (%)	85.4	89
Recall (%)	86	88

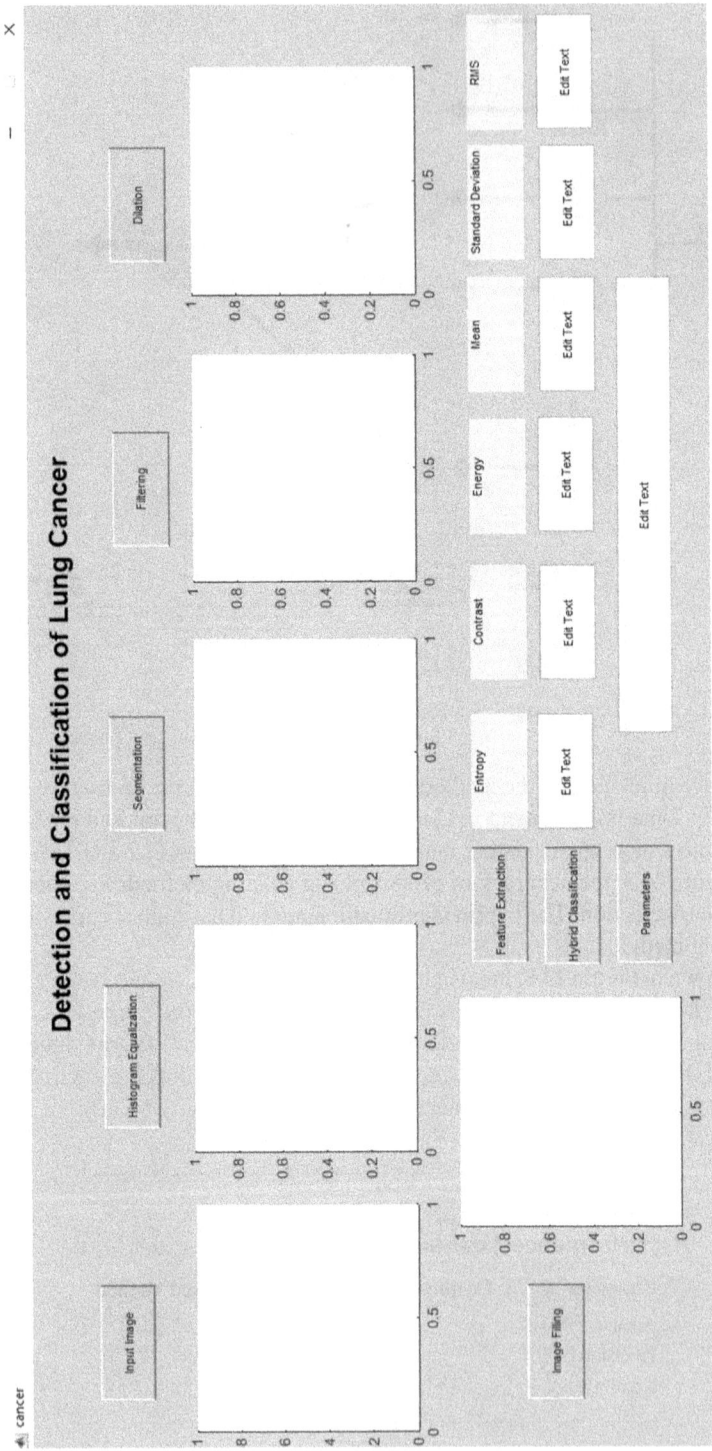

FIGURE 13.5   Default interface.

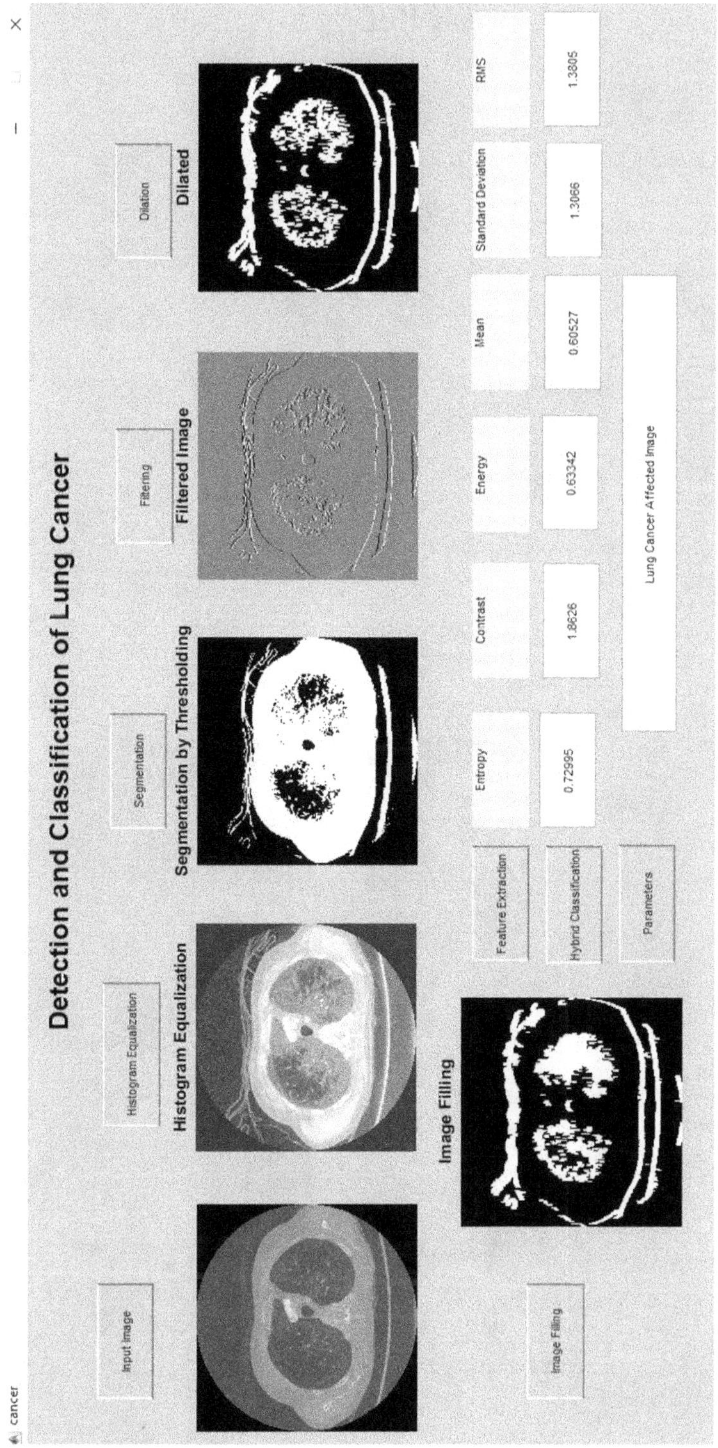

FIGURE 13.6 Processed interface.

Accuracy

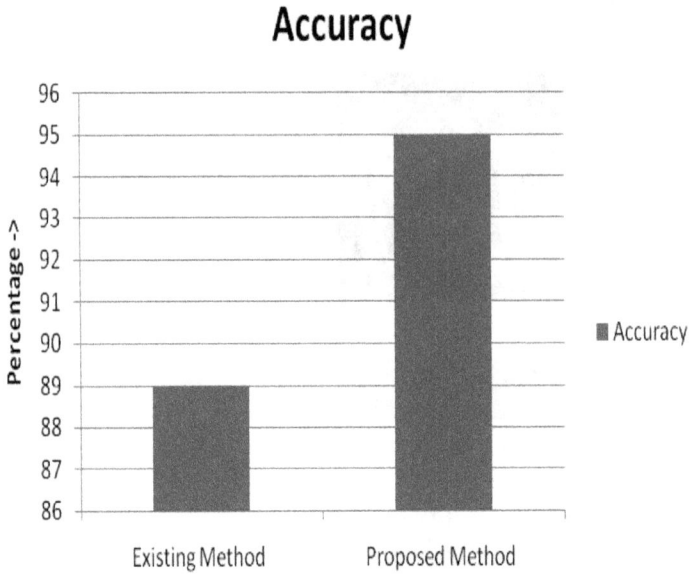

FIGURE 13.7　Accuracy analysis.

Performance Analysis

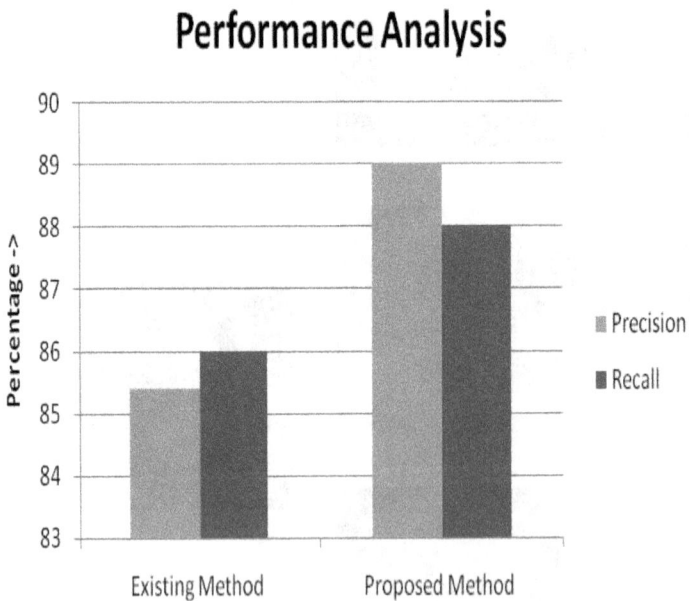

FIGURE 13.8　Precision and recall analysis.

TABLE 13.3
Comparative Analysis

Work	Database	Accuracy (%)	Recall (%)	Precision (%)
Nascimento et al. [24]	LIDC	92.77	85.64	86.73
Krewer et al. [25]	LIDC-IDRI	90.92	85.70	84.90
Gupta and Tiwari [26]	Private	90	86.67	88.72
Hua et al. [27]	LIDC	N/A	73.30	74.94
da Silva [28]	LIDC-IDRI	82.3	79.4	N/A
Song, Qing Zeng [7]	LIDC-IDRI	84.15	83.96	N/A
Hybrid (this paper)	LIDC-IDRI	95	88	89

13.6 CONCLUSION

Cancer patients are increasing day by day due to people's lifestyle. In both men and women, lung cancer has become the second most common cancer which causes death. From early reports on cancer it has been seen that more fatalities occur due to lung cancer than from other cancers. Lung cancer occurs due to smoking (or second-hand smoke) or, less often, exposure to other ecological dynamics, which is why it can be prevented. In this research work, the approach of segmentation, feature extraction and hybrid classification is applied for locating and characterizing cancer. The performance of the proposed method is analyzed in terms of accuracy, precision and recall. It is analyzed that the proposed technique has high performance compared to existing methods. This method successfully predicts the cancer with 95% accuracy.

REFERENCES

1. Rao, R. M., & Arora, M. K. (2004). "Overview of Image Processing." In: *Advanced Image Processing Techniques for Remotely Sensed Hyperspectral Data*. Springer, Berlin, Heidelberg.
2. Ranu, G. (2016). "A Survey of Digital Image Processing." *International Journal of Research in Engineering, Technology and Science*, VI(Special Issue).
3. Mokhled, A. (2012). Lung Cancer Detection using Image Processing Techniques." *Leonardo Electronic Journal of Practices and Technologies*, 11.
4. Gindi, A. M., Al Attiatalla, T. A., & Sami, M.M. (2014) "A Comparative Study for Comparing Two Feature Extraction Methods and Two Classifiers in Classification of Earlystage Lung Cancer Diagnosis of chest x-ray images." *Journal of American Science*, 10(6):13–22.
5. Bariqi, A., Alhadi, B., & Devvi, S. (2016). "Image Processing Based Detection of Lung Cancer on CT Scan Images." The Asian Mathematical Conference.
6. Xiuhua, G., Tao, S., & Zhigang, L. (2011) "Prediction Models for Malignant Pulmonary Nodules Based on Texture Features of CT Image." In *Theory and Applications of CT Imaging and Analysis*. DOI: 10.5772/14766.
7. Song, Q. Z., Zhao, L., Luo, X. K., & Dou, X. C. (2017). "Using Deep Learning for Classification of Lung Nodules on Computed Tomography Images." *Journal of Healthcare Engineering*.2017:1–7. 10.1155/2017/8314740.

8. Aggarwal, T., Furqan, A., & Kalra, K. (2015). "Feature Extraction and LDA Based Classification of Lung Nodules in Chest CT Scan Images." 2015 International Conference On Advances In Computing, Communications And Informatics (ICACCI). DOI: 10.1109/ICACCI.2015.7275773.

9. Rendon-Gonzalez, E., & Ponomaryov, V. (2016). "Automatic Lung Nodule Segmentation and Classification in CT Images Based on SVM." 2016 9th International Kharkiv Symposium on Physics and Engineering of Microwaves, Millimeter and Submillimeter Waves (MSMW). DOI: 10.1109/MSMW.2016.7537995.

10. Ignatious, S., &Joseph, R. (2015) "Computer Aided Lung Cancer Detection System." 2015 Global Conference on Communication Technologies (GCCT). DOI: 10.1109/GCCT.2015.7342723.

11. Roy, T., Sirohi, N., & Patle, A. (2015). "Classification of Lung Image and Nodule Detection Using Fuzzy Inference System." International Conference On Computing, Communication & Automation. DOI: 10.1109/CCAA.2015.7148560.

12. Sangamithraa, P., & Govindaraju, S. (2016) "Lung Tumour Detection and Classification Using EK-Mean Clustering." 2016 International Conference on Wireless Communications, Signal Processing and Networking (Wispnet). DOI: 10.1109/WiSPNET.2016.7566533.

13. Jin, X., Zhang, Y., & Jin, Q. (2016) "Pulmonary Nodule Detection Based on CT Images Using Convolution Neural Network." 2016 9th International Symposium on Computational Intelligence and Design (ISCID). DOI: 10.1109/ISCID.2016.1053.

14. Punithavathy, K., Ramya, M. M., & Poobal, S. (2015). "Analysis of Statistical Texture Features for Automatic Lung Cancer Detection in PET/CT Images." International Conference on Robotics, Automation, Control and Embedded Systems.

15. Dev, C., Kumar, K., Palathil, A., Anjali, T., & Panicker, V. (2019). "Machine Learning Based Approach for Detection of Lung Cancer in DICOM CT Image." In: *Ambient Communications and Computer Systems*, Springer, Singapore.

16. Anifah, L., Haryanto, R. H., Zaimah, P., Puput, W. R., & Adam, R. M. (2017). "Cancer Lungs Detection on CT Scan Image Using Artificial Neural Network Backpropagation Based Gray Level Coocurrence Matrices Feature." *International Conference on Advanced Computer Science and Information Systems (ICACSIS)*, 327–332.

17. Song, J., Hu, M., Wang, J., Zhou, M., Sun, L., Song, Q., Li, Q., Sun, Z., & Wang, Y.(2019). "ALK Positive Lung Cancer Identification and Targeted Drugs Evaluation using Microscopic Hyperspectral Imaging Technique." *Infrared Physics & Technology*, 96: 267–275.

18. Haralick, R. M. (1979)."Statistical and Structural Approaches to Texture." *Proceedings of the IEEE*, 67, no. 5:786–804.

19. Rahman, A., & Tasnim, S. (2014). "Ensemble Classifiers and Their Applications: A Review." *International Journal of Computer Trends and Technology*, 10. 10.14445/22312803/IJCTT-V10P107.

20. Zhou, Z.-H. (2012). *Ensemble Methods: Foundations and Algorithms*. Chapman and Hall/CRC.

21. Dietterich, T. G. (2002). "Ensemble learning." *The Handbook of Brain Theory and Neural Networks*, 2: 110–125.

22. Berk, R. A. (2006). "An Introduction To Ensemble Methods for Data Analysis." *Sociological Methods &Research*, 34, no. 3: 263–295.

23. Vendt, B. (2019). LIDC-IDRI, https://wiki.cancerimagingarchive.net/display/Public/LIDC-IDRI. [Accessed on 05 September, 2019.]

24. Nascimento, L. B., de Paiva, A. C., & Silva, A. C. (2012). "Lung Nodules Classification in CT Images Using Shannon and Simpson Diversity Indices and SVM." In: *Machine Learning and Data Mining in Pattern Recognition*, Springer, 454–466.

25. Krewer, H., Geiger, B., & Hall, L. O. et al. (2013). "Effect of Texture Features in Computer Aided Diagnosis of Pulmonary Nodules in Lowdose Computed Tomography." In: Proceedings of the IEEE International Conference on Systems, Man, and Cybernetics (SMC). Manchester, UK, 3887–3891.
26. Gupta, B. & Tiwari, S. (2014). "Lung Cancer Detection Using Curvelet Transform and Neural Network." *International Journal of Computer Applications*, 86: 1.
27. Hua, K. L., Hsu, C. H., Hidayati, S. C., Cheng, W. H., & Chen, Y. J. (2014). "Computer-Aided Classification of Lung Nodules on Computed Tomography Images via Deep Learning Technique." *Onco Targets and Therapy*, 8: 2015–2022.
28. da Silva, G. L. F., Silva, A. C., de Paiva, A. C., & Gattass, M. (2018). "Classification of Malignancy of Lung Nodules in CT Images Using Convolutional Neural Network." *Medical & Biological Engineering & Computing*, 56(11): 2125–2136. DOI: 10.1007/s11517-018-1841-0.

14 A Revisit on the Progress of Intelligent Robotic Systems (IRS) over the Past Three Decades

Saumyadip Hazra, Abhimanyu Kumar and Souvik Ganguli
Thapar Institute of Engineering and Technology

CONTENTS

14.1 Introduction 185
14.2 Literature Review 188
    14.2.1 Modelling of Intelligent Robotic Systems 188
    14.2.2 Control Aspects of Intelligent Robotic Systems 189
    14.2.3 Fuzzy-Based Applications of Intelligent Robotic Systems 190
    14.2.4 Artificial Intelligence Based Applications of Robots 190
    14.2.5 Medical and Surgery-Based Applications 191
    14.2.6 Swarm Robots 192
    14.2.7 Communication between Robots and Navigation-Based Systems ... 192
    14.2.8 Miscellaneous Areas of IRS 192
    14.2.9 Discussions 193
14.3 Conclusion 193
References 194

14.1 INTRODUCTION

During the late 1960s, our world witnessed a revolution in robotics as well as industrial automation almost simultaneously. The first generation of robots consisted of hardly any computing facilities or sensory capabilities. In the second generation, robots with very limited computational abilities and feedback mechanism were developed. From the third generation onwards, robots with multiple sensing capabilities and decision-making abilities were designed. This outcome resulted from the development of classical control theory to some advancement with adaptive control, self-organizing control and other intelligent control methods [1].

Later on, these robots found applications in diverse disciplines due to the increase in development of hardware and software topologies in all interactive levels (viz. organization, coordination and execution) in intelligent robotic systems. Stellakis et al. [2]

applied fuzzy logic to formulate the organization model of intelligent robotic systems. A method to emulate human skills in assembly operation is given in Reference [3]. Toshio et al. [4] introduced a new paradigm to the field of engineering with the combination of cellular and microrobotic systems. With the cellular robotic system, many attractive topics, viz. swarm intelligence, communications and robot mechanisms, were covered. Micromechanisms, their control and intelligence were deliberated upon with respect to the microrobots. Kawamura et al. [5] paved the way to develop robots with functionalities such as interpretation of fuzzy command, recognition of objects, face tracking, and task planning and learning in order to feed individuals with physical disabilities. Fukuda et al. [6] suggested methodologies to develop adaptation, learning and evolution skills when subjected to an unknown environment. Further, the processing of perception, decision-making and action-taking simultaneously were also synthesized in the intelligent robot system. Akbarzadeh et al. [7] made a fusion of genetic algorithms (GA), genetic programming (GP), and neural networks (NN) with fuzzy schemes to increase the intelligence artificially of the autonomous robots. These hybridizations led to additional reasoning, adaptability and learning capabilities.

Distributed artificial intelligence (DAI) in robotic systems is categorized as distributed problem-solving (DPS) and multi-agent systems (MASs). The information management issues of several units working together are dealt with by DPS to achieve a common objective, MAS focuses on the collective behaviour management of many independent entities or agents [8]. Figure 14.1 shows the Venn diagram for the frameworks of robotic software systems and MASs. Tzafestas et al. [9] presented both the idea of human machine interfacing (HMI) and graphical human machine interfacing (GHMI) with virtual reality system facilities. Further, the class of natural language interfaces was also revisited. Ultimately, the force sensing/tactile-based HMIs were taken up for implementation on several robotic systems and applications. Chen et al. [10] employed robots for gas tungsten arc welding purposes with the help of vision-sensing and neural-control techniques in a real-time environment. Martinoli et al. [11] investigated discrete-time modelling of the dynamics of distributed manipulators at both microscopic as well as macroscopic levels with the help of a swarm of autonomous robots having reactive controllers. Walking machines started developing in the late 20th century, overcoming the limitations of the wheeled systems in transportation technologies [12].

Liu et al. [13] consolidated the controller design techniques of intelligent robotic systems. Various intelligent mechanisms, such as neural networks, evolutionary computation, fuzzy logic, etc., that applied to the design of control scheme were considered in this study for implementation of several physical systems. Madhavan et al. [14] felt the lack of progress in intelligence-based robotics in the manufacturing sector, health-care units and search-and-rescue operations due to the absence of any performance metrics and standards. Hence, they formulated certain benchmarks and standards applicable to intelligent robotic systems. Robotic vision is another important aspect of intelligent robot system. It is usually a very complex task, hindered by sensing and environmental uncertainties. Chen et al. [15] provided a broad survey over the developments in robot vision over the past two decades. Surgery in medical systems is an equally important area where the applications of intelligent robots can be used. Najarian et al. [16] reviewed the various robotic and computer-aided

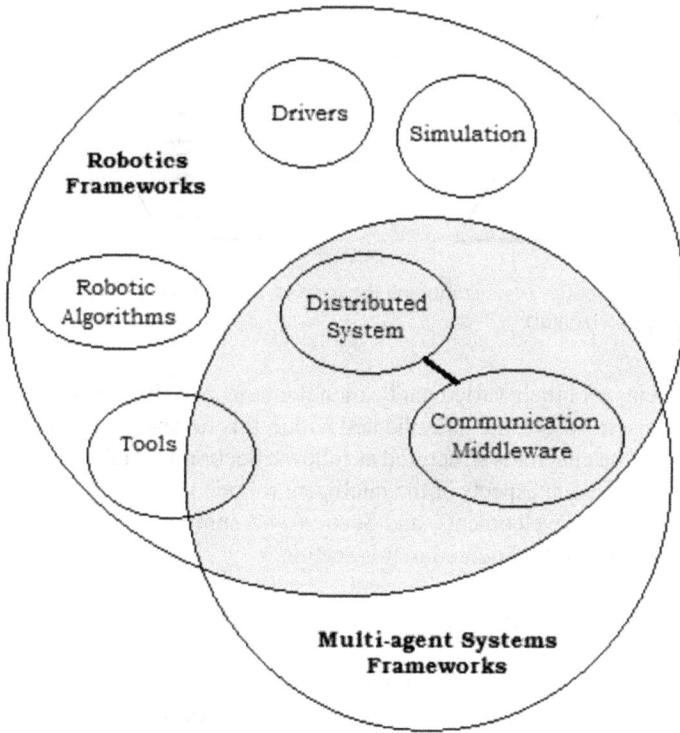

FIGURE 14.1   Venn diagram of robotics software and multi-agent frameworks.

systems that were developed in medical and surgical applications. Desrochers [17] provided a deep insight into the space exploration capabilities of an intelligent robot.

Robots are often needed to communicate amongst themselves to achieve several cooperative tasks. Sabattini et al. [18] solved this problem of communication in a multi-robot system by applying a decentralized control strategy to establish global connectivity. Mobility decreases with age amongst elderly people. Intelligent robot-assisted devices can help them to lead a smooth life, overcoming the adversities [19]. Ahmadzadeh et al. [20] identified the characteristics and application areas of modular robotic systems (MRS). They also developed a new MITE framework constituting of module, information, task and environment based on more than hundred domain features supplemented by a mapping scheme to characterize the properties as well as the applications of MRS.

Figure 14.2 shows the layered robotics software architecture for certain examples. Objects placed in similar layers usually have the same process. MASs are typically placed on the top layer. Liu et al. [21] suggested the technical issues associated with multi-agent robotic systems. Further they addressed on the learning and adaptation skills of autonomous robots in a decentralized environment. Behera et al. [22] illustrated the basic principles for implementing the smart robotic systems (SRS), along with the development of advanced algorithms. Thus, various aspects of intelligent

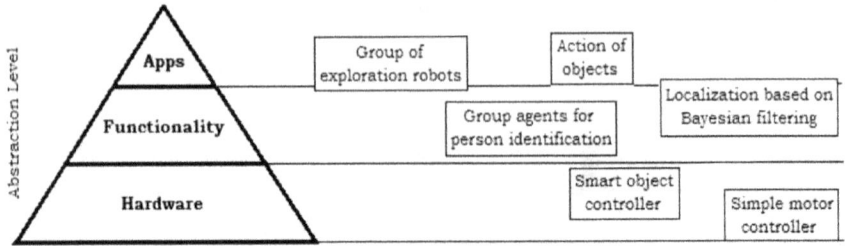

FIGURE 14.2 Robotics organization on the basis of abstraction level (vertical) and degree of intelligence (horizontal).

robotic systems and their varied application domains have been developed over the past three decades which are consolidated within this review.

The rest of the chapter is structured as follows: Section 14.2 takes up the literature review of the different aspects of the intelligent robotic systems followed by discussions on futuristic developments, and Section 14.3 summarizes the chapter. Future scope of work is also highlighted in this section.

14.2 LITERATURE REVIEW

The literature section is classified into the following subsections. The first subsection addresses the modelling aspects of the intelligent robotic systems. The second subsection deals with the control aspects of the robotic mechanisms. The third and the fourth subsections respectively emphasize the fuzzy and other artificial intelligence (AI) based applications in a robotic environment. The fifth subsection narrates the medical and surgical-based applications of the robots. The sixth subsection discusses swarm robots. The seventh subsection deliberates on the communication between robots and between robot and operator as well as robotic systems dealing with navigation-based methods. The final subsection takes into account some miscellaneous applications such as space exploration, robot vision system and robotic-assisted systems for the welfare of humans.

14.2.1 MODELLING OF INTELLIGENT ROBOTIC SYSTEMS

Valvanis et al. [23] presented the hardware and software design for the three modelling levels of intelligent modelling systems. All the three hierarchical levels were modelled differently, and microprocessor and discrete logic-based techniques were proposed for the hardware. Later, an algorithm in the assembly and organization level of robots with different constraints was utilized for the optimization of planning procedure for each system in order to make them capable for doing the requested jobs [24]. A probabilistic approach was presented in Reference [25], in which the hierarchical level of performing tasks was employed. In order to mathematically interpret the system functions, probability and entropy functions were used. Valvanis et al. [26] proposed a general-purpose method for modelling robotic assemblies and intelligent robotic systems, which considered both the pre-defined and fuzzy commands.

Yang et al. [27] decomposed the uncertain robotic model into repetitive and non-repetitive. The Lyapunov model was used for developing iterative control methods for robots having structured certainty and uncertainty. Martinoli et al. [11] presented discrete-time methodology for incremental modelling at the microscopic and macroscopic levels for manipulation using the swarm autonomous robotic systems and reviewed the then existing methodologies by proposing a few changes. Hu et al. [28] designed the team-formation mechanism of the multi-robot system and used it for modelling purposes on the basis of discrete event system specification (DEV) formalism and proved the process of simulation based on the physical mapping of DEVs. Stojanov et al. [29] discussed the overview of the research done in robotics and how curiosity was associated with it through a one-dimensional and reductionist approach. The modelling of curiosity in context of cognitive architecture was also presented. Wurm et al. [30] presented an approach consisting of 3D modelling algorithms using a probabilistic estimation and also reviewed all the models existing then for 3D modelling. The results proved that the probabilistic approach was able to model the robotic systems. Collins et al. [31] illustrated basic principles and algorithms developed for the realization of intelligent robotics. The book also dealt with stability and validations for the kinematics and dynamics of the robot.

14.2.2 Control Aspects of Intelligent Robotic Systems

Kazerooni et al. [32] described the dynamics of extender robotic systems which are worn by human beings for performing various tasks. Humans exchange both power and signal with the extenders, and an arbitrary relationship has been derived between the human and load force. Dubowsky et al. [33] discussed the dynamics and control issues for the free-flying and free-floating–based robots used for different tasks and presented three methods for the planning and controlling of these robots. Sugano et al. [34] studied the stability of the manipulators for vehicles, assuming that they have relation with the vehicle speed, and presented the concept of stability and region of stable operation using ZMP criteria. Tarn et al. [35] applied event-based planning and showed that the representations in path-based and time-based feedback systems have non-linear nature in order to obtain stability. The results were also verified using experiments. Vemuri et al. [36] investigated the problems faced in the fault detection of rigid-link robotic systems, including all the uncertainties, and established the robustness and stability. The use of neural networks had been applied in the robust fault detection and stability of system. Silva [37] reviewed the techniques used for integrating intelligence in the control systems of robots. Also, many concepts related to approximations were explored and presented a generalized model for knowledge-based decision-making. Fukuda et al. [38] studied some of the intelligent techniques for the control of robotic systems and introduced a network of robotic systems and discussed major schemes for the different hierarchical levels of the robotic systems. Tarn et al. [39] presented new control and planning for intelligent robotic systems significantly related to their motion by the combination of closed event-based plan with the non-linear feedback and also presented stability criteria. Jacak [40] presented a method for realization of intelligent robotic systems, including some of the important tools, and described the necessary actions for planning, coordination and control of robots. Choi et al. [41] proposed an adaptive iterative learning control (AILC) scheme for the stability of the robotic system including a balanced conventional adaptive control and iterative control and complemented the disadvantage of each method. Behal et al. [42] described the use of Lyapunov-based systems for the design of a non-linear controller which can help to solve the problems faced by the robot in manipulating the environments. Antonelli et al. [43] proposed a controller based on prioritized task kinematics to provide error stability and convergence. For prioritizing the data and its stability, Lyapunov-based stability was used base on the kinematics algorithm. Short et al. [44] described a method to develop design philosophy and reference architecture for advanced robotic architecture and also described a prototype using the same method which was used for describing the controller architecture. Krishnan et al. [19] reviewed articles on assistive devices which focused on self-transfer and mobility systems. Also, they covered the advances made in the above-mentioned field along with its limitations. Na et al. [45] studied the parameter estimation and controlling of the non-linear robots by minimizing the errors. The equation for parameter estimation was based on the auxiliary filtered variables and investigated robustness against disturbances.

14.2.3 Fuzzy-Based Applications of Intelligent Robotic Systems

Stellakis et al. [2] presented a fuzzy-logic methodology for modelling an intelligent robotic system at the organizational level. Fuzzy relations have been used for connecting every event with a command, along with an approximate reasoning process. The most possible plan has been chosen and considered as optimal for a particular application and then implemented. Kawamura et al. [5] studied the imparting of intelligence to the robotic system in order to make intelligent decisions using fuzzy interpretations, face tracking, object recognition and task planning, and an intelligent system has been designed. Fukuda et al. [46] dealt with the fuzzy logic based robotic systems, including intelligence, and proposed a system with structural intelligence so that a fuzzy controller can be used to make the robot perceive the environment. Castillo et al. [47] described a novel approach for the control of robotic systems, using the hybrid of fuzzy and neural networks. Chang et al. [48] addressed the problems faced in the adaptive fuzzy-logic–based design for a robust controller for uncertain non-linear robotic systems. The solution was reached by solving a linear matrix inequality in order to guarantee the transient and asymptotic performance. Lian [49] developed fuzzy radial basis function neural network controller to solve the problems faced by self-organizing fuzzy controller (SFOC) by eliminating the dynamic coupling effects. Simulation results proved that it was capable of producing better results.

14.2.4 Artificial Intelligence Based Applications of Robots

Tzafestas et al. [50] studied the underlying concepts and the issues of blackboard architecture for realization of an integrated knowledge-based intelligent system. Blackboard architecture uses multiple problem-solving agents. Beni et al. [51] studied the critical problems of distributed robotic systems and concluded that algorithms to solve complex problems may be developed using the hybrid algorithms consisting of the applications in the humans. Kosuge et al. [3] presented human skills to be represented in assembly tasks in order to use them in intelligent robotic systems. Two levels of control strategies have been extracted from human operators, and a transition-state model was generated which was further analyzed. Nakamura et al. [52] proposed a statistical fusion method with the help of geometry of uncertainties for robotic systems, which includes multiple sensors. The result showed that the optical fusion is similar response to the Bayesian inference, the Kalman filter theory and weighted least square. Fukuda et al. [6] presented the methods for the learning, adaptation and evolution of the robotic systems, which include the processing of decision-making, perception and action. Tzafestas et al. [9] discussed the issues faced by the human machine interaction faced in robotics where HMI, natural language interfaces (NLI), GHMI with the virtual reality (VR) technology have been presented along with the application of HMI in various robotic systems. Carbone [12] presented a survey consisting of the solutions existing for walking machines and problems associated with their design and then discussed their feasibility with respect to LARM in Italy. Sanfeliu et al. [53] introduced the concept of Network Robot Systems (NRS) and how they were understood in the USA, Japan and Europe. Also, the projects in those countries and a summary of related papers were presented. Stanescu et al. [54] presented the history of paradigm shifts which ultimately resulted in the advancements of robotics. A generic design and architecture for the development of a complex robotic structure was also presented. Iñigo-Blasco et al. [55] identified and analyzed the different types of frameworks to compare them with MAS frameworks. The frameworks consist of characteristics and act as basic building steps for multi-agent robotic systems (MARS). Lueth et al. [56] included the works done in the fields of distributed robotic system design, sensory navigation, multi-robotic control and navigation systems for robots. Fukuda [57] presented the trends in the field of intelligent robotic systems related to soft computing methods.

14.2.5 MEDICAL AND SURGERY-BASED APPLICATIONS

Hackwood et al. [58] presented the applications of cellular robots in which the design components were also explored along with the design of different cellular robotic systems. They also studied how these robotic systems can be used to organize different human movements. Trivedi et al. [59] developed robotic systems which can use the multisensory information for performing various inspection and manipulation tasks and an approach to integrate vision, proximity, touch and range with the system. The sensory information was presented in the form of an optimization problem where the error was minimized. Schorr et al. [60] presented object distribution architecture in order to improve the performance of surgical robotic systems, and they showed its usefulness and flexibility in integrating the existing software to the surgical robots to achieve complex tasks. Numerous design methodologies and algorithms for the development of the robotic systems are currently available. Different types of computer-assisted robotic systems in the field of medicine and surgery have been developed. Bogatyrenko et al. [61] developed a motion-tracking device for the heart surface using a reconstruction algorithm. It combined the physical stochastic methods for obtaining accurate results and used automatic-initialization for heart motion.

14.2.6 SWARM ROBOTS

Beni et al. [62] defined intelligent robotic systems with the help of unpredictability of improbable behaviour. They also analyzed two types of unpredictability which may result in a non-trivial swarm intelligence and presented engineering problems of swam intelligence with CRS. Lerman et al. [63] presented an analytical model in a group of robots consisting of a series of differential equations. The results were computationally efficient, and many positive conclusions were drawn in the field of macroscopic analysis. Lerman et al. [64] reviewed the methods for analysis and macroscopic modelling of swarm robotic system and compared the results obtained from the probabilistic sensor and numeric-based simulations with the models' prediction.

14.2.7 COMMUNICATION BETWEEN ROBOTS AND NAVIGATION-BASED SYSTEMS

Balch et al. [65] studied the communication between robots and measured performance for three types of communication. The simulations helped in the determination of parameters for reactive control systems. Hu et al. [66] described the building of an internet-based robotic system capable of teleoperation in which the control can be done remotely to navigate it and make it do some useful tasks. Jia et al. [67] proposed a platform for the design of telecare robotic system that uses Cobra robotics for the communication and can be controlled remotely using web browser. Parker et al. [68] addressed the issues important for decentralized robotic systems. Some of the major topics covered communications, reconfigurable robots, biological inspirations architectures, sensing, etc. Sabattini et al. [16] presented a decentralized control technique for connectivity maintenance using a gradient-based technique which included algebraic connectivity estimation. The results obtained through simulation proved the efficacy of the method.

14.2.8 MISCELLANEOUS AREAS OF IRS

Madhavan et al. [14] identified the importance of the use of the standards and benchmarking for the performance evaluation of intelligent robotic systems and also presented a survey consisting of the attempts made by the researchers for the development of the same. Chen et al. [15] provided a survey on active vision in the area of robotics. Further, the paper described the problems which arose from various aspects and finally proposed a set of solutions to solve them. Xu et al. [69] summarized all the modelling concepts of the robotic systems and derived all the kinematic and dynamic equations related to it. The paper also presented planning approaches and ground verifications. Coradeschi et al. [70] reviewed the research done in the field of symbol grounding in the 21st century up to 2012 for the application in intelligent robotic systems. Shah et al. [71] used the transformation of velocity of kinematic constraints such as natural orthogonal complement and decoupled NOC in order to obtain the lowest-order dynamic equation and proved that it had many benefits. Xu et al. [72] first presented the advances and the trends in the field of robot navigation, including localization in complex environments. The next part consisted of sensing approaches and the final section addressed problem in robot dynamics. Liu et al. [21] addressed decentralized robots in regard to learning and adaptation of multi-agent robotic systems. The book also included the architecture of autonomous robots. Bien et al. [73] proposed a new type of society in which all the members would have experienced welfare and the systems would be completely high-tech in all facilities. Desrochers [17] presented the work done in assembly in space, along with the algorithms and design and also addressed problems associated with the control of the robots under zero gravity. Asama et al. [74] presented the work conducted in robotic systems by researchers across the globe. Katic et al. [75] presented the design and algorithms behind the decision-making, task-performing, etc. of intelligent robots and also highlighted the importance of cognitive architecture behind it.

14.2.9 Discussions

The modelling and control aspects of the intelligent robotic systems and their developments over the past few decades are narrated. With the advent of fuzzy logic, neural networks and AI-based techniques, improvement in control schemes, adaptability and learning features of robotic systems have increased tremendously. The improvisations of the intelligent robotic system (IRS) in medical examination and surgery assistance are also described. Swarm robots are an integral part of industrial automation. Their chronological developments have been deliberated in this chapter as well. Communication in a multi-robot system as well as the communication between robot and its operator connected via internet is also discussed. The navigation-based applications of IRS are also taken into consideration. Finally, robotic applications for space exploration, robotic vision systems, computer-assisted systems for the elderly and people with physical disabilities, and benchmarking and standardisation for IRS etc. are also dealt with in the miscellaneous category.

Internet of things (IoT) based robotic applications may be thought of in the near future. With the use of IoT, cloud storage applications for robotic systems can be a useful feature. Though evolutionary computation, GA and GP are applied in the intelligent robotic systems, yet nature-inspired metaheuristic algorithms have not been popularly used except for robotic path and trajectory planning based applications. Intelligent controllers can thus be designed by these algorithms for the robotic systems.

14.3 CONCLUSION

The chronological developments of intelligent robotic systems with certain categories have been considered in this chapter. Mostly the study is carried out for quite a long span of 30 years. Third-generation robots have been addressed. Several application areas, such as surgical assists, space exploration, swarm and path or trajectory planning, robot vision, etc., have been thoroughly deliberated in this chapter. The review is performed based on some of the significant works contributed in books, journal articles and as well as conference papers. The modelling aspects as well as works on intelligent control techniques have been presented. The use of robots in surgical applications has also been reported. The use of swarm robots and their communication aspects are also deliberated. Some navigation-based applications are also provided in this chapter. Some miscellaneous application fields are also explored. Some future propositions are highlighted at the end for the research to move on.

REFERENCES

1. Valavanis, K. P., & Saridis, G. N. (2012). *Intelligent robotic systems: theory, design and applications* (Vol. 182). Springer Science & Business Media.
2. Stellakis, H. M., & Valavanis, K. P. (1991). Fuzzy logic-based formulation of the organizer of intelligent robotic systems. *Journal of Intelligent and Robotic Systems*, 4(1), 1–24.
3. Kosuge, K., Fukuda, T., & Asada, H. (1991). Acquisition of human skills for robotic systems. In *Proceedings of the 1991 IEEE International Symposium on Intelligent Control* (pp. 469–474). IEEE.
4. Toshio, F., Tom, H., & Tsuyoshi, U. (1994). *Cellular robotics and micro robotic systems* (Vol. 10). World Scientific.
5. Kawamura, K., Bagchi, S., Iskarous, M., & Bishay, M. (1995). Intelligent robotic systems in service of the disabled. *IEEE Transactions on Rehabilitation Engineering*, 3(1), 14–21.
6. Fukuda, T., & Kubota, N. (1999). Intelligent robotic systems: Adaptation, learning, and evolution. *Artificial life and Robotics*, 3(1), 32–38.
7. Akbarzadeh-T, M. R., Kumbla, K., Tunstel, E., & Jamshidi, M. (2000). Soft computing for autonomous robotic systems. *Computers & Electrical Engineering*, 26(1), 5–32.
8. Stone, P., & Veloso, M. (2000). Multiagent systems: A survey from a machine learning perspective. *Autonomous Robots*, 8(3), 345–383.
9. Tzafestas, S. G., & Tzafestas, E. S. (2001). Human–machine interaction in intelligent robotic systems: A unifying consideration with implementation examples. *Journal of Intelligent and Robotic Systems*, 32(2), 119–141.
10. Chen, S. B., Zhang, Y., Qiu, T., & Lin, T. (2003). Robotic welding systems with vision-sensing and self-learning neuron control of arc welding dynamic process. *Journal of Intelligent and Robotic Systems*, 36(2), 191–208.
11. Martinoli, A., Easton, K., & Agassounon, W. (2004). Modeling swarm robotic systems: A case study in collaborative distributed manipulation. *The International Journal of Robotics Research*, 23(4–5), 415–436.
12. Carbone, G., & Ceccarelli, M. (2005). *Legged robotic systems* (pp. 557–561). INTECH Open Access Publisher.
13. Liu, D., Wang, L., & Tan, K. C. (Eds.). (2009). *Design and control of intelligent robotic systems* (Vol. 177). Springer.
14. Madhavan, R., Lakaemper, R., & Kalmár-Nagy, T. (2009). Benchmarking and standardization of intelligent robotic systems. In *2009 International Conference on Advanced Robotics*, 1–7. IEEE.
15. Chen, S., Li, Y., & Kwok, N. M. (2011). Active vision in robotic systems: A survey of recent developments. *The International Journal of Robotics Research*, 30(11), 1343–1377.
16. Najarian, S., Fallahnezhad, M., & Afshari, E. (2011). Advances in medical robotic systems with specific applications in surgery: A review. *Journal of Medical Engineering & Technology*, 35(1), 19–33.

17. Desrochers, A. A. (Ed.). (2012). *Intelligent robotic systems for space exploration* (Vol. 168). Springer Science & Business Media.
18. Sabattini, L., Chopra, N., & Secchi, C. (2013). Decentralized connectivity maintenance for cooperative control of mobile robotic systems. *The International Journal of Robotics Research*, *32*(12), 1411–1423.
19. Krishnan, R. H., & Pugazhenthi, S. (2014). Mobility assistive devices and self-transfer robotic systems for elderly, a review. *Intelligent Service Robotics*, *7*(1), 37–49.
20. Ahmadzadeh, H., Masehian, E., & Asadpour, M. (2016). Modular robotic systems: characteristics and applications. *Journal of Intelligent & Robotic Systems*, *81*(3-4), 317–357.
21. Liu, J., & Wu, J. (2018). *Multiagent robotic systems*. CRC press.
22. Behera, L., Kumar, S., Patchaikani, P. K., Nair, R. R., & Dutta, S. (2020). *Intelligent control of robotic systems*. CRC Press.
23. Valavanis, K. P., & Yuan, P. H. (1989). Hardware and software for intelligent robotic systems. *Journal of Intelligent and Robotic Systems*, *1*(4), 343–373.
24. Valavanis, K. P., & Carelo, S. J. (1990). An efficient planning technique for robotic assemblies and intelligent robotic systems. *Journal of Intelligent and Robotic Systems*, *3*(4), 321–347.
25. Valavanis, K. P., & Saridis, G. N. (1991). Probabilistic modeling of intelligent robotic systems. *IEEE Transactions on Robotics and Automation*, *7*(1), 164–171.
26. Valavanis, K. P., & Stellakis, K. M. (1991). A general organizer model for robotic assemblies and intelligent robotic systems. *IEEE Transactions on Systems, Man, and Cybernetics*, *21*(2), 302–317.
27. Yang, S., Fan, X., & Luo, A. (2002). Adaptive robust iterative learning control for uncertain robotic systems. In *Proceedings of the 4th World Congress on Intelligent Control and Automation (Cat. No. 02EX527)* (Vol. 2, pp. 964–968). IEEE.
28. Hu, X., & Zeigler, B. P. (2004). Model continuity to support software development for distributed robotic systems: A team formation example. *Journal of Intelligent and Robotic Systems*, *39*(1), 71–87.
29. Stojanov, G., Kulakov, A., & Clauzel, D. (2006). On curiosity in intelligent robotic systems. In *AAAI Fall Symposium: Interaction and Emergent Phenomena in Societies of Agents* (pp. 44–51).
30. Wurm, K. M., Hornung, A., Bennewitz, M., Stachniss, C., & Burgard, W. (2010). OctoMap: A probabilistic, flexible, and compact 3D map representation for robotic systems. In *Proceedings of the ICRA 2010 Workshop on Best Practice in 3D Perception and Modeling for Mobile Manipulation* (Vol. 2, pp. 1–8).
31. Collins, H., Collins, L., & Mim, M. S. (2019). Intelligent robotic systems research for human-machine interaction. https://dc.uwm.edu/uwsurca/2019/Posters/24.
32. Kazerooni, H., & Mahoney, S. L. (1991). Dynamics and control of robotic systems worn by humans. *Journal of Dynamic Systems, Measurement, and Control*, *113*(3), 379–387.
33. Dubowsky, S., & Papadopoulos, E. (1993). The kinematics, dynamics, and control of free-flying and free-floating space robotic systems. *IEEE Transactions on Robotics and Automation*, *9*(5), 531–543.
34. Sugano, S., Huang, Q., & Kato, I. (1993). Stability criteria in controlling mobile robotic systems. In *Proceedings of 1993 IEEE/RSJ International Conference on Intelligent Robots and Systems (IROS '93)* (Vol. 2, pp. 832–838). IEEE.
35. Tarn, T. J., Xi, N., & Bejczy, A. K. (1996). Path-based approach to integrated planning and control for robotic systems. *Automatica*, *32*(12), 1675–1687.
36. Vemuri, A. T., & Polycarpou, M. M. (1997). Neural-network-based robust fault diagnosis in robotic systems. *IEEE Transactions on Neural Networks*, *8*(6), 1410–1420.
37. de Silva, C. W. (1997). Intelligent control of robotic systems with application in industrial processes. *Robotics and Autonomous Systems*, *21*(3), 221–237.

38. Fukuda, T., & Shimojima, K. (1998). Intelligent robotic systems based on soft computing—Adaptation, learning and evolution. In *Computational Intelligence: Soft Computing and Fuzzy-Neuro Integration with Applications* (pp. 450–481). Springer, Berlin, Heidelberg.

39. Tarn, T. J., & Xi, N. (1998). Event-Based Planning and Control for Robotic Systems: Theory and Implementation. In *Essays on Mathematical Robotics* (pp. 31–59). Springer, New York, NY.

40. Jacak, W. (1999). *Intelligent robotic systems: design, planning, and control* (Vol. 14). Springer Science & Business Media.

41. Choi, J. Y., & Lee, J. S. (2000). Adaptive iterative learning control of uncertain robotic systems. *IEE Proceedings-Control Theory and Applications, 147*(2), 217–223.

42 Behal, A., Dixon, W., Dawson, D. M., & Xian, B. (2009). *Lyapunov-based control of robotic systems*. CRC Press.

43. Antonelli, G., Indiveri, G., & Chiaverini, S. (2009). Prioritized closed-loop inverse kinematic algorithms for redundant robotic systems with velocity saturations. In *2009 IEEE/RSJ International Conference on Intelligent Robots and Systems* (pp. 5892–5897). IEEE.

44. Short, M., & Burn, K. (2011). A generic controller architecture for intelligent robotic systems. *Robotics and Computer-Integrated Manufacturing, 27*(2), 292–305.

45. Na, J., Mahyuddin, M. N., Herrmann, G., Ren, X., & Barber, P. (2015). Robust adaptive finite-time parameter estimation and control for robotic systems. *International Journal of Robust and Nonlinear Control, 25*(16), 3045–3071.

46. Fukuda, T., & Kubota, N. (1999). An intelligent robotic system based on a fuzzy approach. *Proceedings of the IEEE, 87*(9), 1448–1470.

47. Castillo, O., & Melin, P. (2003). Intelligent adaptive model-based control of robotic dynamic systems with a hybrid fuzzy-neural approach. *Applied Soft Computing, 3*(4), 363–378.

48. Chang, Y. C. (2005). Intelligent robust control for uncertain nonlinear time-varying systems and its application to robotic systems. *IEEE Transactions on Systems, Man, and Cybernetics, Part B (Cybernetics), 35*(6), 1108–1119.

49. Lian, R. J. (2011). Intelligent controller for robotic motion control. *IEEE Transactions on Industrial Electronics, 58*(11), 5220–5230.

50. Tzafestas, S., & Tzafestas, E. (1991). The blackboard architecture in knowledge-based robotic systems. In *Expert Systems and Robotics* (pp. 285–317). Springer, Berlin, Heidelberg.

51. Beni, G., & Wang, J. (1991, April). Theoretical problems for the realization of distributed robotic systems. In *Proceedings 1991 IEEE International Conference on Robotics and Automation* (pp. 1914–1920). IEEE.

52. Nakamura, Y., & Xu, Y. (1995). *Geometrical fusion method for multisensor robotic systems* (pp. 241–259). Ablex Publishing Corporation, Norwood, NJ.

53. Sanfeliu, A., Hagita, N., & Saffiotti, A. (2008). Network robot systems. *Robotics and autonomous systems, 56*(10), 793–797.

54. Stanescu, A. M., Nita, A., Moisescu, M. A., & Sacala, I. S. (2008). From industrial robotics towards intelligent robotic systems. In *2008 4th International IEEE Conference Intelligent Systems* (Vol. 1, pp. 6–73). IEEE.

55. Iñigo-Blasco, P., Diaz-del-Rio, F., Romero-Ternero, M. C., Cagigas-Muñiz, D., & Vicente-Diaz, S. (2012). Robotics software frameworks for multi-agent robotic systems development. *Robotics and Autonomous Systems, 60*(6), 803–821.

56. Lueth, T., Dillmann, R., Dario, P., & Wörn, H. (2012). *Distributed autonomous robotic systems 3*. Springer Science & Business Media.

57. Fukuda, T. (Ed.). (2013). *Soft computing for intelligent robotic systems* (Vol. 21). Physica.

58. Hackwood, S., & Wang, J. (1988). The engineering of cellular robotic systems. In *Proceedings IEEE International Symposium on Intelligent Control 1988* (pp. 70–75). IEEE.
59. Trivedi, M. M., Abidi, M. A., Eason, R. O., & Gonzalez, R. C. (1990). Developing robotic systems with multiple sensors. *IEEE Transactions on Systems, Man, and Cybernetics, 20*(6), 1285–1300.
60. Schorr, O., Hata, N., Bzostek, A., Kumar, R., Burghart, C., Taylor, R. H., & Kikinis, R. (2000). Distributed modular computer-integrated surgical robotic systems: Architecture for intelligent object distribution. In *International Conference on Medical Image Computing and Computer-Assisted Intervention* (pp. 979–987). Springer, Berlin, Heidelberg.
61. Bogatyrenko, E., Pompey, P., & Hanebeck, U. D. (2011). Efficient physics-based tracking of heart surface motion for beating heart surgery robotic systems. *International Journal of Computer Assisted Radiology and Surgery, 6*(3), 387–399.
62. Beni, G., & Wang, J. (1993). Swarm intelligence in cellular robotic systems. In *Robots and biological systems: towards a new bionics?* (pp. 703–712). Springer, Berlin, Heidelberg.
63. Lerman, K., Galstyan, A., Martinoli, A., & Ijspeert, A. (2001). A macroscopic analytical model of collaboration in distributed robotic systems. *Artificial Life, 7*(4), 375–393.
64. Lerman, K., Martinoli, A., & Galstyan, A. (2004). A review of probabilistic macroscopic models for swarm robotic systems. In *International workshop on swarm robotics* (pp. 143–152). Springer, Berlin, Heidelberg.
65. Balch, T., & Arkin, R. C. (1994). Communication in reactive multiagent robotic systems. *Autonomous robots, 1*(1), 27–52.
66. Hu, H., Yu, L., Tsui, P. W., & Zhou, Q. (2001). Internet-based robotic systems for teleoperation. *Assembly Automation.*
67. Jia, S., Hada, Y., Ye, G., & Takase, K. (2002, May). Distributed telecare robotic systems using CORBA as a communication architecture. In *Proceedings 2002 IEEE International Conference on Robotics and Automation (Cat. No. 02CH37292)* (Vol. 2, pp. 2202–2207). IEEE.
68. Parker, L. E., Bekey, G., & Barhen, J. (Eds.). (2012). *Distributed autonomous robotic systems 4.* Springer Science & Business Media.
69. Xu, W., Liang, B., & Xu, Y. (2011). Survey of modeling, planning, and ground verification of space robotic systems. *ActaAstronautica, 68*(11–12), 1629–1649.
70. Coradeschi, S., Loutfi, A., & Wrede, B. (2013). A short review of symbol grounding in robotic and intelligent systems. *KI-KünstlicheIntelligenz, 27*(2), 129–136.
71. Shah, S. V., Saha, S. K., & Dutt, J. K. (2013). Dynamics of tree-type robotic systems. In *Dynamics of tree-type robotic systems* (pp. 73–88). Springer, Dordrecht.
72. Xu, X., & Liu, Y. (2017). Recent advances in intelligent robotic systems. *CAAI Transactions on Intelligence Technology, 2*(4), 141–141.
73. Bien, Z. Z., Park, K. H., Kim, D. J., & Jung, J. W. (2004). 4 welfare-oriented service robotic systems: Intelligent Sweet Home & KARES II. In *Advances in rehabilitation robotics* (pp. 57–94). Springer, Berlin, Heidelberg.
74. Asama, H., Fukuda, T., Arai, T., & Endo, I. (Eds.). (2013). *Distributed autonomous robotic systems 2.* Springer Science & Business Media.
75. Katic, D., & Vukobratovic, M. (2013). *Intelligent control of robotic systems* (Vol. 25). Springer Science & Business Media.

15 Intelligent Robotic Systems

Shivang Tyagi and Nthatisi Magaret Hlapisi
Lovely Professional University

CONTENTS

15.1 Introduction 200
   15.1.1 Background and Motivation 200
   15.1.2 Problem Formulation 201
   15.1.3 Objectives and Goals of This Project 202
15.2 Design Procedure..202
   15.2.1 Materials/Parts..202
      15.2.1.1 Raspberry Pi 202
      15.2.1.2 Actuator..203
      15.2.1.3 PiCamera..203
      15.2.1.4 Battery..203
      15.2.1.5 Gears 204
      15.2.1.6 Motor Drivers...204
   15.2.2 Design Calculations for the project 204
      15.2.2.1 Design of Battery Pack 204
      15.2.2.2 Design of Snake's Body 206
      15.2.2.3 Design of First Link of Snake............................207
   15.2.3 Working Principle..208
      15.2.3.1 Process Flow Chart..208
15.3 Testing...209
   15.3.1 Obstacle-Avoidance Testing 209
   15.3.2 Raspberry Piand PiCamera (Streaming Videos) 209
      15.3.2.1 Algorithm/Aim 209
   15.3.3 Raspberry Pi and Motors (Locomotion Testing)...............210
   15.3.4 Raspberry Pi and PIR Sensors (Human Detection)..........210
15.4 Results...210
   15.4.1 Obstacle-Avoidance Expected Result 210
   15.4.2 Raspberry Pi and PiCamera (Video streaming) 211
   15.4.3 Raspberry Pi and Motors (Locomotion testing)...............213
   15.4.4 Raspberry Pi and PIR Sensors (Human Detection)..........213
   15.4.5 Summary of Results 213
15.5 Discussion 214
   15.5.1 Discussion of Obstacle Avoidance.....................................214
   15.5.2 Discussion of Video Streaming 215

     15.5.3 Discussion of Locomotion Testing 216
     15.5.4 Discussion of Human Detection 216
     15.5.5 Limitations... 217
15.6 Conclusion 217
     15.6.1 Conclusions.. 217
     15.6.2 Future Scope 219
References... 219

15.1 INTRODUCTION

15.1.1 BACKGROUND AND MOTIVATION

This chapter is about intelligent robotic systems for security purposes. An intelligent snake robot is described as an active contour model that uses energy to copy the locomotion of a biological snake. Initially the snake robot was designed by Hirose [1]. A snake robot is hyper-redundant like a normal snake; it moves via having internal changes within its body. Snakes have at least five forms of locomotion. These include side-winding locomotion, slide-pushing locomotion, concertina movement, lateral undulation and rectilinear locomotion [2].

Of the forms listed above, most snake robots, including our own design, try to do the side-winding locomotion [3].This choice is determined by the fact that this type of locomotion offers the most acceleration using the least energy. Choosing to mimic the side-winding locomotion over other types of locomotion does not try to discredit the value of other types of locomotion, especially because when synthesizing side-winding locomotion for a snake robot, the programme generated combines partly rectilinear locomotion and lateral undulation to produce a slithering motion or, explained in a better way, an oscillating motion for the robot [4].

Extinction of animals is disastrous but escalating. We decided to name our snake robot the Liophis Cursor Snake, in honour of an extinct snake. We hope that giving this name to our robot creates awareness about animal extinction and how serious it is.

We were motivated to choose this project because bio-inspired robots are incredible at solving problems that trouble human beings [5].Snake robots prove to be particularly useful in search-and-rescue operations during catastrophes such as

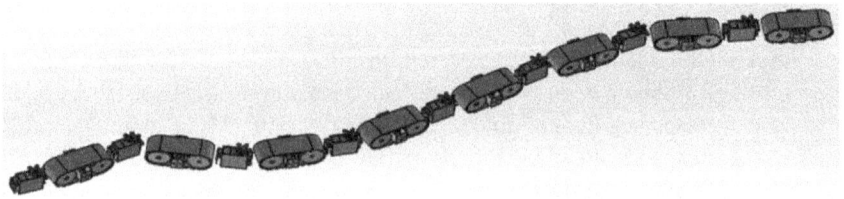

FIGURE 15.1 Snake robot design.

FIGURE 15.2   A wheeled snake robot.

earthquakes. Almost all previous snake designs use wheels as locomotors, which is a problem when navigating through uneven terrain. In this project we propose double-sided timing belts as locomotors [6].

15.1.2   PROBLEM FORMULATION

The problems we took note of when considering the already available designs of snake robots are mainly the difficulty to navigate in uneven terrain and also the high expense of some designs.

Most intelligent snake robots that have difficulty moving in uneven terrain are designed with wheels. So, while wheels are easy to design and implement in snakes, they are not durable, and they get stuck when the snake is moving in even terrain. This poses a particular problem for search-and-rescue teams because such operations need speed and a large degree of freedom. These two factors are largely compromised by these type of snake designs.

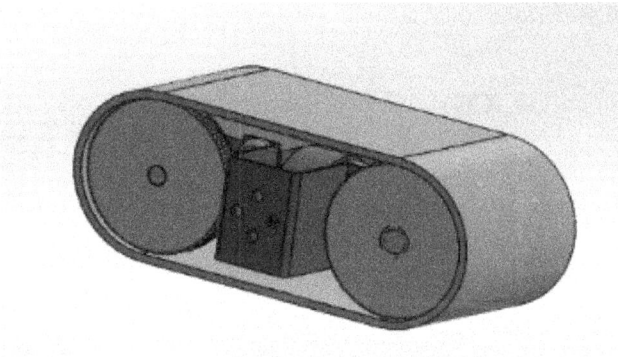

FIGURE 15.3   The double-sided timing belt used in the Liophis Cursor Snake.

The second group of designs entail snake robots that are very complex and expensive to apply.

Our proposed snake robot (Liophis Cursor Snake) is inexpensive, hence feasible for any search-and-rescue team, even in developing countries. Our design uses double-sided timing belts as locomotors. These are more durable and offer a higher degree of freedom than other snake designs [5].

15.1.3 OBJECTIVES AND GOALS OF THIS PROJECT

The main focus of this chapter is to design, implement and showcase a prototype of a hyper-redundant snake robot that can be implemented in search-and-rescue operations in earthquake-stricken areas. This prototype will have distinct locomotors and special sensors. Because this prototype uses efficient yet affordable designs, it is well designed to help search-and-salvage groups throughout the world, including developing and immature nations.

One of the main tasks of the Liophis is to be able to inspect victims under the rubble of buildings destroyed by earthquakes. It should be able to navigate through many obstacles, avoiding them, and taking a live video stream of the environment for people remotely controlling it.

PROJECT GOALS:
The subsequent goals were agreed upon:

1. Range: The human must be detected even if they are metres away from the snake.
2. Video: A high-quality live stream must be at a decent frame rate.
3. Size: The size of the snake must enable it to navigate easily.
4. Power: It must run on low power consumption so that it can run remotely using batteries.
5. Locomotion: The snake should be able to avoid obstacles and move in an uneven terrain.
6. Tests: The design should be built and tested.

15.2 DESIGN PROCEDURE

15.2.1 MATERIALS/PARTS

The Liophis is made of a combination of aluminium and fibre sheets. The body of the snake is made of aluminium because it is light weight. The fibre sheet is used to cover the snake body.

15.2.1.1 Raspberry Pi

Raspberry Pi is a mini board computer. Raspberry Pi is used to teach basic computer skills in collages and developing counties. Raspberry Pi has many models, but the snake uses Raspberry Pi 3B+ because of its advanced features. Raspberry Pi 3B+ has a metal heat spreader on the top for better thermal performance. The heat is dissipated across the whole board for better thermal performance. Raspberry Pi 3B+ does use substantially more power than the previous Raspberry Pi according to the Raspberry Pi foundation.

15.2.1.2 Actuator

The snake uses an actuator to give it motion. The snake uses locomotion by performing two types of motion: angular and rotatory motion. As name suggests, angular motion gives the curved line shape by changing the angle on joints. And rotatory motion provides the rotatory motion to links so that it is comfortable to move forward and backward in specified environment.

We are using two types of actuator:

- Servomotor
- DC gear motor

Servomotor: A DC servomotor is used in feedback control systems. It has high-speed response and low rotor inertia. It is also known as control motor. The response of a servomotor should be fast, and inertia should be low. It is a separately excited or permanent magnet motor. Adjusting the voltage applied to armature can control the speed of the servomotor. The servomotor is used to provide the angular motion by which we get curved line shape of the snake. This shape helps the snake to move in complex places.

DC dual gear motor: A direct current motor rotating electrical device that is used to convert direct current (electrical energy) into mechanical energy. In the middle of motor there is an iron shaft, covered with a coil of wire. The shaft has two fixed north and south magnets pole which causes both an attractive and repulsive force, in turn, producing torque. The dual-shaft DC motor with gear box gives high torque and RPM at low voltage. It is generally appropriate for a light-weight robot running on low voltage. The DC dual-shaft motor is used to provide rotatory motion, makes the model less complex and makes it different to the existing system.

15.2.1.3 PiCamera

A PiCamera is a portable light-weight camera that supports Raspberry Pi. It is a high-quality pixel camera used for various operations, such as object detection, image processing and video streaming. A PiCamera is used to do live streaming onto a web page.

15.2.1.4 Battery

A lithium polymer battery is basically a lithium-molecule using a polymer electrolyte as opposed to a liquid electrolyte. High-conductivity semisolid polymers structure this electrolyte. We use 3300+ mAh to provide a sufficient power supply to devices.

15.2.1.5   Gears

A gear is a rotating machine part having teeth which mesh with other teeth to transmit torque. A gear is a transmission element which transmits torque from one point to another. In the snake we use gears to transmit power to a dual-sided timing belt which makes the snake move.

15.2.1.6   Motor Drivers

A motor driver, or a motor controller, is an integrated circuit (IC) chip. It acts as an interface between the motor and the control circuit in embedded-circuit and autonomous robots [7]. In the project we are using two types of driver:

1. **L298N:** L298N is a dual H-Bridge motor driver. It can be used to drive speed and direction. It can be controlled manually by switches or, much more efficiently, by using a microcontroller such as Arduino. L298N is used to control eight dual-shaft DC motors. We use this motor driver because of its high operating-current rating.
2. **PCA9685:** PCA9685 is a 16-pin motor driver used to control the servomotor by Pulse width Modulation (PWM) The servomotor driver can control 16 servomotors simultaneously. It is a double-wired interface sequential convention to connect low-speed devices such as microcontrollers and other comparative peripherals in embedded systems. This motor driver is used to control 16 channels simultaneously. In the project we need to control 10 servomotors.

15.2.2   Design Calculations for the Project

15.2.2.1   Design of Battery Pack

(1) If we use eight DC gear motors continuously with 3300 mAh capacity of battery, how long will it run?

Operating voltage: 4 – 9V
Rated no-load speed: 165 RPM
Torque: 1.9Kgf.cm
No-load current: 4.54V to 52.5mA
Stall current: 6V to700mA

Solution:

(1) According to reading no-load current at 4.54V to 52.5mA, with eight motors, one of which is 52.5mA.

Load current = 52.5*8
= 420mA

Number of hours = capacity of battery (mAh)/load current (mA)
= 3300mAh/420mA
= 7.85hr

So, the motor will run for **7.85hr**.

(2) According to reading max-load current at 4.54v to 700mA,witheight, one of which is 700mA.

Load current = 700*8 mA
= 5600mA

Number of hours = capacity of battery (mAh)/load current (mA)
= 3300mAh/5600mA
= 0.58hr

So, the motor will run for **0.58hr**.

(2) If we use eight MG995s (servomotors) continuously, with 3300 mAh capacity of battery, How long will it run?

Provided Readings

No-load current: 6V–9.1mA
Max-load current: 6V–450 mA
Operating voltage: 4.8 V–7.2 V
Speed: 0.16 s/60° (6V)
Couple: 10 kgf cm (6V)

Solution:

(1) According to reading no-load current at 6 V– 9.1mA,with eight motors, one of which is 9.1mA.

Load current = 9.1*8 mA
= 72.8mA

Number of hours = capacity of battery (mAh)/load current (mA)
= 3300mAh/72.8mA
= 45.3hr

So the motor will run for **45.3hr**.

(2) According to reading max-load current at 6V–450mA, with eight motors, one of which is 450mA.

Load current = 450*8 mA
= 3600mA

Number of hours = capacity of battery (mAh)/load current (mA)
= 3300mAh/3600mA
= 0.91hr

So the motor will run for **0.91hr**.

15.2.2.2 Design of Snake's Body

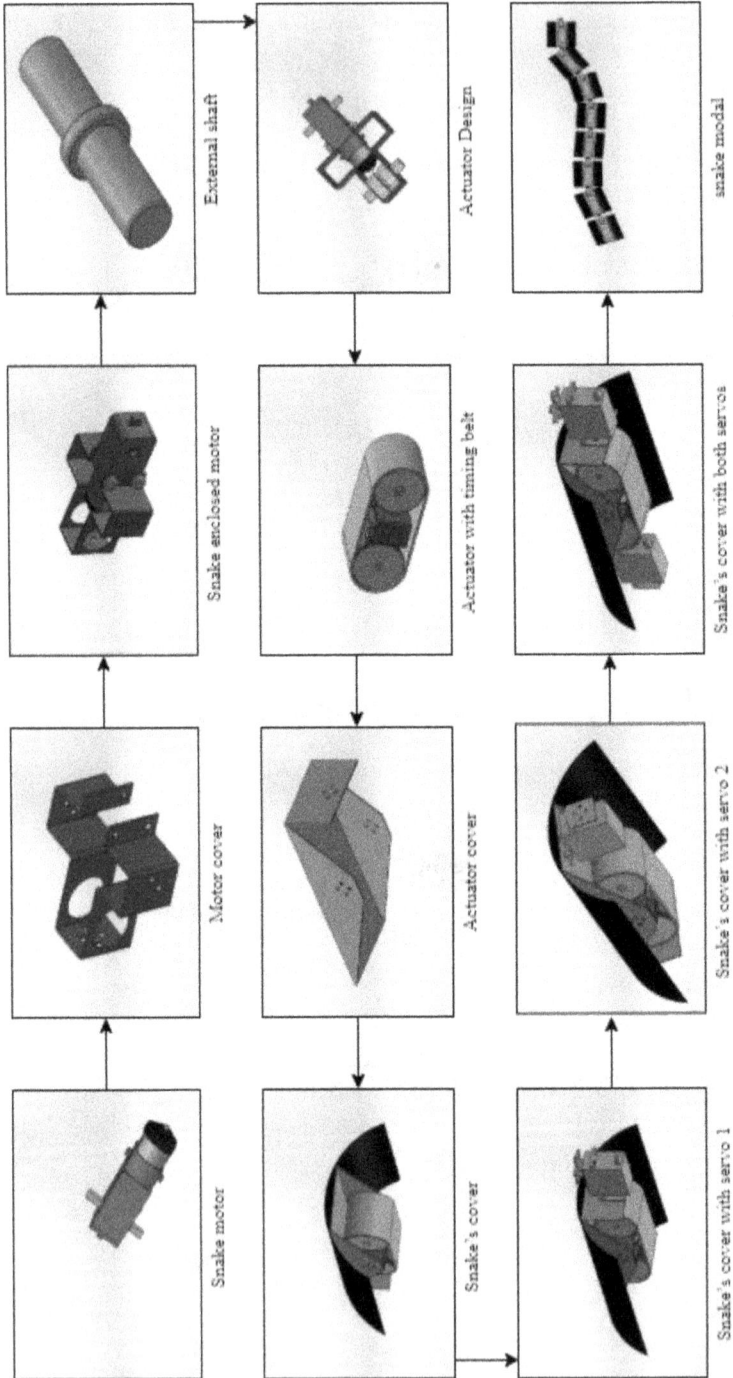

FIGURE 15.4 Assembly of snake's body.

15.2.2.3 Design of First Link of Snake

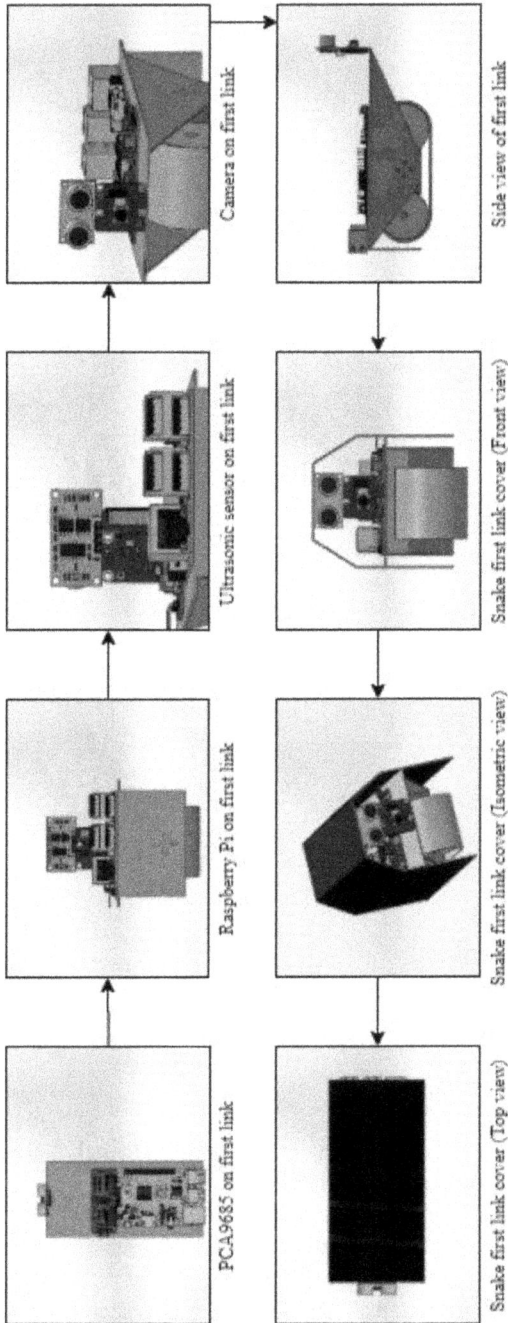

FIGURE 15.5 Assembly of snake's body first link.

15.2.3 WORKING PRINCIPLE

15.2.3.1 Process Flow Chart

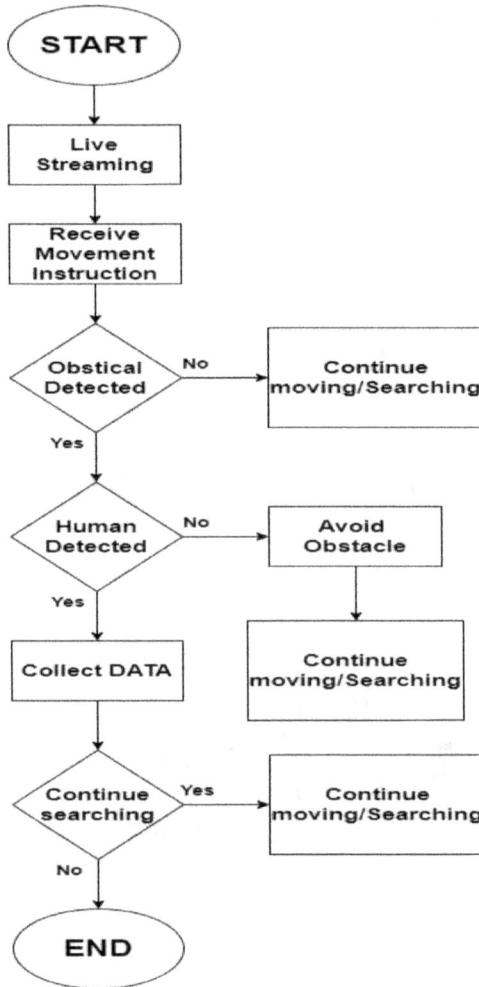

FIGURE 15.6 Process flow chart.

A flow chart is a representation of the sequential steps of how our snake is going to operate. We are using it to represent the workflow in the system.

Figure 15.6 shows that when the snake is turned on, it starts to take live-stream videos and send them to a webpage that can be viewed anywhere. It achieves this by interfacing the Raspberry Pi with PiCamera.

The controller sends movement instructions to the robot, such as forward, left or right. The instructions are sent as PWM signals to the servomotors of the robot. This moves the robot according to the send instruction.

While the robot is moving, its ultrasonic sensor is always on to detect any obstacle in the path. If no obstacle is detected, the robot continues to slither according to the given instruction.

On the other hand, if the obstacle is detected, the robot stops for a while for a passive infrared sensor (PIR) sensor to gather information from the surroundings. The PIR sensor is used to find out if the detected obstacle is a person. If the PIR does not note a higher temperature, then the obstacle is not a person, and the robot avoids and continues to move until a new instruction is given.

If the PIR sensor identifies the detected obstacle as a person, the PiCamera is immediately instructed to take pictures while already taking a video. The images will be used for further analysis; this could be done through OpenCV or any other computer vision software.

Throughout the whole process in applications such as search-and-rescue operations the robot's movement is controlled by the operator through the PiCamera video, and the robot follows the movement instruction while avoiding obstacles in its path. The essence of search-and-rescue operations is to locate injured victims, so the robot is also equipped with a PIR sensor so that it can locate victims appropriately.

15.3 TESTING

15.3.1 OBSTACLE-AVOIDANCE TESTING

In this test, we check the robot's ability to detect and avoid obstacles in its environment. Obstacle avoidance is a crucial task that enables robots to avoid any collisions with objects in their workspace.

ALGORITHIM APPLIED AND TESTED:

The snake uses the ultrasonic HC-SR04 sensor which can detect objects within a 2-cm–400-cm radius with 0.3 resolution. This intelligent snake robot will detect and avoid objects within a 35-cm range.

The robot detects objects within the 35-cm range and turns left/right depending upon which side has no obstacle. It moves around the obstacle detected. The robot is able to move because it can rotate the ultrasonic sensor from 0 to180 degrees with the servomotor installed on the head of the snake.

15.3.2 RASPBERRY PIAND PICAMERA (STREAMING VIDEOS)

Raspberry Pi is a small computer capable of video and media functions. The Raspberry Pi has a Camera Serial Interface (CSI-2) connector which is used to connect the camera module directly to the GPU (Broadcom Video Core 4 Graphics Processing Unit) using the CSI-2 protocol. The camera can deliver H.264-quality video.

15.3.2.1 ALGORITHM/AIM

The objective is to capture a video and stream it on a web server using any device that has a web browser and it is connected to the same IP address as our Pi. We are basically streaming a high-quality video using a Raspberry Pi (which has limited hardware resources). We will transfer a 6-Mbps H.264 video stream of 1280×20 pixels at 25 frames/second.

15.3.3 RASPBERRY PI AND MOTORS (LOCOMOTION TESTING)

Locomotion refers to how our robotic snake will move through the environment and how it will resemble the movement of a biological snake.

Motors are actuators/devices that convert electrical energy into mechanical energy. They operate on the principle of electromagnetism. In this project we are going to use servomotors.

Servomotors are DC motors that allow for precision control in terms of angular position/acceleration or even velocity. They have a revolution from 0to 180 degrees; 90 degrees is often referred to as the neutral position of the motor.

Raspberry Pi controls servomotors by sending PWM output to it to set the angles of rotation for the motor. Rotation, or angle, is set by the length of each PWM signal sent.

15.3.4 RASPBERRY PI AND PIR SENSORS (HUMAN DETECTION)

Human detection is the essence of search-and-rescue work during natural disasters [3]. Victims are holding on by a thin thread as their likelihood of surviving depends on whether they will be found under the rubble of collapsed buildings. Changes of that happening are slim. Search-and-rescue robotic snakes were created to help rescuers to find victims without increasing the risk of further collapsing the building and putting the rescuers in danger. To do all this, robotic snakes need to be able to detect human beings. This is made possible through the PIR sensor.

A PIR is a pyroelectric sensor that is used to detect changes in the temperature in its surrounding. All objects emit heat in the form of infrared heat, so when the robot encounters an object, the sensor detects the change in temperature and converts this change into a voltage output it will pass to the controller.

15.4 RESULTS

15.4.1 OBSTACLE-AVOIDANCE EXPECTED RESULT

Considerations about the testing environment

- All the tests of obstacle avoidance were conducted in the same lab.
- No obstacles in motion were used.
- All obstacles used in the tests had the height higher than that of the robot.
- All the obstacles were placed directly in front of the snake; variations of width and distance from the snake were made.

Theoretical results

The output of the test A (stated in Figure 15.7) will be seen on the desktop of Raspberry Pi as something like shown in Figure 15.11.

TEST A: CAN IT DETECT OBJECTS WITHIN THE 35cm range?

Trial 1: Place the object within 35 cm range of snake
Expected results: the distance of the object is shown on the screen and head motion of snake starts.

Trial 2: Place the object outside 35cm range of snake
Expected result: the object is not detected, no distance of object from the snake is shown

---

TEST B: CAN THE SNAKE AVOID THE OBJECT?
To check if the servo motor can rotate the head of the snake (ultrasonic sensor) when the object is detected.

Trial 1: Place the object within 35cm range of snake and check if the servo rotates the head of the snake.

Expected results: if the object is detected, the head of the snake should turn left or right to look for a better path/avoid the obstacle.

FIGURE 15.7   Tests a and b show how to detect object within the 35-cm range.

Trial 2, obstacle avoidance test was conducted to see if the sensor would take note of obstacles outside the set 35-cm range, and the results shown in Figure 15.12 are expected on the Raspberry Pi terminal.

Test B in 3.2 was conducted to see how the snake robot will navigate around the obstacles. Figure 15.13 shows how it is expected to move.

15.4.2   Raspberry Pi and PiCamera (Video streaming)

Considerations about the testing environment

- All the tests were done in the same lab (to minimize the light noise as much as possible); the lighting was varied by closing windows and turning on and off the lights.

TEST A: CHECK THE WORKING OF PiCamera

Trial 1: Take a picture and show on preview

Trial 2: Take 5 pictures in a row wand save in a file

Trial 3: take a 5 seconds video and store on file.

TEST B: STREAMING REAL TIME VIDEO

Trial 1: Access your video streaming web server at http://<IP ADDRESS>:8000

FIGURE 15.8   Shows how to check the working of piCamera.

TEST A: CHECK OPERATION OF SERVO MOTOR

"For servo motors to rotate, they expect a PWM signal f 50 Hz frequency."

Trial 1: Set the rotation to 0 degrees (Send a PWM signal of .5 ms)
       : Set the rotation to 90 degrees (Send a PWM signal at 1.5 ms)
       : Set the rotation to 180 degrees (Send a PWM signal at 2.5 ms)

TEST B: CHECK THE CODE

Trial 1: Give the robot instructions to turn left, right and forward.

FIGURE 15.9   Shows tests to check operation of servo motor.

TEST A: CHECK THE PIR SENSOR

Trial 1: Connect the PIR output to an LED, it will go, "high", to 3.3V when motion/

human presence is detected.

Trial 2: Adjust the sensitivity of the sensor to an appropriate level

---

FIGURE 15.10   Shows how to check working of PIR sensor.

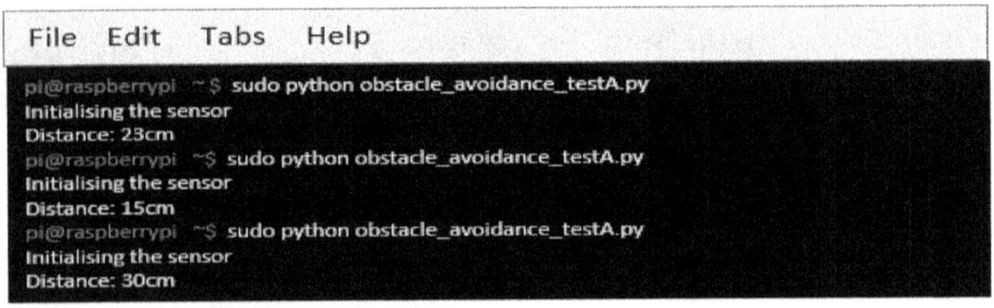

```
 File   Edit   Tabs   Help

 pi@raspberrypi ~$ sudo python obstacle_avoidance_testA.py
 Initialising the sensor
 Distance: 23cm
 pi@raspberrypi ~$ sudo python obstacle_avoidance_testA.py
 Initialising the sensor
 Distance: 15cm
 pi@raspberrypi ~$ sudo python obstacle_avoidance_testA.py
 Initialising the sensor
 Distance: 30cm
```

FIGURE 15.11   Expected output screen of Raspberry Pi terminal window.

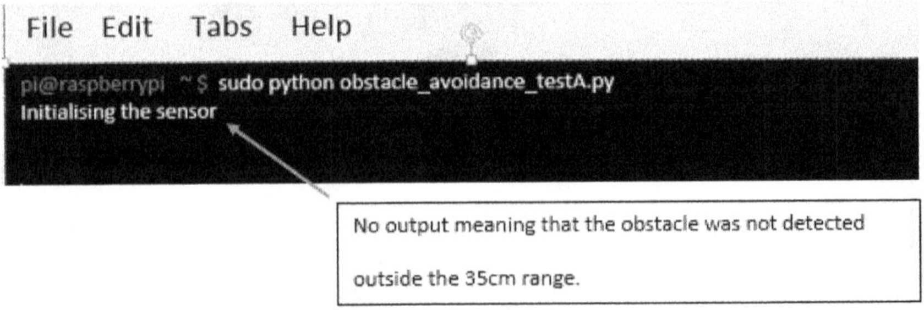

```
 File   Edit   Tabs   Help

 pi@raspberrypi ~$ sudo python obstacle_avoidance_testA.py
 Initialising the sensor
```

No output meaning that the obstacle was not detected

outside the 35cm range.

FIGURE 15.12   No output.

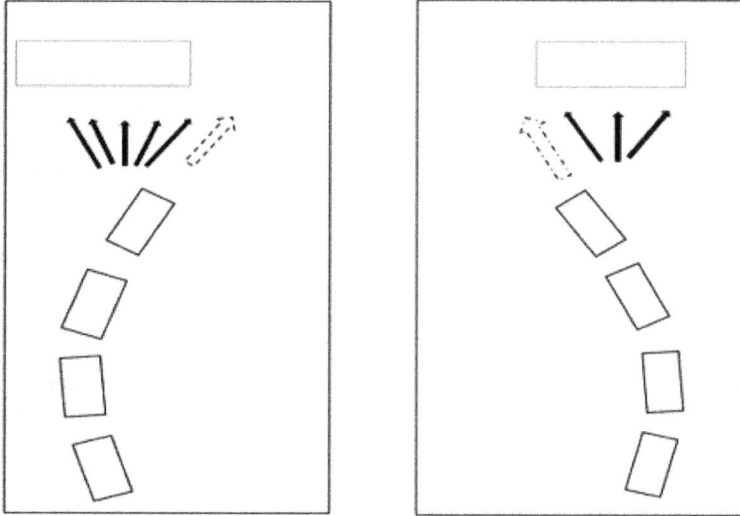

FIGURE 15.13 Turning left and right to avoid obstacle.

- The maximum frame rate was set to 25 frames per second.
- The exposure mode was set to 'on'.

15.4.3 RASPBERRY PI AND MOTORS (LOCOMOTION TESTING)

Considerations about the testing environment

- All the tests were done in the same lab.
- All the tests were done on even terrain.
- All the motors are driven by eternal battery pack and not power from the Pi's power pin.

15.4.4 RASPBERRY PI AND PIR SENSORS (HUMAN DETECTION)

Considerations about the testing environment

- All tests for human detection were done in the same lab at different times of the day to get an average result on the PIR's performance for different ambient temperatures.
- Only human beings were detected or used in the experiments.

15.4.5 SUMMARY OF RESULTS

Table 15.1 presents a summary of the results of tests run on the snake robot to analyze its behaviour and attributes.

TABLE 15.1
Summary of Results

Results	Sensor Used	Performance
Obstacle avoidance	Ultrasonic sensor	If the obstacle is within a 35-cm range, it is detected and avoided by turning the head left or right.
		If the obstacle is outside the 35-cm range, it is not detected.
Video streaming	Raspberry Pi & PiCamera	The video is streamed and viewed on a web browser using any device that is connected to the same IP address as the Pi.
Locomotion testing	Raspberry Pi & servomotors	The motors move the snake forward, left and right depending on the PWM signal they get from the Raspberry Pi.
Human detection	Raspberry Pi & PIR sensor	The PIR sensor detects the presence of a human.

15.5   DISCUSSION

15.5.1   DISCUSSION OF OBSTACLE AVOIDANCE

To recap the obstacle-avoidance mechanism of the snake robot, we note that the snake uses an ultrasonic sensor (hc-sr04) to detect an obstacle, and it avoids the obstacle not by climbing over it but by going around it. The anticipated results when this attribute is tested are that the snake detects obstacle within the 35-cm range; the robot then has to decide between turning left or right to avoid collision with the object.

In this section of the chapter, we explain how exactly the snake robot avoids the obstacle.

How does the ultrasonic sensor enable the snake robot to avoid obstacles?

- *STEP 1*
  - The first step in the obstacle-avoidance mechanism is the ultrasonic sensor detecting the obstacle in the snake's path.

The ultrasonic sensor has a transmitter that sends out sound waves at a very high frequency. The sensor waits to hear the waves come back hitting an object. Based on the time it takes for the waves to travel back, the distance is calculated and the object can be located. That is, we can know if it is in the snake's travel range.

The ultrasonic sensor uses this formula to calculate how far the obstacle is from the snake robot:

$$\text{Distance} = \text{Time} * \text{Speed of sound}$$

When the calculated distance is 35cm or less, the obstacle-avoidance mechanism moves on to step 2.

> STEP 1 = FINDING OUT IF THE OBSTACLE IS WITHIN
> 35cm FROM THE SNAKE ROBOT OR NOT USING
> ULTRASONIC SENSOR.

FIGURE 15.14 Shows step 1of obstacle avoidance.

- *STEP 2*
  - The second step in obstacle avoidance occurs when the ultrasonic sensor does detect that the obstacle is within a 35-cm range of the snake robot.

After the object is detected, what follows is trying to determine which path to use to avoid the obstacle. The Raspberry Pi instructs the servomotor controlling the snake to look left (angle 45 degrees) and then allows the ultrasonic sensor to figure out how far from the left the object is away from the snake. The calculated distance is noted/saved. The servomotor then turns the snake head right by the same angle, and the ultrasonic sensor takes note of the distance. This cycle is repeated twice, once for 45 degrees and the second time for an additional 45 degrees on each side. (During the second cycle, the head is rotated to 0 degrees and on the right side, it is rotated to 180 degrees.)

Raspberry Pi compares the distance from the left and right side. If the object is closer to the left than to the right, the right path is chosen and vice versa. If the distance is the same on both sides, then by default the left path is chosen.

15.5.2 Discussion of Video Streaming

Recapping the video streaming mechanism of the snake robot, we note that the snake uses a PiCamera to take a video of its environment. The anticipated result when this attribute is tested is that the snake is able to stream a video into a webpage that can be accessed from any device that is connected to the same IP address of that of the Pi.

In this section of the chapter, we explain exactly how the snake robot streams the video into a webpage.

How does the PiCamera enable the snake robot to stream a video on to a webpage?

- *STEP 1*
  - The first step in the video streaming mechanism is the PiCamera taking a video of the environment.

After the camera is enabled *and* its IP address noted down by running the command
pi@raspberry: ~ $ifconfig
then step 2 follows.

- *STEP 2*
  - Here the recorded video is streamed on a webpage.

This done by going onto any web browser and typing
http://<Your_Pi_IP_Address>:8000

15.5.3 DISCUSSION OF LOCOMOTION TESTING

Recapping the locomotion mechanism of the snake robot, we note that the snake uses servomotors to navigate through its environment. The anticipated result when this attribute is tested is that the snake is able to move left, right and forward when the servomotors are given different PWM signals.

In this section of the chapter, we explain exactly how snake robot moves through its environment.

How do servomotors enable the snake robot to locomote through its environment?

- *STEP 1*
  - For the snake robot to mimic the movement of a biological snake, a sinusoidal signal is generated and given to the servo motors in the form of PWM signals.

Servomotors are DC motors that allow for a precision control in terms of angular position. The angle of rotation of the servo is set by the width of PWM signal sent by the Raspberry Pi. The servo will rotate between 0 and 90 degrees.

The following section of the code is used to generate the sinusoidal signal for the servo motors imitate the locomotion of a real snake.

- *STEP 2*
  - The snake robot has defined several types of move, such as forward, left and right, because of the code.

15.5.4 DISCUSSION OF HUMAN DETECTION

Recapping on the human detection mechanism of the snake robot, we note that the snake uses PIR sensors to detect the presence of humans in its vicinity. The anticipated results when this attribute is tested are that the LED mounted on the Snake turns on and the PiCamera starts taking pictures when the human is detected.

In this section of the chapter, we explain exactly how snake robot detects human beings.

How does the PIR sensor enable the snake robot to detect humans in its environment?

The PIR sensor is a passive infrared motion detector sensor. It detects the infrared heat that bodies generate. If a human being is present near the PIR sensor, it will output a 5-V signal for a period of a minute to the GPIO pin of the Raspberry Pi that it is connected to.

When the Raspberry Pi receives a high voltage at this pin, it interprets it as that a human is detected by the sensor. It then blinks the red LED on the head of the snake for 15 seconds and instructs the camera to take five consecutive pictures that will be analyzed using image processing.

15.5.5 LIMITATIONS

Obstacle Avoidance:
- The snake avoids collision with obstacles by going around them. This mechanism is slow compared to the one where snake avoids obstacles by climbing over them.

Video Streaming:
- The quality of the streamed video is not very good.

Locomotion:

- The locomotion is not like that of a biological snake yet.
- A biological snake has five types of locomotion; currently, our snake can mimic only one type.

Human Detection:
- It can detect humans only within 6–7 metres from the snake.

15.6 CONCLUSION

15.6.1 CONCLUSIONS

After applying multiple test conditions in this chapter, a reliable, mobile, obstacle-avoiding robot that has the ability to stream high-quality live video to a webpage and also to detect the presence of people was designed.

The prototype developed was given numerous tests to try to optimize its attributes. The tests included having the robot stream videos in different light settings so as to adjust the code so that the snake robot takes quality HD videos that are easy to analyze regardless of the light conditions. This was achieved. The snake robot can stream videos to a web page that can be accessed through any device that has a web browser and is connected to the same IP address as the Raspberry Pi. On the web browser the following line of code is used to stream the video taken by the PiCamera:

http://<IP_ADDRESS>:8000.

Another goal was for the snake robot to be able to detect the presence of people in its vicinity. To achieve this, our robot was equipped with a PIR motion sensor. This was based on the principle that any living thing that has an absolute temperature above zero gives out heat in the form of infrared waves. The PIR sensor senses the motion around it. When a human is around, the infrared waves will change the ambient temperature and the PIR sensor will detect that. After the Raspberry Pi (which is the brain of this robot) detects the presence of a person through the PIR sensor, it immediately instructs the PiCamera to take five consecutive pictures of the human so that more analysis can be done.

The other round of tests was done to strengthen the robot's obstacle-avoiding mechanism. This attribute, however, still needs more modifications as all the tests were done with obstacles that are not in motion. This leaves the robot vulnerable to confusion when it is faced with an obstacle that is in motion. Also, we plan to improve this mechanism by enabling the robot to climb over obstacles. Currently avoids obstacles only by going around them, and it consumes a lot of time deciding which path is best to follow to avoid the obstacle.

In terms of locomotion, our snake mimics the side-winding locomotion of a biological snake. This is achieved by generating a sinusoidal signal through an algorithm. The generated sinusoidal signal is given to servomotors (actuators) through PWM signals. Out of the five types of locomotion of natural snakes, we chose to mimic the side-winding locomotion because it gives a snake high acceleration using minimum energy. In terms of hardware, the servomotors were our best bet because of the precision control they give to control angular position. However, in the future we plan to include other types of locomotion to increase the degree of flexibility of the snake and to make it more hyper-redundant.

In regard to limitations, we have an average list of them. When an object is detected, the normal procedure we created allows the PiCamera to simultaneously take pictures while recording videos; in this instance the quality of the video taken reduces. Even with this problem, we decided to still allow the PiCamera to take pictures and videos at the same time when an object is detected because the pictures it took are going to be used in image processing, which is more easily done on pictures than on videos.

Second, for the locomotion, the more servomotors used the better the snake is at the side-winding motion. But we didn't use as many servomotors as required to accurately locomote like a natural snake because we were limited by the battery pack we were using. Also, all the locomotion tests were conducted on even terrain, so our design has not yet proven how it well it can perform in uneven terrain.

One other limitation is found when the robot avoids obstacles. Our snake robot goes around obstacles, and in doing so, it takes a noticeable amount of time to decide which direction it should turn to avoid an obstacle. In scenarios where the only way forward is to climb over the obstacle, our robot fails. It cannot yet climb over obstacles, but this feature will be included in the future design of this snake robot.

Obstacles can be avoided only if they are detected. The accuracy of the obstacle-detecting mechanism determines the precision of the robot when avoiding collisions. In this project we detect obstacles through an ultrasonic sensor, which has the following limitations: It has a limited detection range; it is rigid and therefore inflexible; and it does not offer the most accurate results. To improve the robot's obstacle-detection method, we will detect obstacles through computer vision. This will be implanted with the help of Raspberry Pi, PiCamera and image processing.

Summarizing the future modifications to be done on this project, the list is as follows: For obstacle detection, we plan to rely on computer vision and not just the ultrasonic sensor; this, we believe, will increase the sensitivity of our robot to its environment. For locomotion, we will add more types of snake locomotion such as slide-pushing and rectilinear locomotion. This will be done with the hope of making the snake more hyper-redundant and adept in moving through uneven terrain.

The uniqueness of our snake robot is the design. Unlike other snake robots, which have wheels as locomotors, we will use compressor belts. The prototype of this design seemed likely to perform better in uneven terrain than its wheel-equipped counterparts.

15.6.2   FUTURE SCOPE

- Inclusion of computer vision through Raspberry Pi Image processing with OpenCV. This will be done to improve the obstacle-detection attribute of the snake robot.
- Inclusion of a GPS module to remotely control the snake.
- Inclusion of solar panels on the body of snake so that the snake has a more reliable power supply.
- Use of a no infrared (NoIR) PiCamera instead of the PiCamera V1 5MP that is used in this project. It will give our snake the ability to see in the dark with infrared lighting.

REFERENCES

1. Hirose, Shigeo, and Edwardo F. Funkushima. "Snakes and strings: New robotic components for rescue operations." The International Journal of Robotics Research 23.4–5, (2004): 341 349.
2. Bayraktaroglu, Z. B., F. Butel and V. Pasqui. "Snake-like locomotion: Integration of geometry and kineto-statics". Advanced Robotics, 14.6, (2000): 447–458.
3. Burdick, J., J. Radford and G. S. Chirikjian, "A side-winding locomotion gait for hyper-redundant robots." Advanced Robotics, 9, (1995): 195–216.
4. Ma, Shugen, Naoki Tadokoro and Kousuke Inoue. "Influence of the gradient of a slope on optimal locomotion curves of a snake-like robot." Advanced Robotics, 20.4, (2006): 413–428.
5. Huang, Cheng C., and Chung L. Chang. "Design and implementation of bio-inspired snake bone-armed robot for agricultural irrigation application." IFAC-PapersOnLine, 52.30, (2019): 98–101.
6. Kyriakopoulos, K. J., G. Migadis and K. Sarrigeorgidis. "The NTUA snake: Design. planar kinematics, and motion planning." Journal of Robotic Systems,16.1, (1999): 37–72.
7. Lee, W. H. "Dynamic analysis and distributed control of the tetrobot modular reconfigurable robotic system." Autonomous Robots, 10, (2001): 67–82.
8. Erkmen, I., A. M. Erkmen, F. Matsuno, R. Chatterjee and T. Kamegawa. "Snake robots to the rescue!" IEEE Robotics & Automation Magazine, 9,3 (2002):17–25.

16 Model Order Reduction of Some Critical Systems Using an Intelligent Computing Technique

Souvik Ganguli, Parag Nijhawan,
Manish Kumar Singla, Jyoti Gupta and
Abhimanyu Kumar
Thapar Institute of Engineering and Technology

CONTENTS

16.1 Introduction 221
16.2 Problem Formulation 223
16.3 Proposed Work 224
16.4 Simulations and Discussions 225
16.5 Conclusions.. 235
References.. 235

16.1 INTRODUCTION

The model order reduction (MOR) technique refers to the compression/reduction of the order of ordinary full-order differential or difference equations for quick computation and less storage requirements while at the same time preserving the same characteristics of the full-order system [1–3]. The number of equations formed to study the system increases as its order increases. This results in time-consuming calculations and a higher probability of error as well as a tedious operation. Also, the higher the order of the system, the more hardware components required to implement it, thereby raising the cost. Thus, in order to keep all these factors in check, the order of the system is reduced in such a way that all its essential characteristics remain unchanged. This results in fewer equations, less time consumption and fewer components. There are various techniques by which the models can be reduced, collectively referred to as MOR techniques [4, 5]. Two kinds of systems, namely the fractional-order and time-delay systems, are discussed below to create an appropriate background for the chapter.

Researchers in various engineering fields are very enthusiastic about the investigation of fractional-order (FO) structures, as their mathematical models have proven to be more successful in various phenomena such as electrochemical processes, long-distance lines, dielectric polarization, viscoelastic materials, coloured noise and even chaos [6, 7].

FO systems are basically approximations derived from integer-order (IO) systems. The system dynamics of the FO systems include non-integer differential or integral operators. It should be noted that the FO model reflects more reliable properties than its IO model [8]. The FO system, however, has an infinite dimension. Thus, the infinite-order structures are generalized by finite integer-order models for investigation, simulation and controller design [9].

MOR is an age-old practice in integer-order systems, but limited research has been performed for fractional-order models. From the literature it can be inferred that most of the classical reduction techniques employed for IO linear time-invariant systems have been applied for FO systems. Further, the use of soft computing techniques has not been explored so much except in References [6, 8, 10, 11].

Time-delay systems (TDS) are referred to as after-effect or dead-time systems, hereditary systems or differential-difference equations. They belong to a class of infinite-dimensional functional differential equations (FDE) [12]. Time delays exist in a variety of technical structures, such as aircraft controllers, lasers, modem, medical instruments, etc. These delays may be caused by transport, communication or measurement [13].

It is not easy to apply the principle of continuous-time control theory directly for dealing with time delays. Padé approximations for system analysis are common practices for the expression of time-delay systems [14]. But those lead to a higher-order system representation for which a higher-order controller is required. A higher-order controller may not always be feasible or implementable. The requirement to reduce the order of the finite-dimensional Padé approximate model is therefore necessary. A great deal of research has already developed in the area of reduced-order modelling for time-delay systems. While several avenues of research have been explored for the reduction of TDS orders, soft computing techniques have yet to be built on the path, except in Reference [15].

This led us to formulate a MOR technique for these two critical systems, applying a new optimization method, namely, the Harris Hawks Optimization (HHO) [16]. Based on the hunting tendency of Harris's hawks, it is successfully applied to address the modelling issues of two typical systems, namely, the fractional-order and the time-delay systems. Heuristic methods are involved, hence the need of DC gain matching, stability and non-minimum phase feature are some concerns which have been taken care of with the help of suitable constraints. Thus, this chapter solves a constrained engineering design problem. Plenty of widely cited and new metaheuristic algorithms are employed for comparison.

The rest of the chapter is structured as follows. Section 16.2 provides the problem statement for the model reduction of some critical systems applying the metaheuristic approaches. The challenges associated with the metaheuristic algorithm-based reduction schemes are also dealt with. In Section 3, the methodology of work is detailed, whereas Section 16.4 presents the simulation results with adequate explanations. Section 16.5 concludes the chapter with directions for future study.

16.2   PROBLEM FORMULATION

A common practice of converting any fractional-order (FO) system to an integer-order (IO) system is to use the Oustaloup approximation [17] of the appropriate order. In the continuous-time domain, the transfer function of the Oustaloup approximated model is

$$G_{Oustaloup}(s) = \frac{Y_{k-1}(s)}{X_k(s)} = \frac{\sum_{i=0}^{k-1} q_i s^i}{\sum_{i=0}^{k} p_i s^i} \qquad (16.1)$$

where q_i and p_i denote the coefficients of the numerator and denominator polynomial respectively. Moreover, $Y_{k-1}(s)$ and $X_k(s)$ do not have any common factors. The motive is to determine a reduced system of order r $(r < n)$ so that essential characteristics of the system are retained and is described by

$$G_R(s) = \frac{Y_{r-1}(s)}{X_r(s)} = \frac{\sum_{i=0}^{r-1} f_i s^i}{\sum_{i=0}^{r} e_i s^i} \qquad (16.2)$$

On a similar note, equation (16.1) will also represent the Padé approximated time-delay model of a certain order, which can be reduced heuristically to form the lower-order model as given in equation (16.2). Now the prime challenges faced in performing model order reduction of these systems through metaheuristic algorithms are as follows:

a. matching DC gain
b. ensuring stability
c. retaining the minimum/non-minimum phase characteristics.

These have been appropriately addressed as constraints in deciding the cost function J. Here, the HHO is used to determine the reduced system by minimizing the fitness function:

$$J = \sum_{i=1}^{N} [y(s) - y_R(s)]^2 \qquad (16.3)$$

subject to the constraints that the poles of the reduced system should lie on the left half of the s-plane while the only zero of the system should be on the right half of it. Further, to justify DC gain matching, another constraint $\frac{q_0}{p_0} = \frac{f_0}{e_0}$ must also be satisfied.

In equation (16.3), $y(s)$ and $y_R(s)$ represent respectively the pseudo random binary sequence (PRBS) response of the parent and reduced systems with a selected number of N samples, in the continuous-time domain. Due to simplicity of under-standing and design, the reduced system is considered to be of typical second-order described by

$$G_R(s) = \frac{f_0 + f_1 s}{e_0 + e_1 s + e_2 s^2} \tag{16.4}$$

Further, e_2 is assumed to be unity for both the test systems, so that only four deci-sion variables need to be optimized, using the proposed technique detailed in the next section.

16.3   PROPOSED WORK

The methodology involves handling the order reduction of two typical systems, viz. the fractional-order and the time-delay systems using a heuristic technique which is not very common in the literature. Usually a fractional-order system (FOS), whether com-mensurate or non-commensurate, can be expressed as an integer-order system with the help of Oustaloup approximation. Once the FO system is transformed into integer order, then the normal procedure is followed to reduce it by assuming a desired fixed-structure model, preferably a second-order, whose unknown parameters are obtained using an identification philosophy through response matching applying required con-straints. The approach is referred to as the identification procedure because a PRBS is employed as an input signal for the parent and reduced models. The responses of both models are compared to determine the reduced system by setting the sum of squares to be minimized. Normally step response matching suffers from signal biasing and other demerits and is not adopted in this chapter. Suitable constraints like DC gain match-ing, preserving stability and retaining the non-minimum phase characteristics of the parent model successfully converts the optimization problem into a constrained one. Similarly, the modelling of another test system, namely, the time-delay system poses yet another challenge to be performed using metaheuristic approach. The general prac-tice is to apply the so-called Padé approximation of a certain order before reducing it, using any classical technique. Our technique was no exception to it. We applied a third-order Padé approximation to generate a tenth-order system. Subsequently, this tenth-order system is reduced to a second-order model using a constrained HHO method. The same set of constraints are applied to the optimization to identify the parameters of the numerator and the denominator, using the PRBS response-matching methodol-ogy discussed above for the fractional-order test system.

HHO algorithm [16] imitates the co-operative role of Harris's hawks to hunt and catch prey. These hawks normally employ the mechanism of a 'surprise pounce' to seize the prey. They continuously alter their direction and speed to grasp the prey. The prey in this case is the rabbit. HHO mathematically emulates this procedure of hunting by the Harris's hawks in order to find an optimized solution for the reduced-order model. Once again this is a population-based technique, which is used to opti-mize the test systems, subject to the necessary constraints.

About 11 recently developed metaheuristic algorithms, namely, ant lion optimi-zation (ALO) [18], dragonfly algorithm (DA) [19], grey wolf optimizer (GWO) [20], moth flame optimization (MFO) [21], sine-cosine algorithm (SCA) [22], multi-verse optimizer (MVO) [23], grasshopper optimization algorithm (GOA) [24], whale optimization algorithm (WOA) [25], equilibrium optimizer (EO) [26], salp swarm algorithm (SSA) [27], and marine predator algorithm (MPA) [28] are used to compare with the proposed method. The population size and maximum iterations have been kept the same while applying all algorithms. The reduced-order models, as well as their best fitness values, are also tabulated. The statistical results are reported for the first test system. Since multiple algorithms are compared, Kruskal-Wallis test [29] is also performed for validation of significance of results of the FO system. Moreover, another non-parametric test, namely the Wilcoxon test [30], is carried out to validate the outcomes. Further, an amendment, viz. the Holm-Bonferroni correction [31], is incorporated in Wilcoxon to calculate the correct p-values. A 95% confidence interval is considered for all the non-parametric tests. Besides this, the frequency-domain parameters of the obtained systems are also compared with the parent model several times. The step and Bode responses of the reduced model from our technique are plotted along with the original system to show a close match between the two. These responses have been shown for the fractional-order test system. Apart from this, five popular benchmark error indices are evaluated and compared with the metaheuristic algorithms. Our method has definitely shown an edge over the existing techniques. On an overall assessment, the suggested method proves to be very effective and competitive for both the test systems.

16.4 SIMULATIONS AND DISCUSSIONS

To test the proposed methodology, two test systems are considered here. The first one represents an FO system, while the second one denotes a time-delay system. The model order reduction of both test systems is carried out applying the HHO technique. The population size and iterations are set at 50 and 100, respectively, for the both test systems. Three constraints are taken up for the study to meet the DC gain, non-minimum phase and stability requirements. A handful number of algorithms are used for comparison.

Test system 1: FO systems have been found to be effective in describing several physical processes. The FOS [6] considered is

$$G_{FOS}(s) = \frac{1}{0.8s^{2.2} + 0.5s^{0.9} + 1} \tag{5}$$

The test system of equation (16.1) is first transformed to integer order in the continuous-time domain by the Oustaloup approximation [17]. The resulting transfer function is thus represented by

$G_{Oustaloup}(s) =$

$$\frac{s^8 + 1128s^7 + 2.506e05s^6 + 7.814e06s^5 + 5.322e07s^4 + 5.223e07s^3 + 1.119e07s^2 + 3.368e05s + 1995}{3.185s^{10} + 3181s^9 + 4.354e05s^8 + 1.115e07s^7 + 4.972e07s^6 + 6.811e07s^5 + 8.711e07s^4 + 5.927e07s^3 + 1.147e07s^2 + 3.388e05s + 1997}$$

$$(16.6)$$

Applying the suggested technique, the lower-order approximated model is described by

$$G_r(s) = \frac{-0.097965s + 1.3711}{s^2 + 0.3881s + 1.3718} \tag{16.7}$$

The non-minimum phase feature of the Oustaloup approximated model is found to be preserved in the second-order system. Table 16.1 lists the second-order models obtained from all methods. The fitness function value is provided in the table for the purpose of comparison. The minimum quoted value of the fitness function is high-lighted using boldface.

In Table 16.1 our method yields the least fitness function value, proving the effectiveness of the approach. Since the methods are heuristic in nature, they give rise to

TABLE 16.1

Reduced-Order Systems of the Higher-Order FO System

Methods	Lower-Order Models in Continuous Time	Best Fitness Value
Proposed	$\dfrac{-0.097965s + 1.3711}{s^2 + 0.3881s + 1.3718}$	**0.0165**
ALO	$\dfrac{-0.00012962s + 1.2983}{s^2 + 0.39162s + 1.2996}$	0.0311
DA	$\dfrac{-0.11796s + 1.3872}{s^2 + 0.38435s + 1.3886}$	0.0168
GWO	$\dfrac{-0.160096s + 11.6015}{s^2 + 12.9997s + 11.6138}$	1.5277
GOA	$\dfrac{-0.155056s + 14.9302}{s^2 + 14.9976s + 14.9451}$	1.7893
WOA	$\dfrac{-0.10832s + 1.3808}{s^2 + 0.39011s + 1.3821}$	0.0167
MFO	$\dfrac{-0.16903s + 15}{s^2 + 15s + 15}$	1.8052
MVO	$\dfrac{-0.17053s + 3.0287}{s^2 + 2.1097s + 3.0301}$	1.2215
SCA	$\dfrac{-0.170115s + 15}{s^2 + 15s + 15}$	1.8052
SSA	$\dfrac{-0.383437s + 14.8328}{s^2 + 14.5916s + 14.8477}$	1.7887
EO	$\dfrac{-0.040596s + 1.4409}{s^2 + 0.52507s + 1.4521}$	0.0601
MPA	$\dfrac{-0.046847s + 1.3093}{s^2 + 0.37485s + 1.3101}$	0.0173

TABLE 16.2

Statistical Analysis of Error Function Value

Methods	MIN	Max	Avg	Std Dev
Proposed	**0.0165**	**0.0169**	**0.0167**	**1.3097e-04**
ALO	0.0311	1.8588	1.4531	0.6803
DA	0.0168	2.0720	1.7008	0.6206
GWO	1.5277	1.8588	1.7726	0.0915
GOA	1.7893	2.4364	1.8885	0.1852
WOA	0.0167	0.3042	0.1329	0.1294
MFO	1.8052	2.0720	2.0163	0.1004
MVO	1.2215	1.8588	1.7495	0.1816
SCA	1.8052	2.0720	2.0148	0.1031
SSA	1.7887	1.8646	1.8332	0.0255
EO	0.0601	1.8588	1.6281	0.4521
MPA	0.0173	0.3042	0.2213	0.1279

different results each time they are run on a personal computer. Hence, statistical analysis of the fitness function needs to be carried out. Fifty test runs are conducted to get meaningful statistical measures. The minimum and maximum value of error, as well as its average value and standard deviation, are summarized in Table 16.2. The best results reported in each column of the table are bolded for the ease of the readers to make appropriate interpretations.

In Table 16.2, our heuristic technique outperforms other algorithms in all the columns under study. Only the methods such as DA, WOA and MPA are nearly close with respect to the minimum error value. To check the statistical relevance of these outcomes, some other tests are also performed on the algorithms. The Kruskal-Wallis test [29] provides a very quick measure of the mean ranks of the different multivariate data as drawn in Figure 16.1. It is clearly evident from the figure that HHO, marked as Group 1 on the y-axis, is at least significant in numerical terms with respect to the nine algorithms out of eleven of them used for comparison.

Further, the Wilcoxon rank-sum test [30] is conducted on the data samples to create a second line of defence for the results. The outcomes of the Wilcoxon test are provided in Table 16.3. A 95% confidence interval is set for the results. This means that the p-value quoted in the Table will be considered to be insignificant if the p-value is greater than 0.05.

All the p-values given in Table 16.4 are clearly less than 0.05. Thus, the results quoted in the table are significant. Some corrections with the Holm–Bonferroni test [31] are improvised with the Wilcoxon test results to further correct them. The amended p-values and the h-values are given below in Table 16.4. Analogous considerations for testing the significance interval, viz. 5%, are considered to determine the corrected p-values. A p-value less than 0.05 is thus considered, while the corresponding h-value is taken as 1. Zero in h-value represents insignificant results.

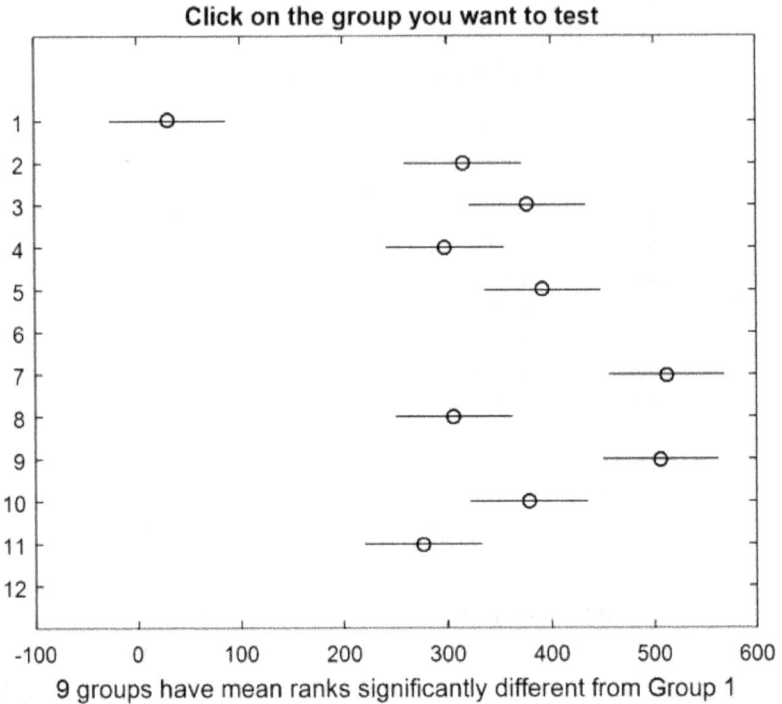

FIGURE 16.1   Kruskal-Wallis test comparison for FO system.

TABLE 16.3
p-Values for HHO vis-à-vis other Methods Applying Wilcoxon Test

Algorithms	ALO	DA	GWO	GOA	WOA	MFO
HHO	5.3936e-18	2.1335e-17	5.5096e-18	5.5096e-18	8.1545e-14	6.5591e-19
	MVO	SCA	SSA	EO	MPA	-
HHO	5.6691e-18	6.5108e-19	5.2032e-18	5.7933e-18	1.0824e-18	-

TABLE 16.4
p-Value Corrections with Holm-Bonferroni Test

Algorithm	Amended p-Values
Proposed (HHO)	$10^{-13} \times$ [0.0004 0.0004 0.0004 0.0003 0.8154 0.0001 0.0003 0.0001 0.0004 0.0002 0.0001]
h-values	[1 1 1 1 1 1 1 1 1 1 1]

TABLE 16.5

Time-Domain Specifications of the Reduced Test Systems for FO System

Test System	Methods	Rise Time (secs)	Settling Time (secs)	Overshoot (%)	Amplitude	Peak Time (secs)
Original		1.0463	26.5272	54.7385	1.5458	2.8241
Reduced	Proposed	0.9926	19.7444	59.2886	1.5911	2.8996
	ALO	1.0439	19.9149	57.8019	1.5765	2.8222
	DA	0.9821	19.6294	59.4534	1.5929	2.8756
	GWO	2.2876	4.1541	0	0.9982	7.5875
	GOA	2.0559	3.7299	0	0.9983	6.8221
	WOA	0.9886	19.6045	59.1027	1.5895	2.8331
	MFO	2.0482	3.7170	0	0.9993	6.7963
	MVO	1.0686	3.4737	9.1730	1.0912	2.3138
	SCA	2.0482	3.7171	0	0.9993	6.7963
	SSA	2.0059	3.6538	0	0.9972	5.8644
	EO	1.0220	14.1279	49.4537	1.4830	2.6312
	MPA	1.0299	20.1407	58.8938	1.5879	2.7028

All the p-values listed in Table 16.4 are less than 0.05. Thus, h-values are reported as unity, implying that the results produced by our approach are totally significant with respect to this non-parametric test. Moreover, the time-domain parameters of the reduced systems are quantified in Table 16.5.

Our method performed with fair satisfaction in terms of rise time, settling time, overshoot amplitude and peak time. Several other methods, such as ALO, DA, WOA and MPA, produce similar results. The method, however, outperforms GWO, GOA, MFO, MVO, SCA, SSA and EO algorithms. The frequency-domain parameters are also calculated and enumerated in Table 16.6. The gain margin and phase margin as well as gain crossover frequency and phase crossover frequency are provided in this table.

The frequency-domain parameters obtained with the help of the proposed technique are also found to be closer to the original higher-order system. Only DA and WOA produce results close to those of the parent system. The rest of the nine algorithms deviate quite largely from the actual system results. The step responses of both the original and reduced models are plotted in Figure 16.2.

The frequency responses of the original and reduced model are plotted in Figure 16.3.

Both the magnitude and the phase part of the frequency response, though similar in nature, show a bigger deviation as the frequency increases. In addition to this, the benchmark error indices for control systems are calculated and the data is provided in Table 16.7. A sufficient number of methods are used for comparison as well.

DA performs slightly better the HHO method in terms of IAE and ITAE and equals it with respect to ITSE as seen in Table 16.7. Our method turns out to be best for ISE and H_{inf} norm results. Very similar results are also provided by WOA and MPA techniques.

TABLE 16.6

Frequency-Domain Parameters of the Reduced Test Systems for Fractional-Order Model

Test System	Methods	Gain Margin (dB)	Phase Margin (degrees)	Gain Crossover Frequency (rad/sec)	Phase Crossover Frequency (rad/sec)
Original		1.7249	8.5671	1.8704	1.6208
Reduced	Proposed	3.8365	19.8116	2.5738	1.6139
	ALO	3.0213e+03	28.1248	62.6409	1.5636
	DA	3.2610	18.6246	2.4315	1.6254
	GWO	81.2012	Inf	30.8815	NaN
	GOA	96.7279	Inf	38.1984	NaN
	WOA	3.6048	19.7523	2.5218	1.6194
	MFO	88.7440	-180	36.6901	0
	MVO	12.3811	113.2867	6.3665	1.2778
	SCA	88.1780	-180	36.5742	0
	SSA	38.0648	Inf	24.0730	NaN
	EO	12.9359	33.4743	4.4824	1.6179
	MPA	8.0038	23.5326	3.4336	1.5752

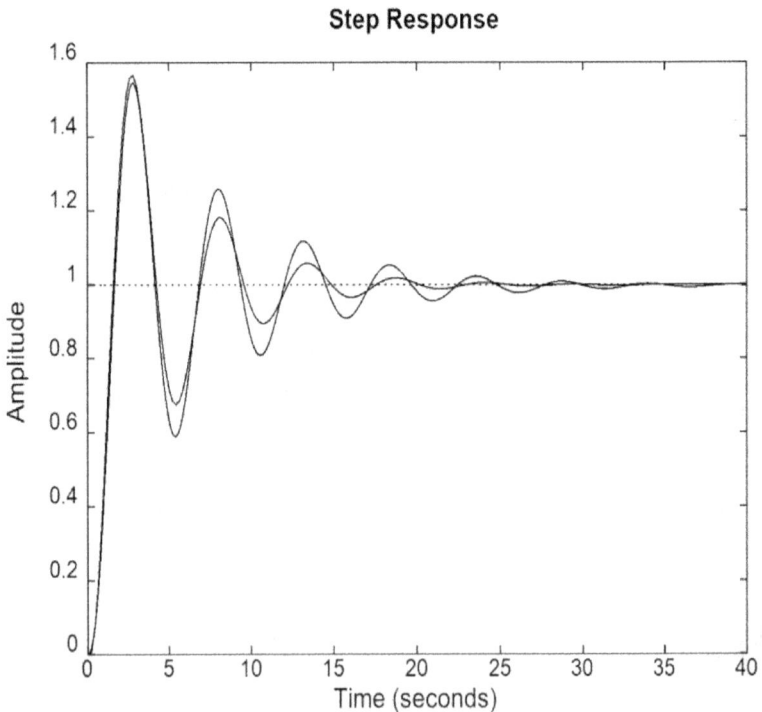

FIGURE 16.2 Time response matching of parent and reduced system applying HHO technique.

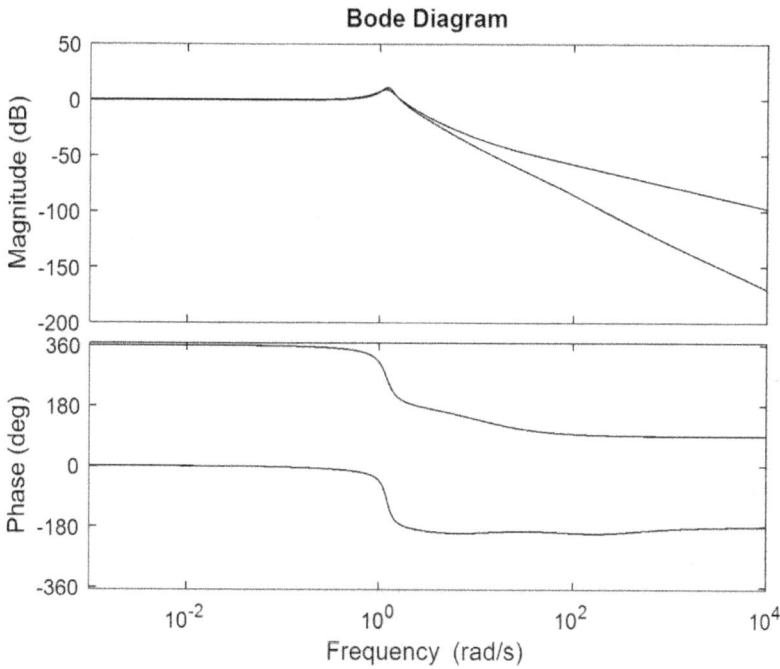

Bode Diagram

FIGURE 16.3 Bode response matching of original and reduced models.

TABLE 16.7
Error Indices Assessment for Reduced Test System of Fractional-Order System

Methods	IAE	ITAE	ISE	ITSE	H_{inf} norm
Proposed	0.0488	0.1119	**6.4733e-04**	**0.0013**	**0.0236**
ALO	0.0652	0.1636	0.0012	0.0030	0.0334
DA	**0.0486**	**0.1089**	6.6265e-04	**0.0013**	0.0240
GWO	0.5334	1.4306	0.0720	0.1955	0.2169
GOA	0.5261	1.3632	0.0738	0.1907	0.2345
WOA	0.0487	0.1133	6.5783e-04	0.0014	0.0238
MFO	0.5258	1.3623	0.0739	0.1907	0.2348
MVO	0.4418	1.3018	0.0488	0.1534	0.1606
SCA	0.5258	1.3624	0.0739	0.1908	0.2348
SSA	0.5285	1.3781	0.0744	0.1938	0.2345
EO	0.1108	0.3104	0.0031	0.0091	0.0435
MPA	0.0551	0.1443	8.6431e-04	0.0023	0.0251

Test system 2: Time delays are usually a part of real-time applications associated with chemical processes and may take place either in state, input or measurement (output). A higher-order time-delay system [15] is taken up whose input-output relation is given by

$$G_{TDS}(s) = \frac{(40000s + 50000)e^{-2s}}{s^7 + 69s^6 + 1764s^5 + 20280s^4 + 102500s^3 + 221375s^2 + 187500s + 50000}$$

(16.8)

The most common practice related to the representation of this time-delay system in MATLAB software is to obtain the Padé approximation of a certain order. A third-order Padé approximated model is thus obtained and is denoted by

$$G_{Pade}(s) =$$

$$\frac{-4000s^4 - 26000s^3 + 24000s^2 - 690000s + 750000}{s^{10} + 75s^9 + 2193s^8 + 31914s^7 + 251675s^6 + 1.167e06s^5 + 3.357e06s^4 + 6.033e06s^3 + 6.433e06s^2 + 3.563e06s + 750000}$$

(16.9)

The second-order approximation of equation (16.9) applying HHO technique is given by

$$G_{rTDS} = \frac{-0.071373s + 0.063676}{s^2 + 0.004968s + 0.063676}$$

(16.10)

The other reduced-order models, generated by the algorithms used for comparison, are shown in Table 16.8. The lower-order models turn out to be non-minimum phase in nature, similar to the Padé approximated higher-order model. The lowest fitness function is also reported in the same table. The lowest value is marked in boldface.

It is evident from this table that the proposed HHO method is better than other techniques in terms of achieving the lowest fitness function value. The time specifications of the reduced systems are reported in Table 16.9 along with the parameters of the parent model. Since the lower-order approximation is non-minimum phase type, the undershoot is expected in each method. Thus, the undershoot percentage is mentioned for each method, as reflected in the table.

The suggested technique performs fairly well in terms of rise time, undershoot and the peak time but shows a very sluggish response (large settling time) and relatively high overshoot. Thus, the need of a controller is felt to reduce the settling time as well as overshoot. Settling time is on the higher side for nearly all the models. ALO, GOA, SSA satisfy the overshoot criteria. The performance of the undershoot remains more or less same for all the models. Nearly competitive rise time is

TABLE 16.8

Second-Order Approximations of the Higher-Order System in Continuous Time

Methods	Second-Order Models in Continuous Time	Fitness Function Value
Proposed	$\dfrac{-0.071373s + 0.063676}{s^2 + 0.004968s + 0.063676}$	**0.013191**
GWO	$\dfrac{-0.089787s + 0.074573}{s^2 + 0.17095s + 0.074573}$	0.018499
ALO	$\dfrac{-0.21924s + 0.16985}{s^2 + 1.582s + 0.16985}$	0.07441
DA	$\dfrac{-0.089643s + 0.074791}{s^2 + 0.16479s + 0.074791}$	0.01825
MFO	$\dfrac{-0.10157s + 0.080438}{s^2 + 0.18385s + 0.080438}$	0.019766
MVO	$\dfrac{-0.069283s + 0.06333}{s^2 + 0.0049844s + 0.06333}$	0.013222
GOA	$\dfrac{-0.26823s + 0.24602}{s^2 + 2.6828s + 0.24602}$	0.09936
WOA	$\dfrac{-0.10508s + 0.083243}{s^2 + 0.3283s + 0.083243}$	0.02492
SCA	$\dfrac{-0.065817s + 0.062803}{s^2 + 0.0148s + 0.062803}$	0.013745
SSA	$\dfrac{-0.16359s + 0.12249}{s^2 + 0.8617s + 0.12249}$	0.048115

provided by the GWO, DA, MFO, MVO, WOA and SCA techniques, while almost similar peak time is obtained by GWO, DA, MFO, MVO, WOA and SCA. The frequency-domain measures are provided in Table 16.10 wherein the gain and phase margins, gain and phase crossover frequencies are reported.

In Table 16.10, the HHO method gives a very mediocre performance. ALO, GOA and SSA perform very similarly in view of phase margin and phase crossover measures. GWO, DA and MFO are similar in view of gain margin, gain and phase crossover frequencies. A few common benchmark performance indices in the systems and control literature are calculated and reported in Table 16.11. The good performance measures in the table are bolded for the understanding of the readers.

In contrast to our model's performance in Table 16.10, the results in Table 16.11 show a considerable improvement. The HHO technique supersedes all other

TABLE 16.9
Quantitative Time-Domain Performance Measures

Test System	Methods	Rise Time (secs)	Settling Time (secs)	Overshoot (%)	Undershoot (%)	Peak Time (secs)
Original		5.3303	11.9231	0	0.5071	18.4030
Reduced	Proposed	3.8804	1.5703e+03	100.6676	3.8290	13.6948
	GWO	4.5945	42.0525	37.0817	4.6628	12.9307
	ALO	19.0069	35.1631	0	6.2031	69.6648
	DA	4.5384	42.1159	38.6885	4.6451	12.1189
	MFO	4.4298	40.4096	35.8680	5.3948	13.0251
	MVO	3.9034	1.5622e+03	100.4240	3.5970	13.7322
	GOA	23.1134	42.0397	0	5.2593	69.0146
	WOA	5.8385	21.3963	11.9605	5.0793	14.3079
	SCA	4.0020	528.1643	94.0394	3.1767	13.7896
	SSA	12.7754	24.3492	0	6.2062	41.7219

algorithms. Only the results of MVO, GWO, DA, MFO and SCA are very close. Thus, the proposed technique is able to cater satisfactorily to the reduced-order modelling of two critical systems, viz. the fractional-order and the time-delay system. With the reduced-order systems developed, the controller synthesis can be thought of in future communications. Improved versions of new algorithms, as well as new hybrid algorithms, can be developed to address the model-reduction problems.

TABLE 16.10
Quantitative Frequency-Domain Performance Measures

Test System	Methods	Gain Margin (dB)	Phase Margin (degrees)	Gain Crossover Frequency (rad/sec)	Phase Crossover Frequency (rad/sec)
Original		1.9791	−180	0.6308	0
Reduced	Proposed	0.0698	−20.6833	0.2610	0.3639
	GWO	1.9047	25.5645	0.4654	0.3577
	ALO	7.2162	−180	1.1814	0
	DA	1.8393	23.5073	0.4608	0.3612
	MFO	1.8100	25.0310	0.4754	0.3707
	MVO	0.0720	−20.1145	0.2606	0.3625
	GOA	10.0030	−180	1.6455	0
	WOA	3.1244	80.4266	0.5859	0.2641
	SCA	0.2249	−16.1244	0.2774	0.3602
	SSA	5.2675	−180	0.8762	0

TABLE 16.11

Benchmark Error Indices for the Lower-Order Approximation of Time-Delay System

Methods	IAE	ITAE	ISE	ITSE	H_{inf} norm
HHO	**0.0444**	**0.1096**	**5.2763e-04**	**0.0013**	**0.0220**
GWO	0.0521	0.1287	7.3994e-04	0.0018	0.0260
ALO	0.1034	0.2883	0.0030	0.0086	0.0523
DA	0.0516	0.1264	7.3001e-04	0.0018	0.0262
MFO	0.0526	0.1221	7.9062e-04	0.0018	0.0279
MVO	0.0445	0.1103	5.2889e-04	**0.0013**	0.0224
GOA	0.1208	0.3503	0.0040	0.0123	0.0607
WOA	0.0604	0.1551	9.9681e-04	0.0026	0.0292
SCA	0.0453	0.1140	5.4979e-04	0.0014	0.0231
SSA	0.0828	0.2181	0.0019	0.0052	0.0428

16.5 CONCLUSION

This chapter considered the reduced-order modelling of two very tough systems, namely, the fractional-order and the time-delay systems, with the help of a new intelligent technique called Harris Hawks Optimization (HHO). The problem is formulated based on a mix of classical and heuristic approaches to obtain the unknowns in a reduced system. A fractional-order model is transformed into integer order by applying the Oustaloup approximation and then reduced to a second-order model with the help of a metaheuristic algorithm applying suitable constraints. On a similar note, the time-delay system is transformed to the integer model using the Padé approximation. Then the Padé approximated model is reduced to a second-order system applying HHO technique formulated with the help of suitable constraints. A sufficient number of algorithms are used for comparison. The models produced by our technique show good competitiveness while being evaluated using several performance measures. Thus, the critical systems like fractional-order and time-delay systems are modelled ably with the aid of constrained intelligent computing tool. New complex systems, viz. unstable and non-minimum phase systems, can now easily be handled with the help of metaheuristic approaches. Even model reduction of time-interval systems can also be taken up in future. Some new versions of HHO, EO and MPA may be thought of in time to come. Use of hybrid computational algorithms may yield further better results. Moreover, the use of additional constraints can be incorporated in the model order to address fulfilling time and frequency-domain criteria.

REFERENCES

1. Rudnyi, B., & Korvink, J. G. (2002). Automatic model reduction for transient simulation of MEMS-based devices. *Sensors Update*, *11*, 3–33.
2. Bai, Z. J. (2002). Krylov subspace techniques for reduced-order modeling of large-scale dynamical systems. *Applied Numerical Mathematics*, *43*, 9–44.

3. Freund, R. W. (2000). Krylov-subspace methods for reduced-order modeling in circuit simulation. *Journal of Computational and Applied Mathematics*, *123*, 395–421.

4. Yang, Y. J., & Yu, C. C. (2004). Extraction of heat-transfer macro models for MEMS devices. *Journal of Micromechanics and Microengineering*, *14*, 587–96.

5. Bechtold, T., Rudnyi, E. B., & Korvink, J. G. (2003). Automatic generation of compact electro-thermal models for semiconductor devices. *IEICE Transactions on Electronics*, *86*, 459–65.

6. Bourouba, B., Ladaci, S., & Chaabi, A. (2018). Reduced-order model approximation of fractional-order systems using differential evolution algorithm. *Journal of Control, Automation and Electrical Systems*, *29*(1), 32–43.

7. Monje, C. A., Chen, Y., Vinagre, B. M., Xue, D., & Feliu-Batlle, V. (2010). *Fractional-order systems and controls: fundamentals and applications*. Springer Science & Business Media.

8. Saxena, S., Yogesh, V., & Arya, P. P. (2016). Reduced-order modeling of commensurate fractional-order systems. In 14th International Conference on Control, Automation, Robotics and Vision (ICARCV), 1–6.

9. Sarkar, P., Shekh, R. R., & Iqbal, A. (2016). A unified approach for reduced order modeling of fractional order system in delta domain. In International Automatic Control Conference (CACS), 257–262.

10. Jain, S., & Hote, Y. V. (2019). Reduced order approximation of incommensurate fractional order systems. In IEEE Conference on Control Technology and Applications (CCTA), 1056–1061.

11. Jain, S., Hote, Y. V., & Saxena, S. (2019). Model order reduction of commensurate fractional-order systems using big bang–big crunch algorithm. *IETE Technical Review*, 1–12. https://doi.org/10.1080/02564602.2019.1653232.

12. Richard, J. P. (2003). Time-delay systems: An overview of some recent advances and open problems. *Automatica*, *39*(10), 1667–1694.

13. Fridman, E. (2014). *Introduction to time-delay systems: Analysis and control*. Springer.

14. Galloway, P. J., & Holt, B. R. (1988). Multivariable time delay approximations for analysis and control. *Computers & Chemical Engineering*, *12*(7), 637–650.

15. Biradar, S., Hote, Y. V., & Saxena, S. (2016). Reduced-order modeling of linear time invariant systems using big bang big crunch optimization and time moment matching method. *Applied Mathematical Modelling*, *40*(15–16), 7225–7244.

16. Heidari, A. A., Mirjalili, S., Faris, H., Aljarah, I., Mafarja, M., & Chen, H. (2019). Harris Hawks optimization: Algorithm and applications. *Future Generation Computer Systems*, *97*, 849–872.

17. Oustaloup, A., Levron, F., Mathieu, B., & Nanot, F. M. (2000). Frequency-band complex noninteger differentiator: Characterization and synthesis. *IEEE Transactions on Circuits and Systems I: Fundamental Theory and Applications*, *47*(1), 25–39.

18. Mirjalili, S. (2015). The ant lion optimizer. *Advances in Engineering Software*, *83*, 80–98.

19. Mirjalili, S. (2016). Dragonfly algorithm: A new meta-heuristic optimization technique for solving single-objective, discrete, and multi-objective problems. *Neural Computing and Applications*, *27*(4), 1053–1073.

20. Mirjalili, S., Mirjalili, S. M., & Lewis, A. (2014). Grey wolf optimizer. *Advances in Engineering Software*, *69*, 46–61.

21. Mirjalili, S. (2015). Moth-flame optimization algorithm: A novel nature-inspired heuristic paradigm. *Knowledge-Based Systems*, *89*, 228–249.

22. Mirjalili, S. (2016). SCA: A sine cosine algorithm for solving optimization problems. *Knowledge-Based Systems*, *96*, 120–133.

23. Mirjalili, S., Mirjalili, S. M., & Hatamlou, A. (2016). Multi-verse optimizer: A nature-inspired algorithm for global optimization. *Neural Computing and Applications, 27*(2), 495–513.

24. Mirjalili, S. Z., Mirjalili, S., Saremi, S., Faris, H., & Aljarah, I. (2018). Grasshopper optimization algorithm for multi-objective optimization problems. *Applied Intelligence, 48*(4), 805–820.

25. Mirjalili, S., & Lewis, A. (2016). The whale optimization algorithm. *Advances in Engineering Software, 95*, 51–67.

26. Faramarzi, A., Heidarinejad, M., Stephens, B., & Mirjalili, S. (2020). Equilibrium optimizer: A novel optimization algorithm. *Knowledge-Based Systems, 191*, 105190.

27. Mirjalili, S., Gandomi, A. H., Mirjalili, S. Z., Saremi, S., Faris, H., & Mirjalili, S. M. (2017). Salp Swarm Algorithm: A bio-inspired optimizer for engineering design problems. *Advances in Engineering Software, 114*, 163–191.

28. Faramarzi, A., Heidarinejad, M., Mirjalili, S., & Gandomi, A. H. (2020). Marine predators algorithm: A nature-inspired metaheuristic. *Expert Systems with Applications.* https://doi.org/10.1016/j.eswa.2020.113377.

29. Breslow, N. (1970). A generalized Kruskal-Wallis test for comparing K samples subject to unequal patterns of censorship. *Biometrika, 57*(3), 579–594.

30. Wilcoxon, F., Katti, S. K., & Wilcox, R. A. (1970). Critical values and probability levels for the Wilcoxon rank sum test and the Wilcoxon signed rank test. *Selected Tables in Mathematical Statistics, 1*, 171–259.

31. Aickin, M., & Gensler, H. (1996). Adjusting for multiple testing when reporting research results: The Bonferroni vs Holm methods. *American Journal of Public Health, 86*(5), 726–728.

17 Advanced Agricultural-Based IoT Technology

Vikalp Joshi
Sara Sae PVT Limited

Manoj Singh Adhikari
Lovely Professional University

CONTENTS

17.1 Introduction 240
17.2 Primary Applications..242
    17.2.1 Sampling for Soil Mapping..242
    17.2.2 Irrigation 243
    17.2.3 Fertilizer 244
    17.2.4 Management of Pests and Crop Disease............................244
    17.2.5 Monitoring Forecasting and Harvesting of Yield.............244
17.3 Advanced Agricultural Practices..245
    17.3.1 Greenhouse Farming 245
    17.3.2 Vertical Farming...245
    17.3.3 Hydroponic Farming..245
    17.3.4 Phenotyping 245
17.4 Wireless Sensors...246
    17.4.1 Acoustic Sensors...246
    17.4.2 Field-Programmable Gate Array Sensors.........................246
    17.4.3 Optical Sensors 246
    17.4.4 Ultrasonic Ranging Sensors 247
    17.4.5 Opto-Electronic Sensors..247
    17.4.6 Airflow Sensors 247
    17.4.7 Electro-Chemical Sensors 247
17.5 IoT-Based Tractors...247
17.6 Harvesting Robots 248
17.7 Communication in Agriculture...248
    17.7.1 Cellular Communication 248
    17.7.2 Zigbee 249
    17.7.3 Bluetooth...249
    17.7.4 LORA 249
    17.7.5 SIGFOX 249

17.8   Cloud Computing...249
17.9   Unmanned Aerial Vehicles...249
          17.9.1 Soil and Worksite Analysis..250
          17.9.2 Planting 251
          17.9.3 Crop Monitoring 251
          17.9.4 Irrigation 251
          17.9.5 Detection of Gap and Plant Counting..251
          17.9.6 Spraying Pesticides/Herbicides...251
          17.9.7 Health Assessment 251
          17.9.8 Identification of Plant Species...251
17.10   Present Challenges and Future Expectations...252
17.11   Conclusion 252
References...252

17.1 INTRODUCTION

Advanced IoT-based technologies offer services to farmers and agriculture indus-tries to check the status of crops from time to time without visiting the worksite. To improve agriculture revenue with fewer resources and labour efforts, the internet of thing (IoT) can be used in agricultural industries. There is need for accurate, pre-cise solutions and advanced technologies due to vastness of the agriculture industry. Advanced agriculture techniques use nanotechnology-based smart sensors and wire-less communication systems. Farmers can observe the status of the worksite by using IoT-based devices. Advanced agriculture technologies are based on smart apparatus and kits for insemination to crop mowing and even during storage space and ship-ping. Advanced agriculture-based IoT technology provides periodic updates through historical trends and graphs. IoT-based advanced agriculture techniques become smart and cost effective. The techniques use embedded-technology–based drones, automotive tractors, harvesters and satellites which are backbone of the IoT-based agriculture system. Many industries such as energy, medical, manufacturing and automotive, already use IoT-based technique to reduce inefficiencies and improve product quality. Figure 17.1 shows the key points of advanced agriculture-based IoT technology, while Figure 17.2 shows challenges in advanced agriculture-based on IoT.

The Food Corporation of India (FCI) monitors and controls IoT-based techniques for regulation of food and environmental safety. Several agriculture commissions and government authorities provide polices and guidelines from time to time. This chapter presents and evaluates the trends of IoT-based agriculture technologies to develop crop quality and production. Some aspects of advanced agriculture-based IoT techniques are as follows:

- Unsolved issue or suggestions regarding quantity of fertilizer soil improvement
- Advanced research and development techniques used in university and research institutions
- Limitation of agriculture industries

Advanced Agricultural-Based IoT Technology   **241**

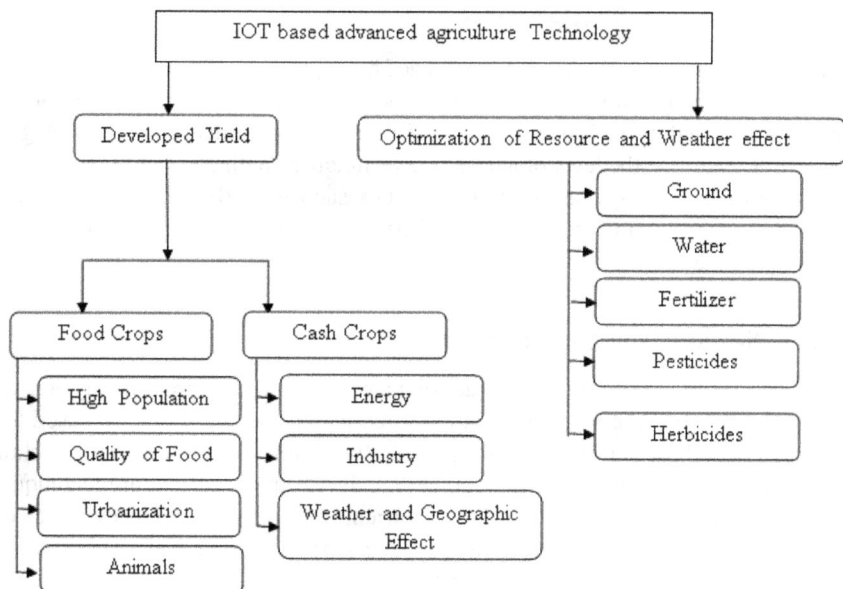

FIGURE 17.1   Key points of advanced agriculture-based IoT technology.

- Roles of advanced agriculture techniques based on IoT (i.e., shortage of processing resources, food resource shortage, food damage, weather and environmental pollution status)

This chapter deals with advanced agriculture techniques to provide smart agriculture technologies through the power of IoT. The next section explores a major application of IoT in the agriculture sector and what can be possible with IoT.

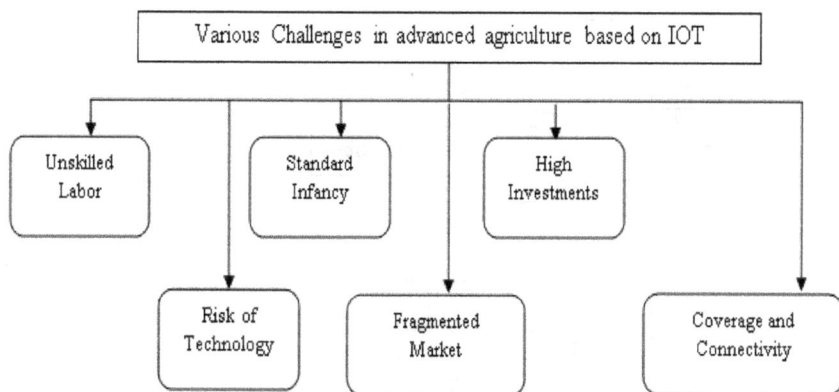

FIGURE 17.2   Challenges in advanced agriculture-based IoT.

17.2   PRIMARY APPLICATIONS

Traditional farming techniques can be changed by implementing high-precision, qual-ity-sensor, communication-based IoT techniques. Wireless-based sensors increase the level of traditional agriculture techniques. Currently, consolidated wireless integra-tion sensors increase the level of agriculture techniques. With constant improvement of agricultural technology, several farming issues, such as yield optimization, drought response, land suitability, pest control and irrigation, can be improved. Figure 17.3 shows a practical orientation of IoT-based smart farming.

17.2.1   Sampling for Soil Mapping

Traditional farming techniques provide soil mapping but they represent only general distribution of soil. The worksite data collected by advanced agriculture-based IoT devices are used for soil-mapping purposes. This technique is necessary to obtain good data samples and the mapping class of soil. Advanced agriculture techniques provide information on the amount of soil nutrients and nutrient deficiency. Based on worksite conditions, a comprehensive soil test is performed on an annual basis. Soil mapping is important for evaluation of land, local planning, agriculture extension, atmosphere protection and many more agriculture-based projects.

Soil is the heart of the plant. To gather worksite information to decide the crop condition at different levels, sampling is necessary. Soil analysis is used to find the amount of nutrients and helps to find nutrient deficiencies. The soil quality, cropping history, fertilizer application and level of irrigation are critical factors for soil analy-sis. Such analysis is useful to determine the biological, physical and chemical status of soil. Generally, soil mapping is useful for selection of seed suitability, sowing time and depth of sowing for various crops.

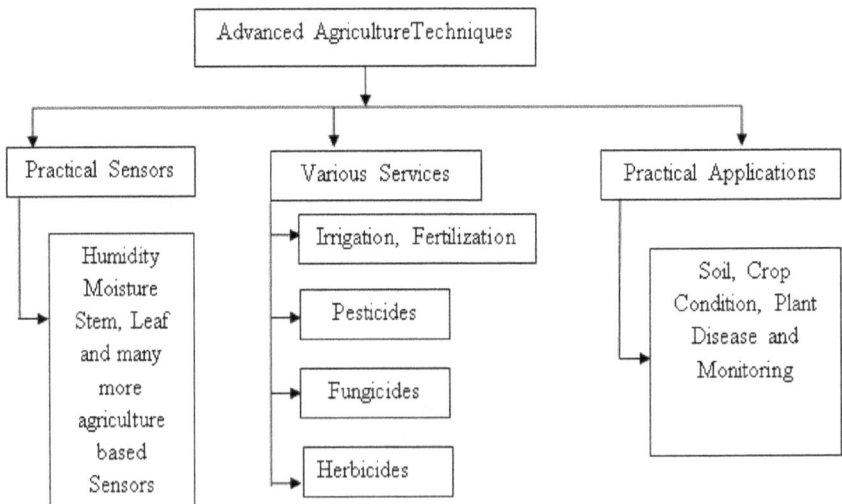

FIGURE 17.3   Practical orientation of IoT-based smart farming.

Currently, several sensors and toolkits are available in market to check the status and quality of soil nutrients. This is helpful for farmers to avoid degradation. Advanced agriculture-based IoT techniques can monitor moisture content, soil texture and rate of absorption, which is useful to minimize acidification, pollution and intersection. Crop productivity depends on several factors, but one important factor is soil moisture. The farmers around the world face problems of dry soil. To overcome this, remote-sensing techniques are used to measure soil moisture. The satellite for Soil Moisture and Ocean Salinity was launched early in 2009. It provides maps of soil moisture in distant regions day by day [1].

Remote-sensing techniques compare worksite data to the soil water deficit index (SWDI) to determine the moisture of the soil [2]. In this, they follow various technique to determine the soil moisture in order to compare with the SWDI acquired from in situ data. The prediction model is based on worksite data and soil mapping. Advanced agriculture-based IoT techniques use real-time cameras and sensors to find the appropriate depth and location for insemination of the seed [3].

Robots record the field data and transmit them to a master control station where the data can be analyzed. Various contactless sensing techniques are planned to decide the flow rate of seed. Such sensors are attached to visible and laser LEDs, transmitter receivers and battery backup [4]. The movement of the seed is captured through the sensors and light detectors, where it signals the conditioning system to convert sensor signals into transmission signal forms.

17.2.2 IRRIGATION

Irrigation is the artificial technique to provide a fixed or variable amount of water to crops. Of all the water on Earth, 3% is freshwater and the remaining 97% saline water in oceans and seas. Two-third of freshwater is in the form of glaciers and frozen forms, and most of the rest is groundwater [5]. In sum up, human life depend on about 0.5%of Earth's water to accomplish all its necessities and to sustain the eco-system. In irrigation, submersible pumps draw water through pipes, canals and other methods to crops at the worksite. Generally the 70% of accessible freshwater is required to drive the agriculture sector [6, 7].

Irrigation improves the quality and quantity of crops and improving 40% of world's food production rate. In several countries, agricultural use of freshwater increase to 75%or more. In Brazil, afore in a number of developing countries, it exceeds 80% [8]. Approximately 80%of farms in United States use irrigation [9]. According to the UN Convention to Combat Desertification, half of the world will be affected by water shortages by 2030. And about 168 countries will affected by desertification [10]. Sprinkler and drip irrigation methods different techniques are used to initiated wastewater issue which were also establish in conventional techniques. A sprinkle irrigation system spreads water in drizzle form by using high output pressure pumps. It spreads water like rainfall by using a small-diameter nozzle in the pipe. Low-level irrigation of a small area is known as micro irrigation.

When water shortage problems occur and irrigation is limited, the quantity and quality of crops are severely affected. The scarcity of water can lead to many problems. Farmers are unable to manage their crop cycles. If irrigation is not regular, then soil nutrients decrease and microbial infections increase. It is not easy to precisely assess the water requirement of crops, considering factors such as precipitation, soil moisture, crop category, irrigation technique, soil variety and crop requirements.

The present state of irrigation is likely to be altered by implementing emergent IoT technologies, which can play a significant role in improving crop efficiency. One such technology is an irrigation management system based on the crop water stress index (CWSI) [11]. To calculate the CWSI, the air temperature and crop canopy at dissimilar time intervals are recorded.

Advanced agriculture-based IoT improves cost efficiency. For the above-mentioned measurement parameters, intelligent software applications use wireless sensors for a monitoring system. For water need assessment in the CWSI model, climate condition statistics and satellite imaging are required. To improve efficiency of water use, use a variable-rate irrigation optimization technique [12].

17.2.3 Fertilizer

A lack of nutrients can be harmful for the plant growth. To improve the growth and fertility of plants, fertilizers (either natural or chemical) are used. Fertilizer includes three micro nutrients – phosphorous, nitrogen and potassium – and some minor nutrients such as magnesium, calcium, sulphur, iron and many more. Potassium, nitrogen and phosphorus are used for root, flower, leaf and fruit development [13]. Approximately one-fourth of the nitrogen in fertilizer is absorbed by the crop. The remaining portion is emitted to the atmosphere. The soil nutrient level can be unbalanced by uneven use of fertilizer.

17.2.4 Management of Pests and Crop Disease

There are several technique to control pests and crop disease. About 20–40% of crop yields are destroyed by pests and diseases, according to the Food and Agriculture Organization (FAO). Agriculture-related chemicals and pesticides have become an essential element of the agriculture industry [14]. Pesticides are not good for humans or animals. So, alternative approaches are being developed. Wireless sensors, robots, drones and IoT-based intelligent systems are four techniques to control plant disease and pests through biological, cultural, chemical and integrated methods. IoT-based techniques provide real-time monitoring and forecasting. [15–20].

17.2.5 Monitoring Forecasting and Harvesting of Yield

Advanced agriculture-based IoT systems continuously monitor and forecast the crop condition from the worksite. Yield monitoring examines a range of aspects related to agricultural yield, such as mass flow of grain, amount of moisture and quantity of harvested grain. Recording the moisture and crop development level is helpful to assess crop quality and what to do ahead. Yield quality depends on several factors. [20–25].

17.3 ADVANCED AGRICULTURAL PRACTICES

To support the high rate of urban population, this section explores the function of IoT in advanced agricultural practices such as vertical farming and hydroponics. Sections 17.4, 17.5 and 17.6 will look at advanced IoT technologies such as sensors robots, tractors and wireless devices. Section 17.7 addresses communication in agriculture, and Section 17.8 describes cloud computing. Section 17.9 shows the advantages of IoT technologies. Section 17.10 indicates the major challenges of IoT technologies. Section 17.11 summarizes the application of IoT-based technologies in agriculture sector.

Humans have been implementing new techniques to improve crops for eras. Primarily we are working to improve the seed quality, fertilizers and pesticides to improve crop production. Table 17.1 demonstrates sensors and their application in the IoT-based agriculture.

17.3.1 GREENHOUSE FARMING

A greenhouse technique uses a framed structure in which vegetable or flower crops can be grown under controlled weather conditions. This type of farming is considered the oldest technique of smart farming. The idea of rising plants in organized surroundings has been established since ancient times. High production of numerous crops in a well-ordered atmosphere depends on several features, such as proper nursing parameters, arrangement of the ventilation system, material to control wind effects, decision support system, etc.

17.3.2 VERTICAL FARMING

Vertical farming enables farmers to develop crops in stacked layers. Increasing food needs require increasing farmable land. However, due to erosion and pollution, one-fourth of arable property has disappeared through the past four decades [26, 27]. Unfortunately, the modern agriculture techniques used in industrial farming are very harmful to soil quality, which can only be rebuilt through nature. A vertical farming system includes buildings, shipping, containers, tunnels, etc.

17.3.3 HYDROPONIC FARMING

To improve the profit of greenhouse farming, agriculture professionals moved ahead an additional step by moving into hydroponic farming. Hydroponic farming is a technique in which plants are grown without soil, using only water that contains a mineral nutrient solution. This is a subset of hydroculture. It provides better use of space and location.

17.3.4 PHENOTYPING

The technique of phenotyping is a reliable approach for producing good quality and quantity of crop. This technique often uses non-invasive technologies and digital techniques. Some advanced methods use wireless and sensing technologies to study additional crop capabilities.

Wireless communication and remote sensing technique are used in modern harvesters, tractors and various types of robot in large-scale agriculture. Tractors with global positioning system (GPS) and geographic information system (GIS) facilities, as well as harvesters and various types of robots, are helpful in precision agriculture. The success rate of precision agriculture depends on accuracy of sample data. Generally, the success rate depends on two factors [28].

17.4 WIRELESS SENSORS

Wireless sensors are connected with transmitters to convert input signals into a radio transmission. Currently various types of apparatus for smart farming are available in the market. To monitor the crop status, wireless sensors play an important role. Wireless sensors are being used individually wherever necessary and applied with tools and heavy equipment to perform various task for particular applications. In agriculture, wireless sensors are generally used in inaccessible locations. The purpose and working procedure of some sensors are discussed below. Table 17.2 gives information about application and specification of wireless sensors.

17.4.1 ACOUSTIC SENSORS

An acoustic sensor is used to sense a physical phenomenon. It converts and transmits mechanical waves by electrical signals. The mechanical wave can be influenced by physical phenomena. The acoustic sensor then converts this wave back into an electrical signal. The difference between input and output amplitude, phase and frequency of electrical signal can be used to measure the presence of the phenomenon. This type of sensor is used in various types of applications in farm management. Generally farm management includes cultivation of soil, seeding, fruit harvesting, etc. The acoustic sensors provide fast response with low cost. The basic concept of acoustic sensors is to measure the level of noise when an input-sensing element is attached to other material. For example, acoustic sensors are used on soil particles to monitor pests and detect seed qualities [29].

17.4.2 FIELD-PROGRAMMABLE GATE ARRAY SENSORS

Field-programmable gate array (FPGA) sensors consist of a programmable array of reconfigurable interconnections. This type of sensor provide reconfiguration flexibility. FPGA sensors measure plant transpiration in real time, humidity and irrigation. The major disadvantages are that FPGA require more power so they are not suitable in real-time monitoring.

17.4.3 OPTICAL SENSORS

An optical sensor converts light waves into electrical signals. Generally it measures the intensity of light. This type of sensor is very helpful in agriculture purpose due to low cost, ease of use and adjustability. Optical sensors are used to measure soil capability with the help of the electromagnetic spectrum concept. The optical sensor reflects light, and the changes that appear in wave reflections help to show the variations in soil density and several additional parameters. The optical sensor is used with microwave scattering and can be used to detect the quality of grove canopies and similar crops.

17.4.4   ULTRASONIC RANGING SENSORS

This type of sensor measures the distance between objects. It transmits acoustic waves of frequency between 25 and 50 KHz, which is greater than the human hearing range. This type of sensor is low cost, can be used in various locations and conditions and is easy to install. Such sensors are useful for monitoring tank spray and distance measurement applications. Camera-based ultrasonic ranging sensors are used for weed detection and crop coverage purposes [30]. Actually, the ultrasonic sensors measure plant height, and the camera shows the forest grass and crop coverage.

17.4.5   OPTO-ELECTRONIC SENSORS

An opto-electronic sensor produces an electrical output signal with respect to input light incident on its active area. This sensor is helpful to detect types of plant, weeds and herbicides [31]. Sensor data and location information are useful for mapping weed distribution and resolution [32].

17.4.6   AIRFLOW SENSORS

Airflow sensors are used to measure air permeability and moisture of soil, to detect soil structure and to differentiate soil category.

17.4.7   ELECTRO-CHEMICAL SENSORS

An electro-chemical sensor is mostly used for measurement of gas. The principle is based on the measured gas interacting with an electrolytic, causing an electrochemical reaction which develops currents. The current is directly dependent upon the concentration of measured gas.

Electro-chemical sensors are typically used to evaluate important soil characteristics, such as pH [33]. This type of sensor replaces the expensive and time-consuming soil chemical analysis process.

17.5   IoT-BASED TRACTORS

Advanced agriculture-based IoT techniques used in modern agriculture create a problem for rural labour because the industry apparatus and equipment, such as tractors and other heavy machinery, work 40 times faster than human beings. [34–38]. To meet the continuously growing demands, agriculture equipment manufacturers such as John Deere and Hello Tractor have started to provide superior solutions.

TABLE 17.1

Sensors and their Application in the IoT-Based Agriculture Sector

S. No.	Sensor	System Sensor To Be Attached	Measurement Variable
1.	XH-M214	Equipment	Yield and moisture
2.	PYCNO	Soil and Weather	Water
3.	Sol Chip Com	Soil	Pollution and Water
4.	MP 406	Soil	Temperature and Moisture
5.	DEX70	Plant	Fruit Size
6.	Wind Sentry 03002	Weather	Wind
7.	SF-4/5	Plant	Fruit Size
8.	Met Station One	Weather	Wind

17.6 HARVESTING ROBOTS

The final stage of the production process is harvesting. It is an important factor which can affect the production considerably. Harvesting robots can work continuously. They provide precise, timely and disease-free output.

17.7 COMMUNICATION IN AGRICULTURE

The backbone of precision agriculture is to provide communication and report field data. To provide field data periodically, telecom providers can play a major role in the agriculture. Table 17.3 represents the sensor based on smart phone used in agriculture applications.

17.7.1 CELLULAR COMMUNICATION

Advanced agriculture-based techniques provide worksite data in real time. To collect the field data, the second-generation to fourth-generation communication may be suitable. The communication depends on bandwidth requirement.

TABLE 17.2

Application and Specification of Wireless Sensors

S. No.	Types of Data	Application	Data Size (Approx.)	Power Consumption
1.	Small	Air, soil temperature, humidity, leaf thickness	100 byte/Sec	Milliampere
2.	Medium	Picture camera, multispectral camera, acoustic camera	10 Mb/Sec	100 Milliampere
3.	Large	Video streaming cameras	10 Mb/Min	50 A

17.7.2 Zigbee

Zigbee is a high-level communication protocol. This type of communication is used in low-power digital transmission systems. Zigbee is used in a large range of uses, particularly as a substitute for prevailing non-standard technologies. Zigbee can be used as personal area network, such as home automation, etc. Generally, Zigbee is used to monitor greenhouse atmosphere and for short-range communication.

17.7.3 Bluetooth

Bluetooth is a wireless technology used over short distances. This is useful when the worksite and data collection centre are not far apart. For data communication between small devices, wireless communication is the standard. The Bluetooth-based moisture and temperature sensors monitor the environment and weather situation of the crop at the worksite. This type of communication is generally used between a fixed device (user handset) and a communication device. A similar type of sensor, known as a sensor node, has been developed which is useful for measurement of light intensity and surrounding temperature.

17.7.4 LORA

This type of communication is used in a wide area network. LORA is based in a spread-spectrum modulation technique. This type of technology is used for long-range purposes and also where low power is required in IoT-based applications.

17.7.5 SIGFOX

SIGFOX is low-power communication system. It is a narrow-band-based technique. SIGFOX and LORA are a low-power low-band communication protocol.

Generally, the primary means of communication in rural areas is mobile phones. Modern advancements in the cellular phone sector have resulted in intense significance increases, making the agriculture sector more attractive, particularly to the small-holding developer in remote areas.

17.8 CLOUD COMPUTING

Cloud computing refers to the storage of programmes and worksite data on a remote internet server instead of on a local computer hard disk. Cloud computing is very useful in precision agriculture. Figure 17.4 represents fluid techniques for smart farming.

17.9 UNMANNED AERIAL VEHICLES

Unmanned aerial vehicles (UAV) are aircraft that have a controller and communication system but do not require a human pilot. In modern times IoT offers remarkable improvement to industries such as poultry, fishing and farming. This type of communication infrastructure and facilities are challenging in developing countries and

TABLE 17.3
Sensor Based on Smart Phone used in Agriculture Applications

S. No	Smart Phone Sensor	Motto	Application in Agriculture Sector
1.	Real-Time Camera	Captures pictures of any object	Plant disease analysis, soil erosion, leaf-area analysis
2.	GPS	Detects location; monitors the latitude and longitude of device	Site information is attached to create alerts; generally used for machine driving and tracking, land management and crop mapping
3.	Microphone	Detects wanted/unwanted sound and convert to electrical signals	Maintenance of machine, detection of bugs
4.	Accelerometer	Measures acceleration forces	Monitoring activities of worker or machine activities
5.	Gyroscope	Measures angular velocity	Movement of equipment.

remote areas, which is one of the difficulties for introducing IoT in the agriculture sector. Drones play a major role for crop monitoring. Some of the key points of drones used in agriculture are given below.

17.9.1 Soil and Worksite Analysis

Soil and worksite analysis enable the farmer to determine the level of nutrients found in the soil. Both analyses are used in fertilizer decisions and affect the profitability

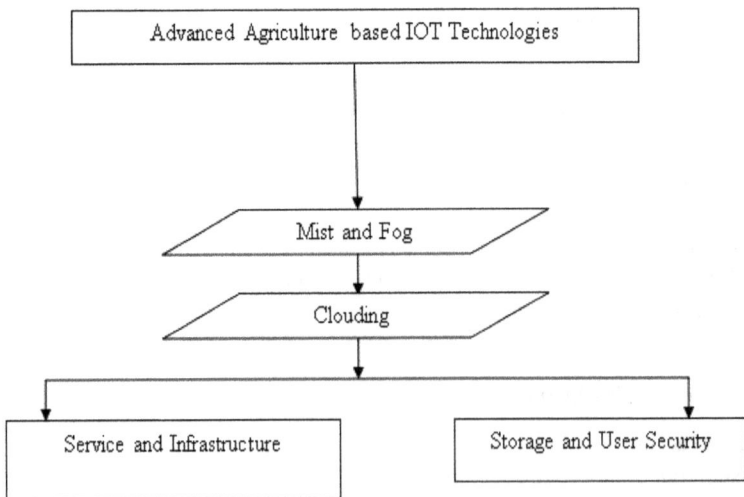

FIGURE 17.4 Fluid techniques for smart farming.

of the farm in the long term. Drones cameras are helpful to produce precise and accurate information. By examining soil conditions before sowing, information from the drone can suggest the category of seed and design of planting which can help to decide the most appropriate crop for a particular plot.

17.9.2 PLANTING

Planting is the process of putting seeds or plants into the ground or a soil container. Currently, a great many acres of property are not used because they are unreachable by humans. So we can use UAV for planting.

17.9.3 CROP MONITORING

Crop monitoring allows the farmer to check land and underwater conditions. Soil parameters, weather conditions and many more factors are used to monitor the crop. Covering a large area is a tough job. Drone cameras provide more accurate and cost effective images in real time.

17.9.4 IRRIGATION

IoT-based drones also used for irrigation purposes. It consists of two aspects. One application
    is associated with the various types of sensors and camera the UAV is equipped with. This captures the image and field data of the crop. The second aspect is that it can be used for water sprinkling purposes at the same time. The technique reduces working time and is very useful in emergencies.

17.9.5 DETECTION OF GAP AND PLANT COUNTING

Precision agriculture significantly increases the 3D data on crop density when creating choices through different types of applications. UAV is a modern technique which is helpful for the detection of gap and plant counting.

17.9.6 SPRAYING PESTICIDES/HERBICIDES

UAVs can be used to spray pesticides. This technique kill pests faster than other pest control technologies. However, apply for such type of applications is more critical.

17.9.7 HEALTH ASSESSMENT

The infrared light sensors in drones can recognize crops that may be infected by bacteria or fungus.

17.9.8 IDENTIFICATION OF PLANT SPECIES

Identification of plant species is carried out by botanists. To monitor plant species, particularly endangered species at risk of disappearing from the Earth, UAVs play major role. UAVs can monitor plant species in inaccessible areas. Another role of UAVs is to detect the forest biomass and fuel with the help of radar and satellites.

17.10   PRESENT CHALLENGES AND FUTURE EXPECTATIONS

In 2015, the UN and international community adopted 'The 2030 Agenda for Sustainable Development', to end starvation by 2030. Recent facts released by the World Health Organisation (WHO) do not show enough improvement in that because approximately 800 million people are facing food shortages. The system hardware (sensors, transmitters and receivers) installed at worst site location where it can be easily damaged. So, periodically maintain it required an IoT-based hardware. These statistics are alarming on their own; another shocking, and possibly more critical, factor is the food quality. There are sources available in our planet and the technologies are in our hand to learn how to use them wisely and accurately. If we do this, then the possibility exists to remove starvation in upcoming decades.

17.11   CONCLUSION

Advanced agriculture technology offers an effective system of doing agriculture. IoT-based agriculture technologies can improve crop growing, quality and production rates to maintain upcoming food demand. The system has the ability to give profitable output to the agriculture industry and to end starvation challenges. A research conducted in 195 countries on 'effect of unhealthy diet' from 1990 to 2017 shows that death rate in one of five per year could be clogged by giving healthier diet. The responsible risk factor for deaths is low grains diet. The conventional technique is in use by farmer in many countries. Advanced agriculture technique is an effective and high-technology system for agriculture.IoT-based agriculture technology has improved production in various ways and reduced the amount of physical work of the farmer. In many regions farmers have kept following conventional agriculture techniques. The blind use of fertilizers and pesticides creates irreversible implications to the environment. The agriculture research organization and farmers need to be run differently and use modern technologies to improve crop quality and quantity.

REFERENCES

1. Prosdocimi, M., Burguet, M., Di Prima, S., Sofia, G., Terol, E., Rodrigo Comino, J., Cerdà, A., and Tarolli, P. (2017). Rainfall simulation and structure-from-motion photogrammetry for the analysis of soil water erosion in Mediterranean vineyards. *Sci. Total Environ.*, 574, 204–215.
2. Han, P., Dong, D., Zhao, X., Jiao, L., and Lang, Y. (2016). A smartphone-based soil color sensor: For soil type classification. *Comput. Electron. Agric.*, 123, 232–241.
3. Santhi, P. V., Kapileswar, N., Chenchela, V. K. R., and Prasad, C. H. V. S. (2017). Sensor and vision based autonomous AGRIBOT for sowing seeds. International Conf. on Energy, Communication, Data Analytics and Soft Computing, Chennai, 1, 100–105.
4. Karimi, H., Navid, H., Besharati, B., Behfar, H., and Eskandari, I., (2017). A practical approach to comparative design of non-contact sensing techniques for seed flow rate detection. *Comput. Electron. Agric.*, 142, 165–172.

5. Romero-Trigueros, C., Nortes P. A., Alarcón, J. J., Hunink, J. E., Parra M., Contreras S., and Droogers P. (2017). Effects of saline reclaimed waters and deficit irrigation on Citrus physiology assessed by UAV remote sensing. *Agric. Water Manag.*, 183, 100–105.

6. Hoffmann, H., Jensen, R., Thomsen, A., Nieto, H., Rasmussen, J., and Friborg, T. (2016). Crop water stress maps for an entire growing season from visible and thermal UAV imagery. *Bio-geosciences*, 13, 6545–6563.

7. Andriamandroso, A. L. H., Lebeau, F., Beckers, Y., Froidmont, E., Dufrasne, I., Heinesch, B., Dumortier, P., Blanchy, G., Blaise, Y., and Bindelle, J. (2017). Development of an open-source algorithm based on inertial measurement units (IMU) of a smartphone to detect cattle grass intake and ruminating behaviors. *Comput. Electron. Agric.*, 139, 126–137.

8. Azam, M. F. M., Rosman, S. H., Mustaffa, M., Mullisi, S. M. S., Wahy, H. Jusoh, M. H., and Ali, M. I. M. (2017 August). Hybrid water pumps system for hilly agricultural site. *IEEE Control Syst. Grad. Res. Colloquium ICSGRC*, 1, 100–105.

9. Hunter, M. C., Smith, R. G., Schipanski, M. E., Atwood, L. W., and Mortensen, D. A. (2017). Agriculture in 2050: Recalibrating targets for sustainable intensification. *Bio. Sci.*, 67 (4), 386–391.

10. Martínez F. J., González- Z, A., Sánchez, N., Gumuzzio, A., and Herrero-J, C.M. (2016). Satellite soil moisture for agricultural drought monitoring: Assessment of the SMOS derived soil water deficit index. *Remote Sensing of Environ.*, 177, 210–220.

11. Vågen, T. G., Winowiecki, L. A., Tondoh, J. E., Desta, L. T., and Gumbricht, T. (2016). Mapping of soil properties and land degradation risk in Africa using MODIS reflectance. *Geoderma*, 263, 216–225.

12. Orlando, F., Movedi, E., Coduto, D., Parisi, S., Brancadoro, L., Pagani, V., Guarneri, T, and Confalonieri, R. (2016). Estimating leaf area index (LAI) in vineyards using the pocket LAI smart-app. *Sensors (Basel)*, 16, 1–12.

13. Moonrungsee, N., Pencharee, S., and Jakmunee, J. (2015). Colorimetric analyzer based on mobile phone camera for determination of available phosphorus in soil. *Talanta*, 136, 204–209.

14. Camacho, H. A. A. (2018). Smartphone-based application for agricultural remote technical assistance and estimation of visible vegetation index to farmer in Colombia: Agro TIC. *Remote Sensor Agric. Ecosyst. Hydrol. SPIE Remote Sens.*, 10783, 100–105.

15. Venkatesan, R., Kathrine, G., Jaspher W, and Ramalakshmi, K. (2018). Internet of things based pest management using natural pesticides for small scale organic gardens. *J. Comput. Theor. Nanoscie.*, 15, 100–104.

16. Wietzke, A., Westphal, C., Gras, P., Kraft, M., Pfohl, K., Karlovsky, P., Pawelzik, E., Tscharntke, T., and Smit, I. (2018). Insect pollination as a key factor for strawberry physiology and marketable fruit quality. *Agric., Ecosyst. & Environ.*, 258, 100–105.

17. Dehaghi, M. A. (2014). Effects of biological and chemical fertilizers nitrogen on yield quality and quantity in cumin. *J. Chem. Health Risks*, 4(2), 55–64.

18. Kou, Z., and Wu, C. (2018). Smartphone based operating behavior modeling of agricultural machinery. *IFAC-Papers On-Line*, 51(17), 521–525.

19. Machado, B. B., Orue, J. P. M., Arruda, M. S., Santos, C. V., Sarath, D. S., Goncalves, W. N, Silva, G. G., Pistori, H., Roel, A. R., and Rodrigues-Jr, J. F. (2016). BioLeaf: A professional mobile application to measure foliar damage caused by insect herbivore. *Comput. Electron. Agric.*, 129, 44–55.

20. Zhang, L., Dabipi, I. K., and Brown, W. L, 2018 Internet of things applications for agriculture. In, *Internet of Things A to Z: Tech. and App.*, Q. Hassan (Ed.). Wiley, Hoboken, NJ.

21. Kong, Q., Chen, H., Mo, Y. L., and Song, G. (2017). Real-time monitoring of water content in sandy soil using shear mode piezo-ceramic transducers and active sensing-A feasibility study. *Sensors*, 2395, 100–105.

22. Pajares, G., Peruzzi, A., and Gonzalez-de-Santos, P. (2013. Sep) Sensors in agriculture and forestry. *Sensors (Basel)*, 1, 12132–12139.

23. Andújar, D., Ribeiro, Á., Fernández-Quintanilla, C., and Dorado, J. (2011). Accuracy and feasibility of optoelectronic sensors for weed mapping in wide row crops. *Sensors*, 1, 2304–2318.

24. Xie, X., Zhang, X., He, B., Liang, D., Zhang, D., and Huang, L. (2016). A system for diagnosis of wheat leaf diseases based on Android smartphone. *Opt. Meas. Tech. Instr.*, 10155, 100–105.

25. Yew, T. K., Yusoff, Y, Sieng, L. K., Lah, H. C., Majid, H., and Shelida, N. (2014). An electrochemical sensor ASIC for agriculture applications. International Convention on Information and Communication Technology, Electronics and Microelectronics (MIPRO), Opatija, 85–90.

26. Maldonado, W., Valeriano, T. T. B., and Souza Rolim, G., (2019). EVAPO: A smartphone application to estimate potential evapotranspiration using cloud gridded meteorological data from NASA-POWER system. *Comput. Electron. Agric.*, 156, 187–192.

27. Bartlett, A. C., Andales, A. A., Arabi, M., and Bauder, T. A. (2015). A smartphone app to extend use of a cloud-based irrigation scheduling tool. *Comput. Electron. Agric.*, 111, 127–130.

28. Freebairn, D., Robinson, B., Mcclymont, D., Raine, S., Schmidt, E., Skowronski, V., and Eberhard, J. (Sep. 2017). Soil water app-monitoring soil water made easy. *Proc.18th Aust. Soc. Agron Conf.*, 24–28, 2017.

29. Ferguson, J. C., Chechetto, R. G., O'Donnell, C. C., Fritz, B. K., Hoffmann, W. C., Coleman, C. E., Chauhan, B. S., Adkins, S. W., Kruger, G. R., and Hewitt, A. J. (2016). Assessing a novel smartphone application – SnapCard, compared to five imaging systems to quantify droplet deposition on artificial collectors. *Comput. Electron. Agric.*, 128, 193–198.

30. Jordan, R., Eudoxie, G., Maharaj, K., Belfon, R., and Bernard, M. (2016). Agri maps: Improving site-specific land management through mobile maps, *Comput. Electron. Agric.*, 123, 100–105.

31. Bueno-Delgado, M. V., Molina-Martínez, J. M., Correoso Campillo, R., and Pavón-Mariño, P. (2016). Ecofert: An Android application for the optimization of fertilizer cost in fertigation. *Comput. Electron. Agric.*, 121, 32–42.

32. Sopegno, A., Calvo, A., Berruto, R., Busato, P., and Bocthis, D. (2016). A web mobile application for agricultural machinery cost analysis. *Comput. Electron. Agric.*, 130, 158–168.

33. Palomino, W., Morales, G., Huaman, S., and Telles, J. (2018). PETEFA: Geographic Information System for Precision Agriculture, IEEE Conf. Electron. Electr. Eng. Comput. INTERCON, 1–4.

34. Herrick, J. E. (2016). The land-potential knowledge system (landpks): Mobile apps and collaboration for optimizing climate change investments. *Ecosyst. Heal. Sustain.*, 2, 100–105.

35. Nakato, G. V., Beed, F., Bouwmeester, H., Ramathani, I., Mpiira, S., Kubiriba, J., and Nanavati, S. (2016). Building agricultural networks of farmers and scientists via mobile phones: case study of banana disease surveillance in Uganda. *Can. J. Plant Pathol.*, 38, 307–316.

36. Masuka, B., Matenda, T., Chipomho, J., Mapope, N., Mupeti, S., and Tatsvarei, S. (2016). Mobile phone use by small-scale farmers: A potential to transform production and marketing in Zimbabwe. *South African J. of Agri Extension (SAJAE)*, 44(2), 121–135.

37. Chung, S., Breshears, L. E., and Yoon, J. Y. (2018). Smart-phone near infrared monitoring of plant stress. *Comput. Electron. Agric.*, 154, 93–98.

38. McGonigle, A. J. S., Wilkes, T. C., Pering, T. D., Willmott, J. R., Cook, J. M., Mims, F. M., and Parisi, A. V. (2018.) Smartphone spectrometers. *Sensors (Switzerland)*, 18(1), 100–105.

18 Machine Learning for Solving a Plethora of Internet of Things Problems

Sparsh Sharma, Abrar Ahmed and Mohd Naseem
Baba Ghulam Shah Badshah University

Surbhi Sharma
Shri Mata Vaishno Devi University

CONTENTS

18.1 Introduction 256
    18.1.1 Contributions of this Article 256
18.2 Categories of Machine Learning 257
    18.2.1 Supervised Learning..258
        18.2.1.1 Classification..258
        18.2.1.2 Regression 258
    18.2.2 Unsupervised Learning 258
        18.2.2.1 Association..259
        18.2.2.2 Clustering..259
        18.2.2.3 Dimensionality Reduction 259
    18.2.3 Semi-Supervised Learning 259
    18.2.4 Reinforcement Learning 260
18.3 Supervised Machine-Learning Algorithms..260
    18.3.1 Linear Regression 260
    18.3.2 Logistic Regression..260
    18.3.3 Classification and Regression Trees 260
    18.3.4 Linear Discriminant Analysis 261
    18.3.5 Naïve Bayes 261
    18.3.6 K-Nearest Neighbour...261
    18.3.7 Support Vector Machine 262
18.4 Unsupervised Machine-Learning Algorithms 262
    18.4.1 K-Means Clustering..262
    18.4.2 Principal Component Analysis 262
    18.4.3 Apriori 262

18.5 Various Applications of Machine Learning 263
    18.5.1 Machine Learning in Intelligent Transportation Systems 263
    18.5.2 Machine Learning in Agriculture 264
    18.5.3 Machine Learning in Health care 264
    18.5.4 Machine Learning in Bioinformatics 265
    18.5.5 Machine Learning in Text Classification 266
    18.5.6 Machine Learning in Data Mining 266
18.6 Implementation and Software Tools for Machine Learning 267
18.7 Conclusion 269
Acknowledgement 270
References 270

18.1 INTRODUCTION

Machine learning (ML) is an essential part of the technology of artificial intelligence (AI). It analyzes enormous amounts of data and comes up with customized prediction, which can help the user to deal logically with the overload of information. Machine learning is applied to enable machines to process information and make decisions by figuring out the pattern without explicit programming. This can be achieved via multiple techniques; one of them is through training machines on a large scale of datasets, called training dataset, which is used to create models to help machines to work with real-time data. ML is a subfield of AI that enables the machine to learn and adapt automatically from past experiences without being explicitly programmed. The goal of ML is to develop machine models or software that can learn from experiences similarly to the way humans do. Machine learning mimics the philosophy of human learning (i.e. learning from expert guidance and experience).

The typical machine learning process involves the following steps as (i) gathering data, (ii) pre-processing data, (iii) choosing a model, (iv) training, (v) testing, (vi) model tuning, (vii) predicting, as pictorially shown in Figure 18.1.

Arthur Samuel [1] describes machine learning as an application of artificial intelligence that provides computers with the ability to learn and improve without human interference. Another important definition of machine learning states that a computer programme is said to be a machine learning programme if that programme can learn and adapt itself based on some tasks and performance measures such that its performance gets enhanced with respect to its learned experience [1]. The sole aim of machine learning is to learn and improve performance from the available data. For example, in the task of image classification and recognition, machine learning can be applied for the classification of images.

18.1.1 CONTRIBUTIONS OF THIS ARTICLE

1. Various categories and types of machine-learning algorithms have been presented, along with the explanation of where and when each type can be well suited to attain maximum performance and accuracy.
2. The plethora of applications supported by machine learning in various fields of research have been explored.

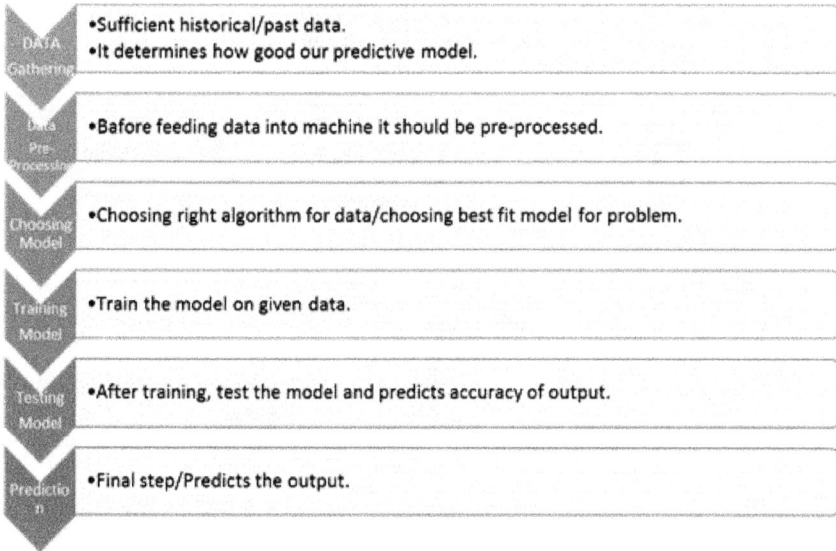

FIGURE 18.1   Machine learning process.

3. Tools, languages and platforms for the implementation of machine learning have also been explored.

18.2 CATEGORIES OF MACHINE LEARNING

In this section, categories and algorithms of machine learning are explored as well as when and where to use each category. Machine learning can be categorized into four types, as pictorially illustrated is Figure 18.2. These types can be further subclassified as shown in Figure 18.3.

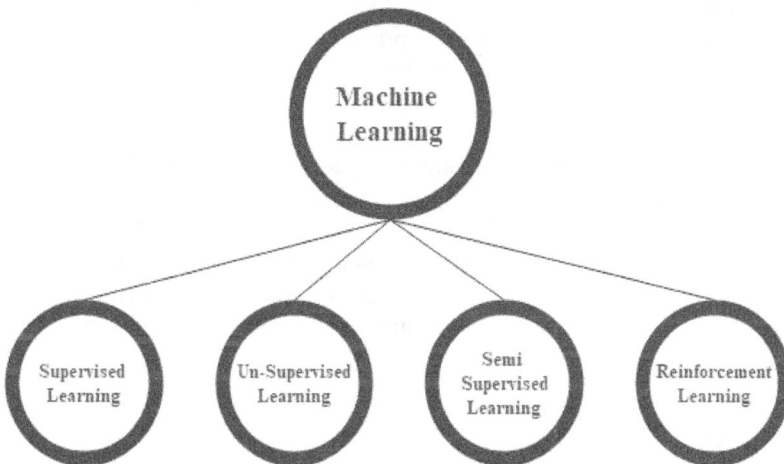

FIGURE 18.2   Categories of machine-learning techniques.

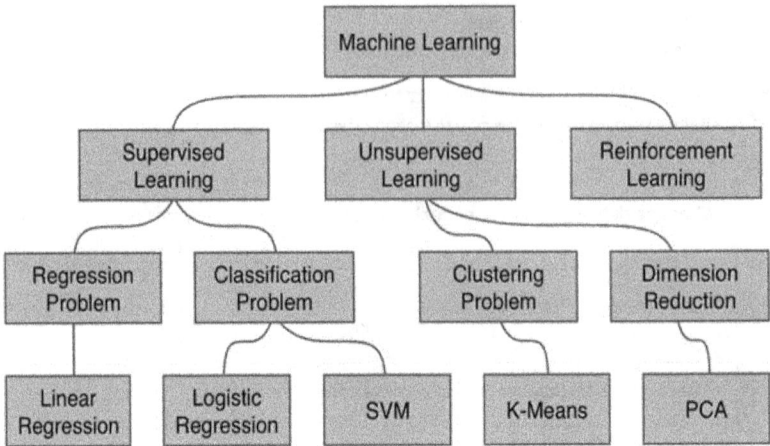

FIGURE 18.3 Machine learning hierarchy.

18.2.1 SUPERVISED LEARNING

Supervised learning (SL) relies on the use of known and labelled data as input. The programmes are trained on a pre-defined set of training examples; this training facilitates its ability to reach accurate results or conclusions where new data is given [2]. Most of the supervised learning applications have the ultimate goal of developing a predictor function f(x), called the mapping function, which facilitates better prediction for output (i.e. y = f(x) where x is the labelled input data and output variable y is predicted for data). Based on the training data, the learning is generalized such that it can respond accurately to all possible inputs [3]. Supervised learning is classified as follows:

18.2.1.1 Classification

In classification, the model is trained to classify or predict something into classes [3]. Classification involves using data to predict whether a student will pass or fail, whether a person is sick or healthy, whether an email is spam or not.

18.2.1.2 Regression

In regression, models or machines are also trained for prediction, but they differ from classification models because in regression it predicts values like weight, price, height, etc. instead of classifying them into different categories [3, 4]. We predict value based on past data. Predictions such as whether it will rain tomorrow, and if there is rain how much rain we will get, can be calculated by measuring weather factors such as humidity, temperature, wind speed and pressure and then seeing how they correlate to rain in the past.

18.2.2 UNSUPERVISED LEARNING

Unsupervised learning (UL) is based on unlabelled input data (i.e. all the input data here have no prior information) [5]. In unsupervised learning, the algorithms identify the hidden patterns in unlabelled data [6]. It is a technique of discovering the labels from the data itself [7]. It is called unsupervised learning (UL) because no supervision for learning is required. The machines learn through observation and find structure in data. In this technique, only the response variable is known [8]. Unsupervised learning has three types:

18.2.2.1 Association

Association is a powerful technique of analyzing or finding a pattern in the dataset [9]. It was proposed by Rakesh Agrawal in the 1990s for market analysis [10]. The aim of association is to find correlation rules of the form (A1.......... An) → Y; it illustrates that it is possible to find B if all values A1 .... An are calculated in a cart. So it is a beneficial technique for using unsupervised learning exploratory data analysis of a variety of resources in market-oriented applications [10, 11].

It is extensively used in market analysis. For example, if a customer visits a book store and buys a pencil, then they are more likely to purchase a sharpener and eraser as well. So indirectly, the purchase of a pencil implies the purchase of an eraser and sharpener as well. These types of machine models are designed for marketing to estimate the sales products based on the selection of one product that gives machines the ability to suggest or select similar products.

18.2.2.2 Clustering

In a given collection of unlabelled datasets, or input data, the models find the pattern, or structure [12]. From given dataset clustering algorithms, data based on similarities is grouped into a number of clusters, (i.e. k number of clusters) [13], so the data from other clusters are dissimilar [11]. The set of inputs is divided into groups. Popular clustering applications apply to fields such as biology, marketing, city planning and recommendation systems.

18.2.2.3 Dimensionality Reduction

Dimensionality relates to the number of characteristics, or input variables, presented in a dataset. If the dataset is noisy or if the number of input variables is large [14], the algorithms do not form effective models [15]. Dimensional reduction is a process of converting a large dataset into smaller data to transmit similar information concisely [16]. Data is compressed and reduced, so storage space and computation time is also reduced [17], and dimensionality reduction gives better computation. Dimensionality reduction is an unsupervised technique that removes redundant features from data.

18.2.3 SEMI-SUPERVISED LEARNING

Semi-supervised learning (SSL) is between unsupervised and supervised and learning [7]. Lots of input data is unlabelled, but labelled data are limited. Semi-supervised problems occur in many applications where labelling is performed by human experts [18]. It is a useful method for regression and prediction. Several real-world machine-learning problems fall under this technique [19]. A photo archive is a good example of where some images are labelled (as cat, person, car) and the majority are unlabelled.

18.2.4 Reinforcement Learning

Reinforcement learning (RL) is learning through interaction with an environment by taking different actions and experiencing many failures and successes when trying to maximize the benefits received. This technique of machine learning is similar to the natural learning process, where a supervisor is not available, but the learning process evolves through trial and error [7, 11]. This learning method is originated from dynamic programming and actor-critic algorithms [20]. In automating the decision-making process, these models are used in applications such as game playing, robotics and medicine [21].

18.3 SUPERVISED MACHINE-LEARNING ALGORITHMS

18.3.1 Linear Regression

Linear regressions is the simplest and powerful machine-learning algorithm. Regression is a predictive modelling technique that finds the relationship between the target variable (independent) and predictor variable (dependent) [22].The equation of regression is $Y = a+bx+e$, where Y is the output function, a→ intercept of fitted line, b→ is the slope, x→ is predictor value and e→ error in model. We can use various techniques to learn the linear regression model for data, such as descent gradient optimization and a linear algebra solution for ordinary least squares. It is a supervised machine-learning algorithm that gains more popularity for prediction. Well-known examples of this technique are predictions of stock prices, housing prices and a person's weight based on height [23].

18.3.2 Logistic Regression

A logistic regression algorithm relies on the classification learning technique, where the dependent variable takes only two possible values [24], as yes or no, 1 or 0, male or female, living or dead, etc. It predicts the numeric variables [12] where the output is an S-curve called a sigmoid curve [3]. Logistic regression is applied in many real-life areas such as image classification; prediction of weather, voting results or student results; computer vision and adaptive object tracking.

18.3.3 Classification and Regression Trees

The classification and regression tree is a popular supervised ML-based decision tree algorithm. CART was introduced by Leo Breiman to refers to a decision tree [25] used for regression and classification. The decision trees is a widely used classification method [26]. Binary representation is used for CART [27] because at each node there are two branches. It is a recursive algorithm that splits the training dataset by doing binary partition. It takes less computational time, but considerable time is taken to build the tree. Decision trees are represented by a set of questions that break the input dataset into smaller and smaller units. CART asks only yes and no questions such as 'Did the student pass?' or 'Is the sex male?' or 'Is this person applicable for the loan?' To find the best partition the CART algorithm searches all the possible values and variables.

18.3.4   LINEAR DISCRIMINANT ANALYSIS

Linear discriminant analysis (LDA) was proposed by R. Fischer in 1936 [28]. Linear discriminant analysis (LDA) depends on searching for a linear combination of variables that isolates two classes [29]. It is a powerful supervised algorithm for dimensionality reduction. LDA is applied when it is assumed that the predictors are distributed normally and the covariance of each class is the same. While applying LDA on a more complex dataset, it loses flexibility because of its linearity [30]. Two different approaches are used for transformation of the dataset and classification of test vectors: a) class dependent and b) class independent. Feature selection and feature extraction are the two methods for dimensionality reduction [31]. LDA is applied in supervised machine learning areas such as speech classification, face recognition, handwriting recognition, heartbeat classification and bioinformatics. [32, 33].

18.3.5   NAÏVE BAYES

Naïve Bayes (NB) is a classification algorithm based on Bayes' theorem [2]. The Bayesian network introduced by Pearl in 1988 is a high-level representation of probability distribution over a set of variables [34]. Naïve Bayes is a probabilistic classification algorithm as it uses probability to make predictions for the purpose of classification [35]. It doesn't require a large amount of data for training to estimating the necessary parameters for classification [36]. The presence of a particular feature in a class is not related to the presence of another feature; this is the assumption of a simple base classifier [37]. It is simple and fast to predict class datasets as well as multiclass prediction. It handles real and discrete data. [35]. Naïve Bayes is a faster, easier to implement and very effective algorithm in ML. NB is usually beneficial for high-dimensional data, where the probability of every feature is independently estimated [34]. It is one of the best and most important algorithms in data mining, as described by Wu et al. [38].

18.3.6   K-NEAREST NEIGHBOUR

In ML, the supervised algorithms play a very important role for labelled data. We have discussed some algorithms earlier in this chapter. The k-nearest neighbour (KNN) algorithm is based on the nearest neighbour rule for classification [5, 36, 39, 40]. We simply say that a decision is made just by examining the labels, based on the nearest voting or k-nearest neighbour voting. In the case of large training sets, it has a practical problem because, for each test instance, the entire set must be searched. This algorithm is mainly used for regression and classification problems and output depends on the type of application, (i.e. classification or regression). It classifies the objects based on closest training example in the featured space [41, 42]. The K-nearest neighbour algorithm is considered as one of the simplest ML algorithms.

This algorithm is applied in many application areas, such as fault classification, process making and sensor development [43].

18.3.7 SUPPORT VECTOR MACHINE

The support vector machine (SVM) algorithm was introduced by Vapnik and his team in the early 1990s. It proved to be a promising and effective technique for data mining [44]. SVM is a supervised algorithm used for both classification and regression. However, it is mostly used in classification problems [45]. It provides excellent results in terms of precision when data can be separated linearly or non-linearly [46]. The SVM algorithm performs a linear categorization by obtaining the hyperplane, which increases the margin between two classes. A hyperplane is determined by the data points known as support vectors [47]. Both linear and non-linear classification can be performed by SVM based on the kernel in use [48]. SVMs are popular for pattern recognition and text classification [49]. For a pattern-classification problem, SVM is considered the most effective algorithm [43] compared with other supervised learning methods, support vector machines may have better generalization under many cases.

18.4 UNSUPERVISED MACHINE-LEARNING ALGORITHMS

18.4.1 K-MEANS CLUSTERING

K-means clustering is a well-known algorithm in unsupervised machine learning for solving clustering problems [50]. Clustering involves the grouping of objects together in such a way that objects that are similar to each other are kept in one cluster [9]. The k-means algorithm is the best known squared error-based clustering algorithm. Due to its simplicity, it is easily implemented in real-world problems [13]. This algorithm is named k-means because k number of distinct clusters are created in it. It is applied in many real-world applications, such as data compression, image analysis, and computer graphics.

18.4.2 PRINCIPAL COMPONENT ANALYSIS

The principal component analysis (PCA) transformation was initially coined by Pearson in 1901 and is still being used in its original generic form and also as a basis for complicated data mining algorithmic techniques. PCA is incredibly useful when we need to visualize high-dimensional data. It's also simple to implement but requires plenty of work beforehand. In PCA the dimension of the data is reduced to make the computations easier and faster [13]. PCA is a very useful technique for pattern recognition, image recognition and compression, text recognition and face recognition [50–52].

18.4.3 APRIORI

The Apriori algorithm is based on the association rule [10]. It is used for finding all frequent item sets [53]. The Apriori algorithm implements a level search using properties of frequent elements and can be further optimized [54]. It is called Apriori because it is based on the fact that the algorithm uses prior knowledge of an item set [55]. The item sets are chosen on the Boolean association rules. It is widely used for market basket analysis, autofill applications, etc.

18.5 VARIOUS APPLICATIONS OF MACHINE LEARNING

Nowadays machine learning is applied everywhere. In the era of technology when everyone directly or indirectly interacts with machines, there are many applications of machine learning, such as image recognition, text classification, news classification, bioinformatics system, data mining, autonomous vehicles, marketing, robotics, speech recognition, cybersecurity, etc. This section looks at various application domains where machine learning is applicable as shown in Figure 18.4.

18.5.1 MACHINE LEARNING IN INTELLIGENT TRANSPORTATION SYSTEMS

Machine learning has a huge list of applications in intelligent transportation systems (ITSs). Machine learning is being used in the ITS sector for reducing traffic accidents, traffic congestion, the waiting time of vehicles in jams, etc. Machine learning can be used to predict traffic patterns in order to make intelligent decisions regarding traffic diversions and controlling the traffic lights. Machine learning can also

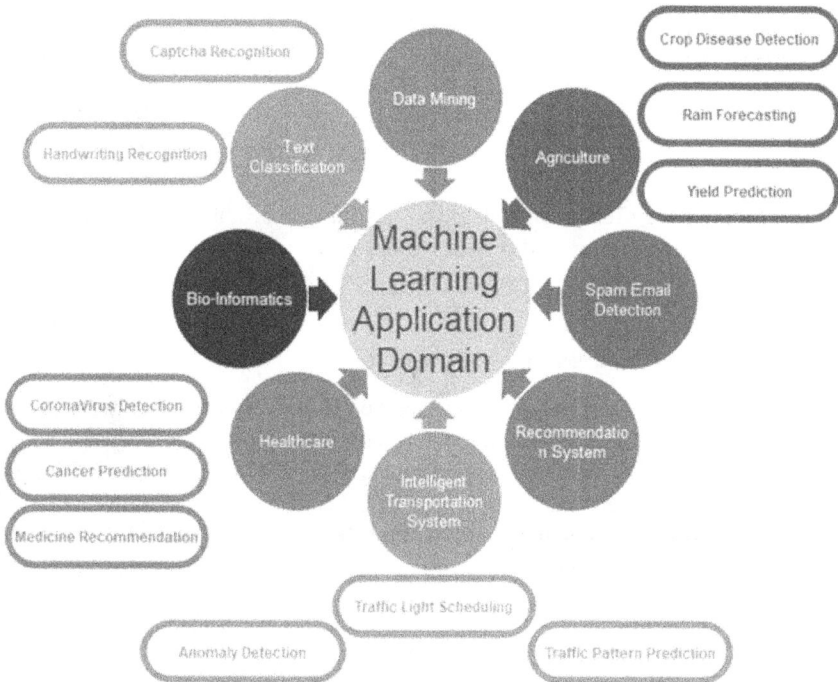

FIGURE 18.4 Various applications supported by machine learning.

TABLE 18.1

Machine Learning in Intelligent Transportation System

Reference	ITS Application	Machine-Learning Model	Dataset Used	Remarks
[56]	Pedestrian Detection	AdaBoost and Support Vector Machine	Self-Captured Images	AdaBoost is used for segmentation of pedestrians. SVM is used for removing false positives.
[57]	Licence Plate Recognition	Support Vector Machine	Self-Captured Images of Real Number Plates	SVM is used for character recognition with 93.54% accuracy.
[58]	Real-Time Pothole Detection	Support Vector Machine	Real Testbed	Performed on real deployment in Australia and achieves 97.5% accuracy in road damage detection.
[59]	Road Marking Detection	Binarized Normed Gradient (BING) and PCA Network	Road Marking Dataset [60]	System achieves 96.8% accuracy in detecting nine classes of road markings
[61]	Intrusion Detection System	Dolphin Swarm Algorithm Optimized SVM	Road Vehicles dataset	Use of multi-cluster head based clustering for load distribution along with anomaly intrusion detection system (IDS).

be used for advance estimation of trip charges for cabs. Table 18.1 shows some of the research work being carried out on various applications of ITS to provide safe, accident-free and convenient travelling.

18.5.2 Machine Learning in Agriculture

Machine Learning can be used for increasing the yield of crops in agriculture. Various machine learning models can be applied in tasks such as identifying diseases and recommending appropriate fertilizers. Similarly, machine learning can be used for predicting and forecasting the yield of a crop at a particular time and with particular soil type. Rain levels can also be predicted with machine learning. Table 18.2 lists some of the research being carried out on machine learning in agriculture.

18.5.3 Machine Learning in Health care

Machine learning has had a remarkable impact in the health-care sector also. Machine learning can be used in health care for tasks such as medication recommendation,

TABLE 18.2
Machine Learning in Agriculture

Reference	ITS Application	Machine-Learning Model	Dataset Used	Remarks
[62]	Wheat Yield Prediction	Self-Organizing Map Models i.e. CP-ANN, SKN, XYF	Real Testbed	Accuracy of 81.65 %, 78.3 %, and 80.92 % was achieved by SKN, CP-ANN, and XYF, respectively.
[63]	Soil Drying Prediction	Classification Trees, K-NN, Boosted Perceptrons	Real Testbed in Urbana, USA	K-NN and boosted perceptron was able to predict the soil dryness with 91–94% accuracy.
[64]	Tomato Disease Detection	Deep Learning models i.e. AlexNet and VGG16 net	PlantVillage	Pre-trained model was used for identification of diseases in tomato crops and achieved 97.29% and 97.49% accuracy for VGG16 and AlexNet respectively.
[65]	Soil Irrigation Decision Support System	Partial Least Square Regression, and Adaptive Neuro Fuzzy Inference System	Real Testbed	Experiment was conducted in southeast Spain.

disease recognition, diagnosis, etc. It can also be used to identify unique disease patterns from the patient's records for further quick and accurate diagnosis. Some of the research work carried out by researchers in health care is presented in Table 18.3.

18.5.4 MACHINE LEARNING IN BIOINFORMATICS

Bioinformatics is a field of study that combines computer science (CS), statistics, mathematics and engineering for analyzing and interpreting biological data [72]. In bioinformatics, one of the most challenging aspects is data. Due to the large volume of patient data, sometimes it's difficult to implement this data as a knowledge for the machine. Working with clean and meaningful data, the machine model accurately predicts disease and recommends prescriptions based on the data [73]. ML covers various applications of bioinformatics, such as genomics, proteomics, microarray and system biology. Machine learning is applied for solving various bioinformatics problems: (1) gene finding, (2) gene expression where clustering algorithms are used, (3) population stratification in which PCA, multi-dimension scaling (MDS) and manifold learning techniques have been adopted and (4) DNA sequencing. Konstantina Kourou et al. [74] applied machine learning for determining cancer prognosis.

TABLE 18.3

Machine Learning in Health Care

Reference	ITS Application	Machine-Learning Model	Dataset Used	Remarks
[66]	Breast Cancer Prediction	Support Vector Machine and Support Vector Machine Ensembles	UCI machine learning repository [67] ACM SIGKDD Cup 2008 [68]	SVM ensembles performed better than the SVM.
[69]	Lung Cancer Prediction	Convolutional Neural Network	Kaggle data science bowel 2017 [70]	Classification performance in the region of low 90s AUC points was achieved.
[71]	Coronavirus (Covid-19) prediction	Support Vector Machine	Dataset containing 150 CT images.	CT images were used for the identification of Covid-19.

18.5.5 MACHINE LEARNING IN TEXT CLASSIFICATION

ML is applied for handwriting recognition and text classification. In the recent advancement of technology, various systems have been developed for recognition of text and numbers. The MNIST dataset is used for the recognition of numbers and letters in a text. Classification-based ML models predict the outcome of the desired text. Natural language processing (NLP), text mining and the techniques of automatic learning work together to classify and discover patterns from the electronic documents automatically. Table 18.4 presents some of the applications of text classifications where machine learning has been applied by researchers.

18.5.6 MACHINE LEARNING IN DATA MINING

Over the past several decades, data mining (DM) has played a key role in knowledge discovery in decision-making. Data mining refers to the technique of extracting features and relationships from larger data. Nowadays, data mining techniques are used to assemble machine learning models that power modern artificial intelligence applications such as search engines, recommendation systems and algorithms and for better prediction. The main feature incorporated in the DM/ML combination is the emphasis on the characteristics and distribution of the data [79]. DM techniques are cost-effective and efficient compared to other statistical data applications as DM beneficial for companies to get knowledge-based information. Kalpana Kushwaha and Pinkeshwar Mishra [80] describe the techniques for data mining using ML. Applications of data mining include qualitative data mining (QDM), spam filtering, fraud detection, health care, bioinformatics, credit risk management and database marketing. Xindong Wu et al. [38] describe the techniques and also discuss the top ten algorithms for data mining in ML. Zhiqiang Ge et al. [43] discuss the supervised,

TABLE 18.4

Machine Learning in Text Classification

Reference	ITS Application	Machine-Learning Model	Dataset Used	Remarks
[75]	Text Classification	Naïve Bayes	Chinese text classification corpus	Multi-class odds ratio and class discriminating measure were the two metrics used for performance evaluation.
[76]	Handwritten Numerical String Recognition	Support Vector Machine	NIST SD19 [77]	SVM was applied for the recognition of numerical strings.
[78]	Captcha Recognition	Neural Network, K-Nearest Neighbour, SVM	Bubble Captcha	Pattern recognition neural network performs well in terms of both accuracy and computational cost.

unsupervised, and semi-supervised algorithms for data mining in the data-processing industry using machine learning techniques.

18.6 IMPLEMENTATION AND SOFTWARE TOOLS FOR MACHINE LEARNING

Here we introduce the most popular and useful ML tools used in industry and in research. The languages used for ML are Python, R, Javascript, Java, Scala. The most popular framework for implementation of ML algorithms is TensorFlow. It is the most vibrant and useful framework for the implementation of deep learning (DL) models, where neural networks (NN) and computer vision (CV) models are implemented. SciKit-learn is a Python library mainly used for data mining and analysis. A wide range of ML algorithms is implemented on it. Various machine learning algorithms based on supervised, unsupervised and semi-supervised learning types are implemented using Scikit-Learn. Keras is a Python library used for implementing deep neural networks, and it runs on Tensorflow or Theano. PyTorch is a powerful deep learning framework for machine learning written in Python. PyTorch has strong tensor computations similar to the NumPy array with strong GPU acceleration. PyTorch is a tool for ML which is already used on Facebook, Google, Twitter, etc. [81]. Orange is a GUI-based data mining environment used for machine learning and data visualization. It comes with a large toolbox and offers interactive data analysis workflow. It's a drag-and-drop tool for building ML models and analyzing data, and it resides inside the Anaconda Navigator. Google's AutoML is a drag-and-drop method in which various machine learning models can be used without any coding knowledge requirement. Pre-trained neural network models are provided by Google, which is available via its APIs for accomplishing and performing certain required tasks. Graphical user interface of AutoML enables the dragging of image sets by the users. Google Colab is a cloud-based service for ML from Google which supports Python. While using Colab, there is no need for a GPU to be installed in our own system. Colab provides all the GPU-based ML frameworks, including TensorFlow, PyTorch, NumPy, Keras, etc. ANNdotNET is another deep learning tool from Microsoft built on .NET platform. It is a Windows desktop application for creating and training ANN models and is written in C#. The application is considered a GUI tool for the CNTK library, with extensions in data. It relies on CNTK and Microsoft Cognitive Toolkit. Table 18.5 provides the overview of some of the software tools widely used in the implementation of machine learning.

TABLE 18.5
Implementation and Software Tools for Machine Learning

Tool	Platform	Programming Language Used	Supported Features & Algorithm	Pros/ Cons	URL
Scikit Learn	Macintosh, Windows, Linux	Python, C++, Python, C	Pre-processing, Classification, Dimensionality reduction, Regression, Model Selection Clustering	**Pros:** Proper documentation of its in-built functions. Supports change in parameters of algorithms during the object calling	http://scikit-learn.org/stable/
PyTorch	Windows, Mac, Linux, OS	CUDA, C++, Python	nn Module Autograd Module Optim Module	**Pros:** Easy front end. Computational graphs can be designed easily	https://pytorch.org/
TensorFlow	Windows, Mac, Linux, OS	CUDA, C++, Python	Dataflow programming library will be offered	**Cons:** A bit difficult to learn	https://www.tensorflow.org/
Weka	Windows, Mac, Linux, OS	Java	Association rules mining Clustering Data preparation Classification Visualization Regression	**Pros:** GUI **Cons:** Not much Documentation available	https://www.cs.waikato.ac.nz/ml/weka/

(continued)

TABLE 18.5

Implementation and Software Tools for Machine Learning *(Continued)*

Tool	Platform	Programming Language Used	Supported Features & Algorithm	Pros/ Cons	URL
Colab	Cloud based	N/A	Has the support of libraries of OpenCV, TensorFlow, PyTorch, and Keras	**Pros**: Use it on browser without specialized hardware	https://colab.research.google.com/
Shogun	Windows, Mac, Linux, OS,UNIX	C++	Classification Regression Online learning Clustering Support vector machines Dimensionality reduction	**Pros:** Supports large datasets. Easy to use and good support	http://shogun-toolbox.org/
Keras.io	Cross-platform	Python	Neural network API will be provided	**Pros:** Ease to use, extendible **Cons:** Requires Theano, tenserflow and CNTK for using it	https://keras.io/
Rapid Miner	Cross-platform	Java	Data loading & transformation data pre-processing & visualization	**Pros:** Ease to use, GUI, No programming required. Functionality can be extended by plugins **Cons**: A bit expensive	https://rapidminer.com/

18.7 CONCLUSION

Machine learning has revolutionized countless real-life domains and is now in use in almost all the industry sectors of the world. The use of machine learning techniques has various advantages, such as improved accuracy, enhanced performance and productivity. Various categories of machine-learning algorithms exist, and it is crucial to know which machine-learning algorithm to use and which should not be used in a certain scenario. In this article, all the four types of ML types.

Machine learning supports a plethora of applications. So various applications where machine learning is useful nowadays were presented here along with the various research work being done in machine learning. Finally, various working environments that are required for the implementation and deployment of machine learning algorithms were examined and discussed. This study on machine learning will provide a kick-start for the budding researcher in understanding the concepts of machine learning.

ACKNOWLEDGEMENT

The authors wish to thank TEQIP-III and World Bank for helping to complete this study.

REFERENCES

1. Samuel, Arthur L. *Some studies in machine learning using the game of checkers. II— Recent progress. IBM Journal of Research and Development.* 1967. **11**(6): p. 601–617.
2. Dey, Ayon. *Machine learning algorithms: a review. International Journal of Computer Science and Information Technologies.* 2016. **7**(3): p. 1174–1179.
3. Singh, A., N. Thakur, and A. Sharma. *A review of supervised machine learning algorithms.* In *2016 3rd International Conference on Computing for Sustainable Global Development (INDIACom).* 2016. IEEE.
4. Park, Byeonghwa, and Jae Kwon Bae. *Using machine learning algorithms for housing price prediction: The case of Fairfax County, Virginia housing data. Expert Systems with Applications.* 2015. **42**(6): p. 2928–2934.
5. Dy, Jennifer G., and Carla E. Brodley. *Feature selection for unsupervised learning. Journal of Machine Learning Research.* 2004. **5**(Aug): p. 845–889.
6. Simon, A., et al., An *overview of machine learning and its applications. International Journal of Electrical Sciences & Engineering (IJESE).* 2016. **1**: p. 22–24.
7. Sandhya, N., and K. R. Charanjeet. *A review on machine learning techniques. International Journal on Recent and Innovation Trends in Computing and Communication.* 2016. **4**(3): 451–458.
8. Amruthnath, N. and T. Gupta. *A research study on unsupervised machine learning algorithms for early fault detection in predictive maintenance.* In *2018 5th International Conference on Industrial Engineering and Applications (ICIEA).* 2018. IEEE.
9. Liu, Y. *A Survey of Machine Learning Based Packet Classification.* In Symposium on Computational Intelligence for Security and Defence Applications (CISDA). 2009.
10. Agrawal, R., T. Imieliński, and A. Swami. *Mining association rules between sets of items in large databases.* In *Proceedings of the 1993 ACM SIGMOD International Conference on Management of data.* 1993.
11. Kavakiotis, I., et al., Machine learning and data mining methods in diabetes research. *Computational and Structural Biotechnology Journal.* 2017. **15**: p. 104–116.
12. Shanthamallu, U.S., et al., *A brief survey of machine learning methods and their sensor and IoT applications.* In *2017 8th International Conference on Information, Intelligence, Systems & Applications (IISA).* 2017. IEEE.
13. Singh, L., S. Singh, and P.K. Dubey. *Applications of clustering algorithms and self-organizing maps as data mining and business intelligence tools on real world data sets.* In *2010 International Conference on Methods and Models in Computer Science (ICM2CS-2010).* 2010. IEEE.
14. Cunningham, John P., and Zoubin Ghahramani. Linear dimensionality reduction: Survey, insights, and generalizations. *The Journal of Machine Learning Research.* 2015. **16**(1)(): p. 2859–2900.

15. Sacha, D., et al., Visual *interaction with dimensionality reduction*: A *structured literature analysis. IEEE Transactions on Visualization and Computer Graphics.* 2016. **23**(1): p. 241–250.
16. Boutsidis, C., et al., Randomized dimensionality reduction for k-means clustering. *IEEE Transactions on Information Theory.* 2014. **61**(2): p. 1045–1062.
17. Xu, X., et al., Review of *classical dimensionality reduction and sample selection methods for large-scale data pro*cessing. *Neurocomputing.* 2019. **328**: p. 5–15.
18. Grandvalet, Y. and Y. Bengio, *Semi-supervised learning by entropy minimization. Advances in Neural Information Processing Systems.* 2005. **17**: p. 529–536.
19. Zhu, X., Z. Ghahramani and J.D. Lafferty. *Semi-supervised learning using Gaussian fields and harmonic functions.* In *Proceedings of the 20th International conference on Machine learning (ICML-03).* 2003.
20. Stefán, Péter, László Monostori, and Ferenc Erdélyi. *Reinforcement learning for solving shortest-path and dynamic scheduling problems. Proceedings of the 3rd International Workshop Emergent Synth. IWES* 1.1 (2001): 83–88.
21. Hammoudeh, A., *A concise introduction to reinforcement learning.* n.d.
22. Gharehchopogh, F.S., et al., *A linear regression approach to prediction of stock market trading volume:* A *case study. International Journal of Managing Value and Supply Chains.* 2013. **4**(3): p. 25.
23. Mokhade, Y.S.I.P.A., *Use of Linear Regression in Machine Learning for Ranking. International Journal for Scientific Research & Development* 2013.
24. Gladence, L.M., et al., A *statistical comparison of logistic regression and different Bayes classification methods for machine learning. ARPN Journal of Engineering and Applied Sciences.* 2015. **10**(14): p. 5947–5953.
25. Wang, W., et al., *An improved algorithm for CART based on the rough set theory.* In *2013 Fourth Global Congress on Intelligent Systems.* 2013. IEEE.
26. Yang, B.H. and S. Li, *Remote sense image classification based on CART algorithm.* In *Advanced Materials Research.* 2014. Trans Tech Publ.
27. Pinem, A.F.A. and E.B. Setiawan. Implementation of classification and regression tree (CART) and fuzzy logic algorithm for intrusion detection system. In *2015 3rd International Conference on Information and Communication Technology (ICoICT).* 2015. IEEE.
28. Dorfer, M., R. Kelz and G.J.A.P.A. Widmer, *Deep linear discriminant analysis.* arXiv preprint arXiv:1511.04707. 2015.
29. Tharwat, A., et al., Linear *discriminant* analysis: A *detailed tutorial. AI Communications.* 2017. **30**(2): p. 169–190.
30. Siddiqi, M.H., et al., *Human facial expression recognition using stepwise linear discriminant analysis and hidden conditional random fields. IEEE Transactions on Image Processing* 2015. **24**(4): p. 1386–1398.
31. Sharma, Alok, and Kuldip K. Paliwal. *Linear discriminant analysis for the small sample size problem: an overview. International Journal of Machine Learning and Cybernetics.* 2015. **6**(3): p. 443–454.
32. Soni, Sneha, and Shailendra Shrivastava. Classification of Indian stock market data using machine learning algorithms. *International Journal on Computer Science and Engineering.* 2010. **2**(9): p. 2942–2946.
33. Toygar, Önsen, and A. C. A. N. Adnan. *Face recognition using PCA, LDA and ICA approaches on colored images. Istanbul University-Journal of Electrical & Electronics Engineering.* 2003. **3**(1): 735–743.
34. Taheri, Sona, and Musa Mammadov, *Learning the Naive Bayes classifier with optimization models. International Journal of Applied Mathematics and Computer Science.* 2013. **23**(4): p. 787–795.

35. Saiyed, Sohana, Nikita Bhatt, and Amit P. Ganatra, *A survey on Naive Bayes based prediction of heart disease using risk factors. International Journal of Innovative and Emerging Research in Engineering.* 2016.

36. Putatunda, S., *Machine learning: An introduction.* In *Advances in Analytics and Applications.* 2019, Springer. p. 3–11.

37. Rish, I., *An empirical study of the Naive Bayes classifier.* In *IJCAI 2001 Workshop on Empirical Methods in Artificial Intelligence.* 2001.

38. Wu, X., et al., Top 10 *algorithms in data mining. Knowledge and Information Systems.* 2008. **14**(1): p. 1–37.

39. Larranaga, P., et al., Machine *learning in bioinformatics. Briefings in Bioinformatics.* 2006. **7**(1): p. 86–112.

40. Jain, R.S., *Study of Different Multi-instance Learning kNN Algorithms. International Journal of Computer Applications Technology and Research.* 2014. **3**(7), p. 460–463.

41. Mutyalamma, G., K. Komali, and G. Pushpa, *Ranking and fraud review detection for mobile apps using KNN algorithm. International Journal of Trend in Scientific Research and Development (IJTSRD).* 2017. **2**(1), p.1347–1351.

42. Taneja, S., et al., *An enhanced k-nearest neighbor algorithm using information gain and clustering.* In *2014 Fourth International Conference on Advanced Computing & Communication Technologies.* 2014. IEEE.

43. Ge, Z., et al., Data *mining and analytics in the process industry*: The *role of machine learning.* 2017. *IEEE Acess.* **5**: p. 20590–20616.

44. Tian, Y., et al., Recent *advances on support vector machines research. Technological and Economic Development of Economy.* 2012. **18**(1): p. 5–33.

45. Tong, Simon, and Daphne Koller. *Support vector machine active learning with applications to text classification. Journal of Machine Learning Research.* 2001. **2**(Nov): p. 45–66.

46. Caragea, C., D. Caragea, and V. Honavar. *Learning support vector machines from distributed data sources.* In *Proceedings of the National Conference on Artificial Intelligence.* 2005. Menlo Park, CA; Cambridge, MA; London; AAAI Press; MIT Press.

47. Khan, M.M.R., et al., *Study and observation of the variation of accuracies of KNN, SVM, LMNN, ENN algorithms on eleven different datasets from UCI machine learning repository.* In *2018 4th International Conference on Electrical Engineering and Information & Communication Technology (iCEEiCT).* 2018. IEEE.

48. Das, K., Behera, A *survey on machine learning*: Concept, *algorithms and applications. International Journal of Innovative Research in Computer and Communication Engineering.* 2017. **5**(2): p. 1301–1309.

49. Elleuch, M., R., Maalej and Kherallah, A *new design based-SVM of the CNN classifier architecture with dropout for offline arabic handwritten recognition.* 2016. *Procedia Computer Science.* **80**: p. 1712–1723.

50. Sehgal, S.,et al., *Data analysis using principal component analysis.* In *2014 International Conference on Medical Imaging, m-Health and Emerging Communication Systems (MedCom).* 2014. IEEE.

51. Fu, Q. *Research and implementation of PCA face recognition algorithm based on matlab.* in *MATEC Web of Conferences.* 2015. EDP Sciences.

52. Zhang, Z., Y. Xu, and C.-L. Liu. *Natural scene character recognition using robust PCA and sparse representation.* In *2016 12th IAPR Workshop on Document Analysis Systems (DAS).* 2016. IEEE.

53. Yabing, Jiao., Research of an *improved Apriori algorithm in data mining association rules.* 2013. *International Journal of Computer and Communication Engineering.* **2**(1): p. 25.

54. Patel, P., *Implementing APRIORI Algorithm on Web serve log*. National Conference on Recent Trends in Engineering and Technology (NCRTET-2011)

55. Bhardwaj, S., et al., Improved *Apriori algorithm for association rules. International Journal of Technical Research and Applications*. 2015. **3**(3): p. 238–240.

56. Guo, L., et al., Pedestrian *detection for intelligent transportation systems combining* AdaBoost *algorithm and support vector machine. Expert Systems with Applications*. 2012. **39**(4): p. 4274–4286.

57. Wen, Y., et al., *An Algorithm for license plate recognition applied to intelligent transportation system. IEEE Transactions on Intelligent Transportation Systems*. 2011. **12**(3): p. 830–845.

58. Anaissi, A., et al., *Smart pothole detection system using vehicle-mounted sensors and machine learning. Journal of Civil Structural Health Monitoring*. 2019. **9**(1): p. 91–102.

59. Chen, T., et al., *Road marking detection and classification using machine learning algorithms*. in *2015 IEEE Intelligent Vehicles Symposium (IV)*. 2015. IEEE.

60. Wu, T. and A. Ranganathan. *A practical system for road marking detection and recognition*. in *2012 IEEE Intelligent Vehicles Symposium*. 2012. IEEE.

61. Sharma, S. and A. Kaul, *Hybrid fuzzy multi-criteria decision making based multi cluster head dolphin swarm optimized IDS for VANET. Vehicular Communications*. 2018. **12**: p. 23–38.

62. Pantazi, X.E., et al., *Wheat yield prediction using machine learning and advanced sensing techniques*. 2016. *Computers and Electronics in Agriculture*. **121**: p. 57–65.

63. Coopersmith, E.J., et al., *Machine learning assessments of soil drying for agricultural planning*. 2014. *Computers and Electronics in Agriculture*. **104**: p. 93–104.

64. Rangarajan, A.K., R. Purushothaman and A.J.P.C.S. Ramesh, *Tomato crop disease classification using pre-trained deep learning algorithm*. 2018. *Procedia Computer Science*. **133**: p. 1040–1047.

65. Navarro-Hellín, H., et al., *A decision support system for managing irrigation in agriculture*. 2016. *Computers and Electronics in Agriculture*. **124**: p. 121–131.

66. Huang, M.-W., et al., SVM and SVM *ensembles in breast cancer prediction. PLOS One*. 2017. **12**(1).

67. Asuncion, A. and D. Newman, D., *UCI Machine Learning Repository*. 2007.

68. Perlich, C., et al., *Breast CANCER IDENTIFICATION: KDD CUP WINNER'S REPort. ACM SIGKDD Explorations Newsletter*. 2008. **10**(2): p. 39–42.

69. Kadir, T. and F. Gleeson, *Lung cancer prediction using machine learning and advanced imaging techniques. Translational Lung Cancer Research*. 2018. **7**(3): p. 304.

70. Kaggle, B., *Kaggle Data Science Bowl 2017*. 2017. Kaggle. n.d.

71. Barstugan, M., U. Ozkaya, U., and S. Ozturk, *Coronavirus (COVID-19) classification using CT images by machine learning methods*. arXiv preprint arXiv:2003.09424. 2020.

72. Bhaskar, H., et al., *Machine learning in bioinformatics: A brief survey and recommendations for practitioners*. 2006. *Computers in Biology and Medicine*. **36**(10): p. 1104–1125.

73. Kourou, K., et al., Machine learning applications in cancer prognosis and prediction. *Computational and Structural Biotechnology Journal*. 2015. **13**: p. 8–17.

74. Ikonomakis, M., S. Kotsiantis, S., and V. Tampakas, *Text classification using machine learning techniques*. 2005. *WSEAS transactions on computers* **4**(8): p. 966–974.

75. Chen, J., et al., *Feature selection for text classification with Naïve Bayes*. 2009. *Expert Systems with Applications* **36**(3): p. 5432–5435.

76. Oliveira, L.S. and R. Sabourin. *Support vector machines for handwritten numerical string recognition*. in *Ninth International Workshop on Frontiers in Handwriting Recognition*. 2004. IEEE.

77. Grother, P., *NIST Special Database 19 Handprinted Forms and Characters, 2nd Edition*. National Institute of Standards and Technology, Tech. Rep. 2016.

78. Bostik, O. and J. Klecka, *Recognition of CAPTCHA characters by supervised machine learning algorithms*. IFAC-PapersOnLine. 2018. **51**(6): p. 208–213.

79. Teng, X. and Y. Gong. *Research on application of machine learning in data mining*. In *IOP conference series: materials science and engineering*. 2018. IOP Publishing.

80. Kalpana, K. and M., P. September, *A Survey on Data Mining using Machine Learning Techniques*. 2016, IJARCCE.

81. Abdullah, A.A. and S. Kanaya. *Machine learning using H2O R package: An application in bioinformatics*. In *Proceedings of the Third International Conference on Computing, Mathematics and Statistics (iCMS2017)*. 2019. Springer.

19 The Fusion of Blockchain and IoT Technologies with Industry 4.0

Arun Kumar Rana
Panipat Institute of Engineering and Technology

Sharad Sharma
Maharishi Markandeshwar (Deemed to be University)

CONTENTS

19.1   Introduction 276
19.2   Related Work 277
19.3   Industry 4.0 278
19.4   Blockchain Transformation of Industries and the Economy 278
19.5   Types of Blockchains 279
    19.5.1   Public Blockchain 279
    19.5.2   Private Blockchain 279
    19.5.3   Consortium Blockchain 279
    19.5.4   Hybrid Blockchain 280
19.6   Benefits of Adopting Industry 4.0 280
    19.6.1   Improved Productivity 280
    19.6.2   Improved Efficiency 281
    19.6.3   Increased Knowledge Sharing 281
19.7   Principles and Goals of Industry 4.0 282
    19.7.1   Interconnection 282
    19.7.2   Data Transparency 282
    19.7.3   Specialized Help 283
    19.7.4   Decentralized Decisions 283
19.8   Blockchain in Cybersecurity 283
19.9   Blockchain-Based Manufacturing Model 283
19.10 Blockchain Impact on Supply Chain Process 284
19.11 Fog Computing Model for Industrial IoT with Evaluation 285
19.12 Open Issues or Challenges 286
    19.12.1 Scalability 286
    19.12.2 Security and Privacy 287
    19.12.3 Energy and Cost-Effectiveness 287
    19.12.4 Resource Limitations 287

19.13 Conclusion 287
References...288

19.1 INTRODUCTION

Industry 4.0 is neither modern technology nor a new market framework. This is the emerging pattern of data sharing and automation in the creation and advancement of new technology in our society. Industry 4.0 is a developing trend of availability and association among parts, machines and people that can make huge innovation and proficiency profits, upgrades in personal satisfaction and maintainable ecological results. Industry 4.0 speaks to the development of conventional plants towards genuine industrial facilities, which are intended to be increasingly effective as an asset the executives and to be exceptionally adaptable to ever-changing innovation prerequisites. Historically people have had space and time for their hand-made instruments to finish all of their jobs [1]. The population evolved over many decades, companies grew, individuals began to rely on computers, apparatuses and other devices to complete their jobs. It is the basis for the growth of companies and projects all around the globe. The emergence of the internet of things (IoT) strengthens conventional past thought and enables many, if not all, environmental artefacts to be linked to the network. It can connect the network to cars, household appliances, and other electronic devices, which in turn gives a smarter life to humans [19]. The device performs detection, position, tracking, and monitoring in real time, and automatically triggers corresponding events. Also, IoT is the main component of industrial IoT (IIoT) aimed at producing smart manufacturing products and setting up smart factories with close customer-business partner relations [2]. Blockchain and IoT with Industry 4.0 uses are shown in Figure 19.1.

Since the 1970s, we've been increasingly digital, getting closer to fully digitizing our culture, whether it's mobile home assistants or smart security systems. Industry 4.0 uses the internet of things to digitally enhance factories, making them 'smart.' In this structure, we allow the development of 'cyber-physical' structures, processes supervised by tightly integrated algorithms and software that replicate physical systems to a virtual network that makes decentralized decisions. With the launch of the internet of things, cyber-physical devices can connect and function together, providing users with real-time device interactions. The technology underpinning smart contracts and the 4.0 industry is known as blockchain [26–28]. A unique identifier known as a hash is passed to the next transaction to connect it. Any new transaction is added to the old block, which gives an unalterable, accurate and objectively verifiable record of the relationship. The blockchain is a transparent element that is available for review [21–24]. The precise contents of each transaction or block remain secured and secret by a private key. In addition to the transactions, the blockchain provides a completely replicated copy of the data forming the chain at each node. A complete copy of the ledger would be given in each node. The permission principle includes all those participants who are interested in setting up a blockchain network who can decide which parts of the ledger other members can see. They can even grant them the right to see anything in good time, if necessary [33–36].

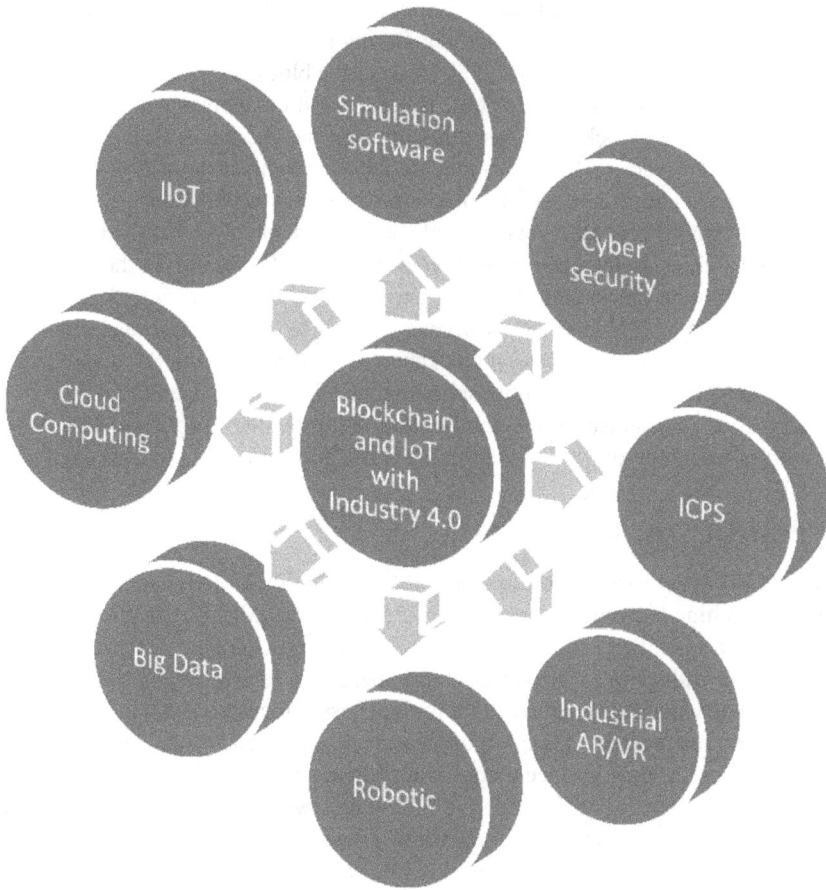

FIGURE 19.1 Blockchain and IoT with Industry 4.0.

19.2 RELATED WORK

The convergence of internet technology and other forward-looking technologies would lead to a paradigm change in industrial development leading to what is known as Industry 4.0. Industry 4.0 gives us an environment focused primarily on the automation of smart services, smart goods and smart factories. This transition provides many possibilities for game-changing ideas in sustainable industrial production. Creating such an automated environment requires internet access along with a large number of artefacts [20]. There have been attempts, for example, to use blockchain in IIoT protection and to promote intelligence variety and capacity approaches. In turn, we viewed numerous centralized overviews and evaluations on IoT, smart urban environments and business blockchain implementations. For examples, the authors of References [2, 3] audited blockchain's usage for IoT and addressed the advantages and impediments of using blockchain in these applications. A few overviews and audits concentrated on the use of blockchain in explicit businesses [5–7, 9] or in applications identifying with edge registering [8] and dispersed-system security administrations [4], etc. Christidis et al. [1] describe blockchains and smart contracts for IoT, and Fernández-Caramés et al. [2] discussed challenges and recommendations for developing blockchain-based IoT applications. Ferrag et al. [3] clarified IoT protection and privacy technologies that are focused on blockchain. Salman et al. [4] proposed blockchain solutions for achieving distributed network security, and Shen and Pena-Mora [5] provide a review of blockchain use cases for Fraga et al. [6] proposed smart cities architecture as well as a survey on blockchain in the automotive sector. A study on blockchain in the oil and gas industry was explored by Lu et al. [7], and Yang et al. [8] identified the convergence of blockchain and edge-computing technology. Xie et al. [9] discussed blockchain use in smart cities. Rana and Sharma [30] introduced the issue of double-spending, using a peer-to-peer network to boost the protection mechanism. Rana and Sharma [31] provide the concept of decentralized digital currency, as well as alternative applications such as property registries.

19.3 INDUSTRY 4.0

Industry 4.0 refers to a new process that focuses heavily on interconnectivity, automation, machine learning and real-time data. Industry 4.0, also often referred to as IIoT or smart manufacturing, integrates smart digital technology, machine learning, and big data with physical production and operations to create a more comprehensive and better-integrated environment for businesses focused on manufacturing and supply chain management [32]. Although every business and organization operating today is different, they all face a common challenge: the need for connectivity and access to systems, partners, goods, and individuals with real-time insights. Industry 4.0 is not only about investing in new innovations and instruments to increase production quality, it's about revolutionizing the way the whole organization functions and develops [39]. This resource will give you an in-depth overview of Industry 4.0 and IIoT topics. Industry 4.0 provides an approach to development that is more systematic, interlinked and holistic. It links physical to digital, enabling better communication and access through agencies, partners, suppliers, goods and individuals. Industry 4.0 empowers business owners to better monitor and understand every aspect of their business and enables them to use instant data to maximize efficiency, optimize processes, and accelerate growth [40].

19.4 BLOCKCHAIN TRANSFORMATION OF INDUSTRIES AND THE ECONOMY

Today, in all parts of society and industry, implementations of blockchain technology are emerging. For example, blockchain can simplify business processes in the finance sector while creating safe, trustworthy records of agreements and transactions. A global consortium of more than 80 institutional members has been set up to establish evidence of the principles and prototypes of financial systems that disrupt the financial sector through the automated real-time execution of financial transactions. In addition, for example, in the case of food supply chains, a blockchain-enabled ecosystem could facilitate an end-to-end service that alleviates interruptions in the incidence of fraudulent products in the supply chain. An efficient and safe transfer of value will take place in the entire process by combining supply chain management with an internet of things (IoT) framework that facilitates automated machine-to-machine communication. A peer-to-peer distributed ledger of time-stamped transactions is the blockchain. For the intent of cryptocurrency, by Bitcoin and other cryptocurrencies, the whole ethos was to decentralize away from central banks. It is, therefore, a movement against centralization and fiat money power. Although central banks are in charge of the ledger with fiat money, the consumer maintains their own copy of the ledger through cryptocurrencies and blockchain technology, and all copies of the ledger are synchronized by what is known as a consensus algorithm [5]. Succeeding in the next industrial age requires manufacturing companies to identify and shape their core value drivers created by digital technologies. Via smart factories and smart supply chains, Industry 4.0 can drive operational efficiencies as well as broaden opportunities through creativity and personalized solutions to maximize customer value. Ultimately, enabled by digitalization, they will lead to entirely new business models and service offerings [21].

19.5 TYPES OF BLOCKCHAINS

Types (Figure 19.2) and function of blockchains are explained below.

19.5.1 PUBLIC BLOCKCHAIN

This is the simplest network and anyone can join any time in public networks blockchain and one of the best examples of this network is a bitcoin. All transactions that occur on public blockchains are completely straightforward, implying that anybody can inspect the transaction subtleties. Public blockchains are proposed to be decentralized, with no one individual or component controlling which trades are recorded in the blockchain or the solicitation in which they are dealt with [2].

19.5.2 PRIVATE BLOCKCHAIN

This is also a simple network and very much similar to the public blockchain network. This technique also has a decentralized peer-to-peer network, and in this technique, a complete network is governed by one organization. Private blockchains are essential for adventures that need to collaborate and exchange data on a public blockchain needn't bother about their complicated business data. These chains are increasingly unified by their demeanour; the elements operating the chain have essential control over the systems of individuals and organizations. Personal blockchains may have a network token drawn in Reference [2].

19.5.3 CONSORTIUM BLOCKCHAIN

Consortium blockchains are often called private blockchain's alternative function. The fundamental distinction between them is that the blockchain's community is

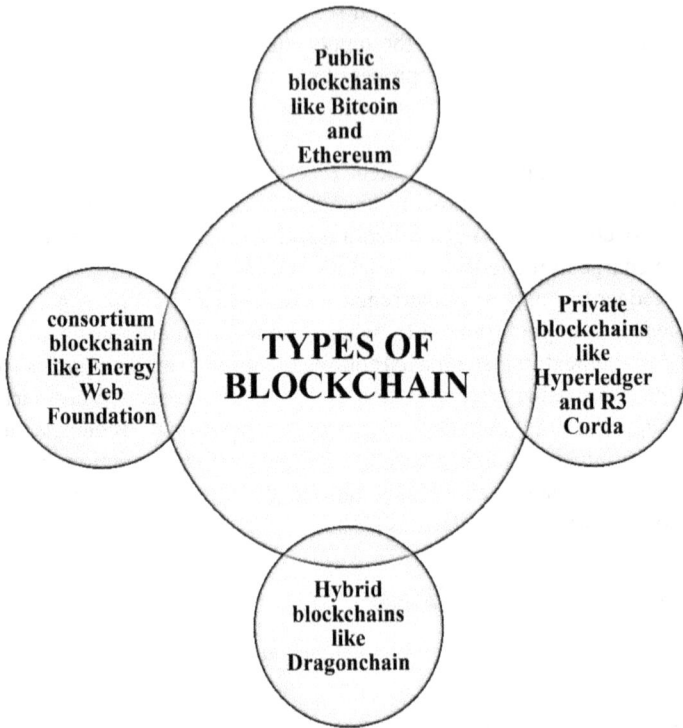

FIGURE 19.2 Types of blockchain.

governed by a collective phenomenon rather than by content alone. This strategy has the same focus points as a private chain, which may be perceived as a specific blockchain's sub-order as opposed to an external chain [2].

19.5.4 HYBRID BLOCKCHAIN

Dragonchain includes a novel spot inside the blockchain condition in that it's a cream blockchain. This suggests it joins the assurance focal points of approval and private blockchain with the security and straightforwardness preferences of an open blockchain. That gives associations enormous versatility to pick what data they have to make open and clear and what data they have to keep hidden [2].

19.6 BENEFITS OF ADOPTING INDUSTRY 4.0

19.6.1 IMPROVED PRODUCTIVITY

In simple terms, innovations from Industry 4.0 allow you to do more with less. In other words, by allocating your resources more cost-effectively and efficiently, you can generate more and faster. Because of improved system control and automated/ semi-automated decision-making, the production lines would also experience less

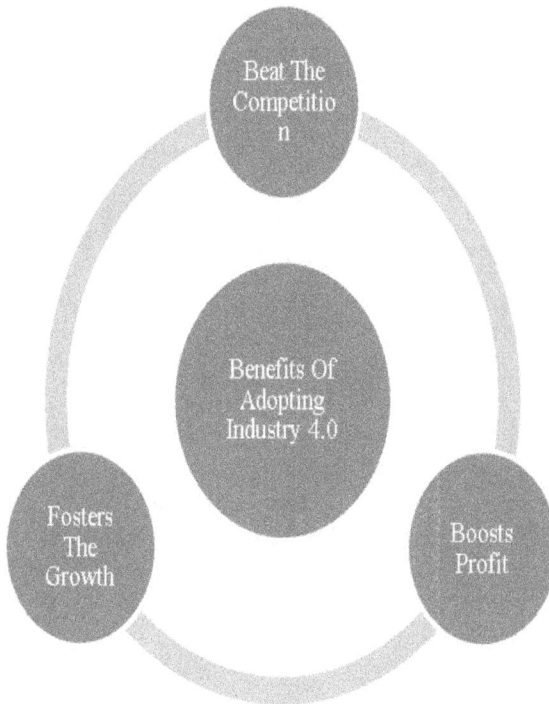

FIGURE 19.3 Benefits of adopting Industry 4.0 [2].

downtime. In truth, as your facility moves closer to being an Industry 4.0 smart plant, the general overall equipment effectiveness (OEE) will increase [43].

19.6.2 Improved Efficiency

As a result of the technologies associated with Industry 4.0, multiple areas of your production line can become more productive. Less computer downtime as well as the ability to make more goods and make them faster are some of these efficiencies mentioned above. Faster batch changeovers, automatic track-and-trace procedures and automated reporting are other examples of improved performance as shown in Figure 19.3. As well as business decision-making and more, new product introductions (NPIs) are also becoming more successful [44].

19.6.3 Increased Knowledge Sharing

Conventional manufacturing plants run in silos. Individual facilities, including individual machines inside a facility, are silos. This results in insufficient cooperation or sharing of information. Industry 4.0 technologies allow connectivity between your manufacturing lines, business processes and departments regardless of location, time zone, network or any other factor. This allows information gained by a sensor on a computer in one plant to be disseminated in the company, for example [43].

19.7 PRINCIPLES AND GOALS OF INDUSTRY 4.0

- Interconnectivity
- Data Transparency
- Specialized help
- Decentralized Decisions

19.7.1 INTERCONNECTION

This refers to the capacity of machinery and related components to communicate and interact through the Internet with individuals as shown in Figure 19.4.

19.7.2 DATA TRANSPARENCY

This theory requires that, through the configuration of digital data into sensor data, information systems should be able to construct virtual copies of the physical world. Raw sensor data must be aggregated with compatible background data in order for this to be accomplished [43].

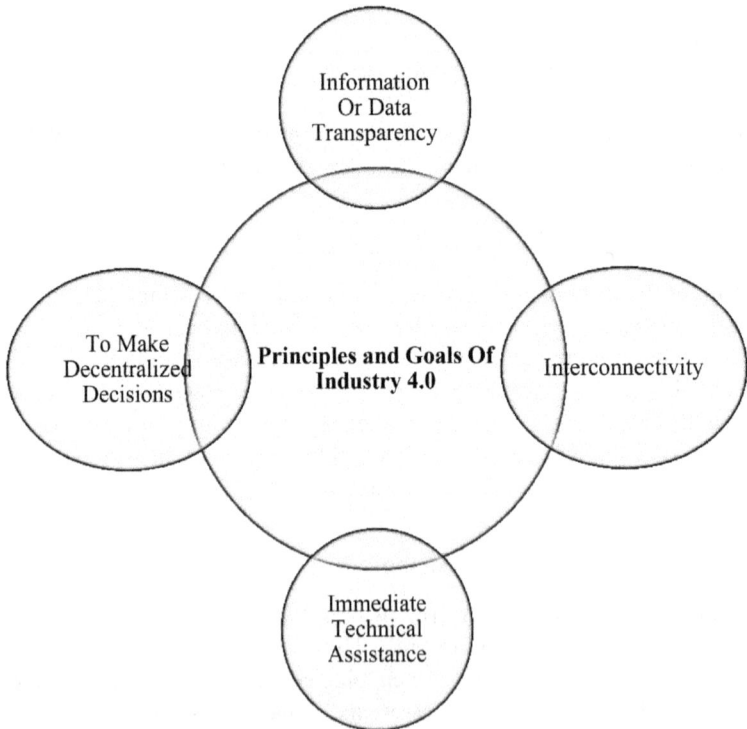

FIGURE 19.4 Principles and goals of Industry 4.0 [43].

19.7.3 SPECIALIZED HELP

This concerns the capacity of systems to help people through systematic aggregation and visualization of data for better decision-making and rapid problem solving. Technical assistance also focuses on the capacity of cyber-enabled systems to support human resources physically by performing different activities that are deemed time-consuming, dangerous and exhausting to individuals [43].

19.7.4 DECENTRALIZED DECISIONS

This concept applies to the ability of cyber-enabled systems to make decisions and execute their dedicated functions independently. Only in the case of intervention or conflict with the intended objectives, which may cause those activities to be performed at other levels, may this be modified. In versatile manufacturing environments, as in industries dealing with mass production, the definition is characterized by greater customization of goods [3].

19.8 BLOCKCHAIN IN CYBERSECURITY

Blockchain is a booming technology that boosts business and industry verticals. Thus, blockchain may theoretically improve cyber-defence through consensus structures that deter malicious practices and identify data manipulation based on its characteristics such as data protection, suitability, organizational stability, immutability, and transparency. Hackers often rely on edge machines such as routers and switches to reach the overall network. In recent days, hackers also find it easier to attack through devices such as smart thermostats, CCTVs and doorbells too [37].

Thus blockchain technology can be used to prevent attacks on those devices. Blockchain technology can provide those IoT devices with the "smarts" to make security decisions without depending on a central authority or administrator. The costs of detecting bad data and rooting out frauds are significant regardless of the industry. Algorithms that are used to fight fraud are largely effective in detecting inconsistencies but are more costly. Data that is filtered and stored through a decentralized blockchain technology tends to be more trustworthy, more securable, and 100% correct as it undergoes a lot of verification process. Thus, the threshold of data veracity is higher when using blockchain technology [41].

19.9 BLOCKCHAIN-BASED MANUFACTURING MODEL

The effect of blockchain is becoming more prevalent as factories across the world become increasingly interconnected. The factory of the future, including equipment manufacturers and distribution firms, covers a whole network of devices, components, goods and value chain participants. Today, more than ever before, producers face the task of exchanging information safely inside and outside factory walls [7]. A manufacturer must perform a systematic evaluation to figure out the best location for blockchain, which starts with defining the current business challenges and potential needs of the company. Subsequently, it will then analyze how it uses the

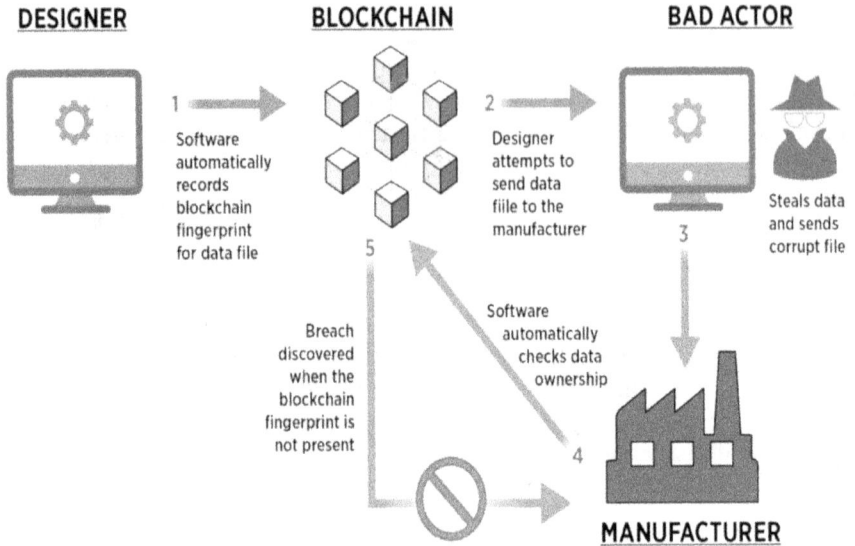

FIGURE 19.5   Blockchain-based manufacturing model.

technology to mitigate the pressure points of the factory and satisfy its needs. Equipped with a good understanding of the opportunities and challenges it faces, the manufacturer may then select the most suitable alternative from the available technology solutions. Essentially, as the name implies, blockchain is a chain of blocks. There is digital information (the block) stored in a public database instead of a physical chain. It is applied to the blockchain when a block stores new data. In order to provide substantial value for industrial businesses, blockchain-powered applications will seamlessly aggregate all of the knowledge and help unlock the full potential of other emerging technologies, such as virtual reality, IoT and 3D printing [36]. Each square of information incorporates a timestamp, cryptographic hash and the past square's hash, which makes a chain of exchanges all connected to a unique square as shown in Figure 19.5. It is clear that the ability of blockchain-powered technologies to build value by enabling companies to solve difficult issues is clear. Blockchains can improve transparency across supply chains, monitor key staff identity and credentials, and create more streamlined functionality of audit and enforcement [42].

19.10   BLOCKCHAIN IMPACT ON SUPPLY CHAIN PROCESS

Many forward-thinking supply chains have already identified and are implementing the benefits of the blockchain into their processes. There is the possibility of a universal operating system for all supply chains, linking the world in a fully transparent and practically incorruptible way, should technology become widespread. The supply chain is a system of partnerships, persons, activities, records and materials that are engaged in bringing an element or administration from a vendor to a customer

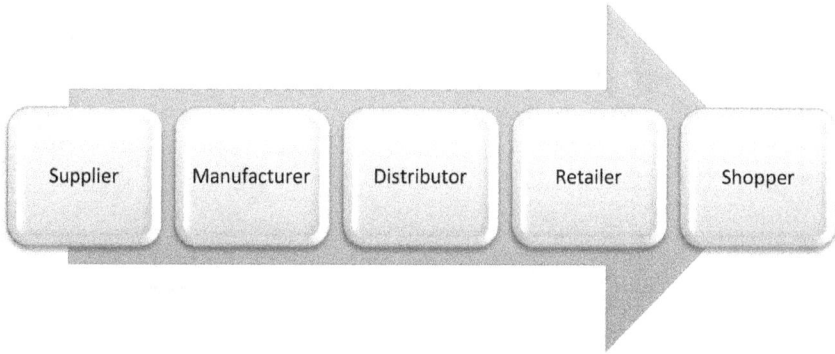

FIGURE 19.6   Flow of supply chain.

(see Figure 19.6). It's hard for clients or purchasers to know the estimation of items because there is a critical absence of straightforwardness in our present framework. Likewise, it's incredibly hard to examine supply chains when there is suspicion of illicit or unscrupulous practices [36]. Accordingly, time delays, human mistakes and the related expenses that influence the supply chains today could be decreased.

19.11   FOG COMPUTING MODEL FOR INDUSTRIAL IoT WITH EVALUATION

In the near future, fog computing offers a hierarchical distributed architecture that facilitates the convergence of technical components and services, such as smart cities, smart grid networks, connected vehicles and smart homes. The deployment of fog systemic networks at the edge is important for the protection of the smart future, to perform intelligent computations, to analyze big data for better predictive analysis, and to detect odd and dangerous incidents. In this evolving process of the fog computing paradigm, the concept of fog computing should therefore be explained. Definitions offer an expanded view of fog computing, but the particular resemblance to the cloud cannot be pointed out. All functionalities of fog with pre-existing structure need to be described and contrasted by the basic definition, so here comes our definition. The fog computing platform focused on a blockchain is suitable for potential industrial IoT as seen in Figure 19.7. The essential requirement is a fast and reliable information exchange focused on low throughput and low latency [38]. The main job is to create a private key on the edge nodes and to register it with the fog node/server to explain hyper ledger and fog computing operation. Cloud system as protection and network solidity, must be assured. Such layers are designed to have service and network features: A consent algorithm is used to connect a block to the chain by the key component. And the more complicated the algorithm is, the higher the consistency, thanks to the ordered form. But the cycle slows down the pace. With the aid of a fog node to preserve speed, the edge collects and processes signals from sensors, controls, and other devices in an urban area network linked via wifi and ethernet in real time [38].

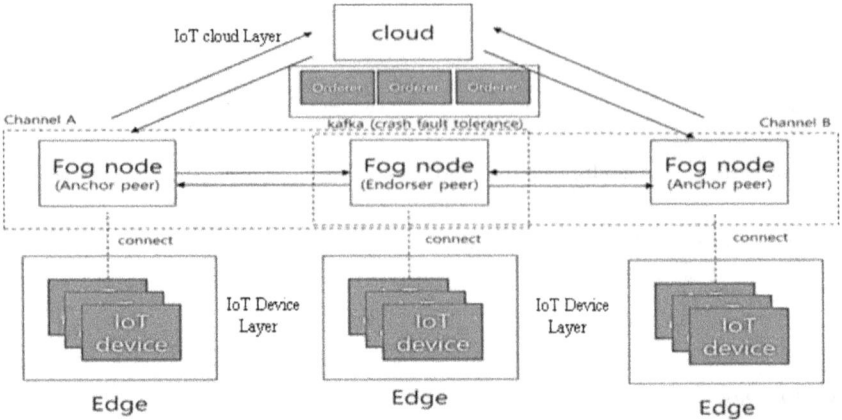

FIGURE 19.7   Blockchain-based Industrial IoT fog computing architecture.

19.12   OPEN ISSUES OR CHALLENGES

Important open issues in the adoption of blockchain in the IoT scenario (shown in Figure 19.8) are discussed below.

19.12.1   SCALABILITY

Adaptability problems with the latest blockchain developments are a major impediment to the widespread identification and organization of blockchains. (For example, VISA may process about 2000 exchanges per second [10], while Bitcoin may process up to 7 exchanges per second [11].) This prompts gigantic overhead power, as opposed to traditional databases. Explicitly in the IIoT case, the problem is exacerbated with the number of sensor hubs increasing and the calculation of knowledge

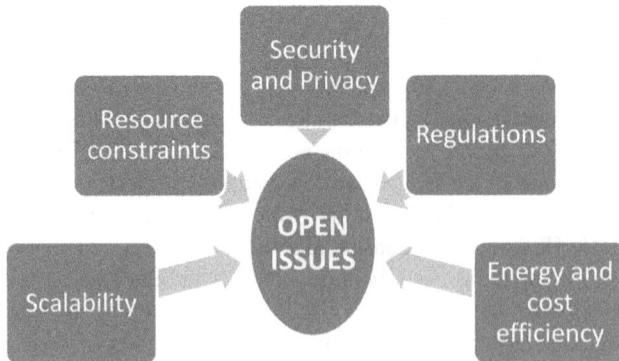

FIGURE 19.8   Open issues or challenges.

generated. Although late lightweight versions such as Ethereum have been created [12], blockchain organizations stay out of extension for asset and capacity compelled IoT gadgets [29].

19.12.2 SECURITY AND PRIVACY

Today, IoT systems are disposed of at different defence ambushes such as replay attacks [13] and tuning in attacks [14]. Restricted by such security challenges, blockchains are taking security flaws into their course of operation. Part of the documented protection flaws in blockchain's development are bugs in the clever agreement programme [15], communication catch [16], etc. Safety spillage is another problem in blockchains, as seen in Reference [17]. (For example, on the blockchain there is a chance of future security breach with trade data being taken care of [18].)

19.12.3 ENERGY AND COST-EFFECTIVENESS

Versatility problems in the existing blockchain progressions are a crucial impediment in blockchains' broad confirmation and affiliation. (For example, VISA may process about 2000 trades every second [10], yet Bitcoin will process up to 7 trades in a single second [11. Though late lightweight versions such as Ethereum have been made [12], blockchain associations avoid augmentation for resource and limit constrained IoT devices [29].

19.12.4 RESOURCE LIMITATIONS

Current blockchain executions require stable system associations. IIoT gadgets don't generally ensure stable system associations; in this way the execution of blockchain innovation is a test right now [25]. The level of reorganization conceivable in the current usage of blockchain coordinated IoT systems is additionally constrained because of the asset constraints of IoT gadgets and systems [3].

19.13 CONCLUSION

It is expected that the use of blockchain technology in IoT-enabled industries will expand and support a wide range of industrial sectors. Tamper-proof and resilient record-keeping methods are said to be the game changers that this technology provides. There have been major investments in blockchain, as seen in numerous case studies reviewed in this paper. The conventional method of centralized data sharing has been deemed safer compared to a decentralized approach, despite scalability concerns, and has been in use in the industry for so long now. With blockchain-promising technologies to fix both security and credibility problems, this is expected to change now. In this chapter, to provide an abstract measure of adoption in practice, we looked at different commercial implementations of blockchain in Industry 4.0 and IIoT. And the problems faced by each of these industries for the implementation of blockchain were further addressed.

REFERENCES

1. Christidis K. and Devetsikiotis M. (2016). "Blockchains and smart contracts for the internet of things," *IEEE Access*, vol. 4, pp. 2292–2303.
2. Fernández-Caramés T. and Fraga-Lamas P. (2018) "A review on the use of Blockchain for the internet of things," *IEEE Access*, vol. 6, pp. 32 979–33 001.
3. Ferrag M.A., Derdour M., Mukherjee M., Derhab A., Maglaras L. and Janicke H. (2018). "Blockchain technologies for the internet of things: Research issues and challenges." *IEEE Internet of Things Journal*, vol. 6, no. 2, pp. 2188–2204.
4. Salman T., Zolanvari M. and Samaka M. (2016) "Security services using Blockchains: A state of the art survey," *IEEE Communications Surveys & Tutorials*, vol. 21, no. 1, pp. 858–880.
5. Shen C. and Pena-Mora F. (2018). Blockchain for cities—a systematic literature review. *IEEE Access*, vol. 6, pp. 76787–76819.
6. Fraga-Lamas P. and Fernández-Caramés T.M. (2019). "A review on Blockchain technologies for an advanced and cyber-resilient automotive industry," *IEEE Access*, vol. 7, pp. 17 578–17 598.
7. Lu H., Huang K., Azimi M. and Guo L. (2019) "Blockchain technology in the oil and gas industry: A review of applications, opportunities, challenges, and risks," *IEEE Access*, vol. 7, pp. 41 426–41 444.
8. Yang R., Yu F.R., Si P., Yang Z. and Zhang Y. (2019). Integrated blockchain and edge computing systems: A survey, some research issues and challenges. *IEEE Communications Surveys & Tutorials*, vol. 21, no. 2, pp. 1508–1532.
9. Xie J., Tang H., Huang T., Yu F.R., Xie R., Liu J. and Liu Y. (2019). A survey of blockchain technology applied to smart cities: Research issues and challenges. *IEEE Communications Surveys & Tutorials*, vol. 21, no. 3, pp. 2794–2830..
10. "Bitcoin and Ethereum vs Visa and PayPal Transactions per second" (2019). http://www.altcointoday.com/bitcoin-ethereum-vs-visa-paypaltransactions-per-second/, [Online; accessed 12-April-2019].
11. Croman K., Decker C., Eyal I., Gencer A.I., Juels A., Kosba, A. and Saxena P. (2016). "On scaling decentralized Blockchains," in International Conference on Financial Cryptography and Data Security. Springer, pp. 106–125.
12. "Ethereum Light Client Protocol." (2019). https://github.com/ethereum/wiki/wiki/Light-client-protocol, [Online; accessed].
13. Lin J., Yu W., Zhang N. and Zhao W. (2017). "A survey on internet of things: Architecture, enabling technologies, security, and privacy, and applications," *IEEE Internet of Things Journal*, vol. 4, no. 5, pp. 1125–1142.
14. Wang H. and Zhao Q. (2016). "An analytical study on eavesdropping attacks in wireless nets of things," *Mobile Information Systems*, vol. 206, pp. 1–11.
15. Li X., Jiang P., Chen T., Luo X. and Wen Q. (2020). A survey on the security of blockchain systems. *Future Generation Computer Systems*, vol. 107, pp. 841–853.
16. Apostolaki M., Zohar A. and Vanbever L. (2017). "Hijacking bitcoin: Routing attacks on cryptocurrencies," in 2017 IEEE Symposium on Security and Privacy (SP). IEEE, pp. 375–392.
17. Conti M., Kumar E.S., Lal C. and Ruj S. (2019). "A survey on security and privacy issues of bitcoin," *IEEE Communications Surveys & Tutorials*, vol. 20, no. 4, pp. 3416–3452.
18. Dorri A., Kanhere S. S. and Jurdak R. (2019). "Mof-bc: A memory-optimized and flexible Blockchain for large scale networks," *Future Generation Computer Systems*, vol. 92, pp. 357–373.

19. Mendling J., Weber I., Aalst V, Brooke J. V. and Dustdar S. (2016). "Blockchains for business process management-challenges and opportunities," *ACM Transactions on Management Information Systems(TMIS)*, vol. 9, no. 1, p. 4.

20. Conoscenti M., Vetro A. and Martin J. C. De. (2016). "Blockchain for the internet of things: A systematic literature review," in 2016 IEEE/ACS 13th International Conference of Computer Systems and Applications (AICCSA). IEEE, 2016, pp. 1–6.

21. Bozic N., Pujolle G. and Secci S. (2016). "A tutorial on blockchain and applications to secure network control-planes," in *2016 3rd Smart Cloud Networks & Systems (SCNS)*. IEEE, pp. 1–8.

22. "Telehash," http://telehash.org, [Online; accessed 12-April-2019].

23. Zhang Q., Zhu J. and Ding Q. (2019, October). OBBC: A Blockchain-Based Data Sharing Scheme for Open Banking. In *CCF China Blockchain Conference* (pp. 1–16). Springer, Singapore.

24. Zou J., Ye B., Qu L., Wang Y., Orgun M.A. and Li L. (2018). A proof-of-trust consensus protocol for enhancing accountability in crowdsourcing services. *IEEE Transactions on Services Computing*, vol. 12, no. 3, pp. 429–445.

25. Biswas K. and Muthukkumarasamy V. (2016). "Securing smart cities using Blockchain technology," *IEEE*, pp. 1392–1393.

26. Rana A. and Salau A. (2019). *Recent Trends in IoT, Its Requisition with IoT Built Engineering: A Review.* Springer, Singapore, ISBN978-981-13-2553-3.

27. Sharma S. and Kumar A. (2019). "Enhanced energy-efficient heterogeneous routing protocols in WSNs for IoT application," *IJEAT*, ISSN: 2249-8958 vol. 9 ISSUE-1.

28. Kumar K., Gupta S. and Rana A. (2018). "Wireless sensor networks: A review on 'Challenges and Opportunities for the Future world-LTE,'" *(AJCS)*, vol. 1, no. 2, pp. 30–34, ISSN: 2456-6616.

29. Sharma S., Kumar A., Krishna R. and Dhawan S. (2019). "Review on Artificial Intelligence with the Internet of Things – Problems, Challenges and Opportunities," 2nd International Conference on Power Energy, Environment and Intelligent Control (PEEIC), Greater Noida, India, pp. 383–387.

30. Rana A.K. and Sharma S. (2020). "Contiki Cooja Security Solution (CCSS) with IPv6 Routing Protocol for Low-Power and Lossy Networks (RPL) in internet of things applications," in *Mobile Radio Communications and 5G Networks* (pp. 251–259). Springer, Singapore.

31. Rana A.K. and Sharma S. (2020). "Industry 4.0 manufacturing based on IoT, cloud computing, and big data: Manufacturing purpose scenario," in *Advances in Communication and Computational Technology* (pp. 1109–1119). Springer, Singapore..

32. Jensen I.J., Selvaraj D.F., and Ranganathan P. (2019). "Blockchain Technology for Networked Swarms of Unmanned Aerial Vehicles (UAVs)," in 2019 IEEE 20th International Symposium on "A World of Wireless, Mobile and Multimedia Networks" (WoWMoM), pp. 1–7. IEEE.

33. Vukolić M. (2015). "The quest for scalable blockchain fabric: Proof-of-work vs. BFT replication." International workshop on open problems in network security, pp. 112–125. Springer, Cham.

34. Premkumar A. (2020). "Application of Blockchain and IoT towards Pharmaceutical Industry," 2020 6th International Conference on Advanced Computing and Communication Systems (ICACCS), Coimbatore, India, pp. 729–733.

35. Khrais L.T. (2020). "The Combination of IoT-Sensors in Appliances and Block-Chain Technology in Smart Cities Energy Solutions," 2020 6th International Conference on Advanced Computing and Communication Systems (ICACCS), Coimbatore, India, pp. 1373–1378.

36. Malibari N.A. (2020). "A Survey on Blockchain-based Applications in Education," 2020 7th International Conference on Computing for Sustainable Global Development (INDIACom), New Delhi, India, pp. 266–270.

37. Anwer M., saad A. and Ashfaque A. (2020). "Security of IoT Using Block chain: A Review," 2020 International Conference on Information Science and Communication Technology (ICISCT), Karachi, Pakistan, pp. 1–5.

38. Jang S., Guejong J., Jeong J. and Sangmin B. (2020). "Fog Computing Architecture Based Blockchain for Industrial IoT," *ICCS*, Faro, Portugal, June 12–14.

39. Okegbile S.D., Maharaj B.T. and Alfa A.S. (2020). Interference characterization in underlay cognitive networks with intra-network and inter-network dependence. *IEEE Transactions on Mobile Computing*, 1536–1233, pp. 1–5.

40. Wu Q., He K. and Chen X. (2020). "Personalized federated learning for intelligent IoT applications: A cloud-edge based framework," *IEEE Computer Graphics and Applications vol. 1, pp. 35–44.*

41. Oksiiuk O. and Dmyrieva I. (2020). "Security and Privacy Issues of Blockchain Technology," 2020 IEEE 15th International Conference on Advanced Trends in Radioelectronics, Telecommunications and Computer Engineering (TCSET), Lviv-Slavske, Ukraine, pp. 1–5.

42. Leng J. (2020). "ManuChain: Combining permissioned blockchain with a holistic optimization model as bi-level intelligence for smart manufacturing," *IEEE Transactions on Systems, Man, and Cybernetics: Systems*, vol. 50, no. 1, pp. 182–192.

43. Habib M.K. and Chimsom C. (2020). "Industry 4.0: Sustainability and Design Principles," 20th International Conference on Research and Education in Mechatronics (REM), Wels, Austria, pp. 1–8.

44. Adamu A.A., Wang D., Salau A.O. and Ajayi O. (2020). An integrated IoT system pathway for smart cities. *International Journal on Emerging Technologies*, vol. 11, no. 1, pp. 01–09.

20 Decentralized Blockchain Technology for Security Enhancement of Internet of Things

Pranav Ratta and Amanpreet Kaur
Chandigarh University

Sparsh Sharma
Baba Ghulam Shah Badshah University

CONTENTS

20.1 Introduction 291
    20.1.1 IoT Architecture...292
20.2 How IoT is Different from Traditional Networks 293
20.3 IoT Enabling Technologies 294
20.4 Cloud Computing vs. Fog Computing and How the Integration of
    Blockchain in IoT Is More Secure than Cloud-based IoT............................295
20.5 Blockchain Technology 296
    20.5.1 Building a Blockchain 296
    20.5.2 Basic Terminology of Blockchain 297
20.6 How Blockchain Can Help in IoT...298
20.7 Factors Affecting IoT Research Technologies...299
    20.7.1 Identification of Things..299
    20.7.2 The Architecture of IoT 300
    20.7.3 Communication Between Devices...300
    20.7.4 Network Technology...301
    20.7.5 Software Service and Algorithms 301
    20.7.6 Hardware 302
20.8 Conclusion 302
References...303

20.1 INTRODUCTION

The internet is used as a global platform for communication, dialogue, and coordination of smart objects, devices, and machines in the new world. The main aim of modern technologies in which smart objects interact with each other is to establish new businesses according to the opportunities of demand in the market [1]. Hence, IoT is built on three pillars, related to the ability of smart objects:

- *Identification*: Anything which is added in the network must be identifiable, which means the device that adds must have an IP address.
- *Communication*: There must be the ability to communicate.
- *Interaction*: The devices which connect to the internet must interact with each other.

20.1.1 IoT Architecture

IoT is the combination of heterogeneous objects, which is not easy because each IoT device has its own architecture [2]. There is no single architecture for IoT which is agreed universally. The basic architecture of IoT has five layers as shown in Figure 20.1.

1. **Perception layer:** This layer contains the sensor which senses or gather information about the environment.

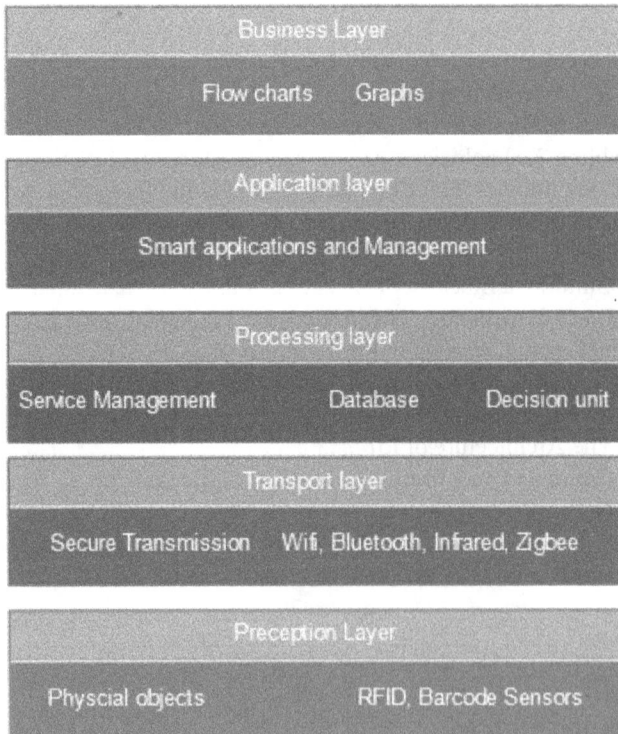

FIGURE 20.1   Layered IoT architecture.

2. **Transport layer:** The data which is gathered in the perception layer is transferred between different layers through a network such as wireless, Bluetooth, radio-frequency identification (RFID), etc.
3. **Processing layer:** A huge amount of data is transferred from the transport layer and maintained in the processing layer. Data can be stored and analyzed in this layer.
4. **Application layer:** This layer is responsible for delivering application-specific service to the user.
5. **Business layer:** Applications, business and profit models, and user privacy are managed in the IoT system. This layer manages the whole IoT system.

20.2 HOW IoT IS DIFFERENT FROM TRADITIONAL NETWORKS

In the field of telecommunication technology, IoT has made huge changes, and a lot of traditional ideas have been broken down by accepting the concept of IoT. The main difference between traditional networks and IoT is the connection between smart objects. The IoT environment is based on the idea of smart objects communicating. The difference between IoT and the internet is shown in tabular form in Table 20.1. The IoT network is the combination of the internet, a wireless sensor network and the smart objects in an intelligent system. The following three categories come under the IoT Infrastructure [3]:

1. *Information items*: The IoT network includes items which have the ability to sense the objects and self-learn with the experience.
2. *Independent networks*: These networks are not dependent on another system and include the following features: self-configuration of the system, self-protection, self-optimization, self-healing.
3. *Intelligent applications*: Their main feature is intelligence. Applications must exhibit the smart features of controlling, processing and exchanging data.

TABLE 20.1
Comparison between Traditional Network and IoT

Parameters	Internet	IoT
Rules for communication	TCP/IP protocol	Lightweight communication protocols
Identification of objects	Can't be identified	Objects can be identified
Functioning	Fixed	Dynamically
Area coverage	Covered wide area	Cover wide area
Networking time	Unlimited	Timing synchronization
Network approach	Determine backbone	Determine backbone
Design of network	Fixed objects	WSN, intelligent environment, smart things
Type of nodes	Active	Active and passive

20.3 IoT ENABLING TECHNOLOGIES

IoT enabling technologies means technologies that differentiate IoT from the usual internet. Following are the factors that categorize enabling technologies into two parts:

1. The basic need is that the technologies must have the feature to make things intelligent. Intelligent means sensing the data, processing the data and taking control action.
2. The second technology includes work on the fundamentals of security and privacy.

There are many enabling technologies in IoT, such as sensors, RFID, M2M, semantic search, cloud computing, semantic data integration, embedded systems, and many more. The main categories of IoT enabling technologies are given in Figure 20.2.

a. **Sensors:** Sensors have the capability to detect the changes and the occurrence of events in a specific quantity, communicate the event or change data to the cloud, detect the external environmental signals, and receive the information back from the cloud. In future research of IoT, sensing technology will be a hot spot [4]. It also has the ability to communicate with other smart objects. Since 2012, sensors have become smaller, and this has affected the popularity of IoT in the market. Now sensors are so small that they can be embedded in many places, such as clothes, mobile phones, watches and many other smart objects.

b. **Wireless Sensor Networks:** Wireless Sensor Networks (WSNs) comprise distributed devices and sensors which have the capability to monitor the environment and the physical conditions. It is a combination of all these elements: end nodes, routers and coordinators. End nodes consist of several sensors that also act as routers. The function of the routers is to transfer data packets from end nodes to the coordinator. The function of the

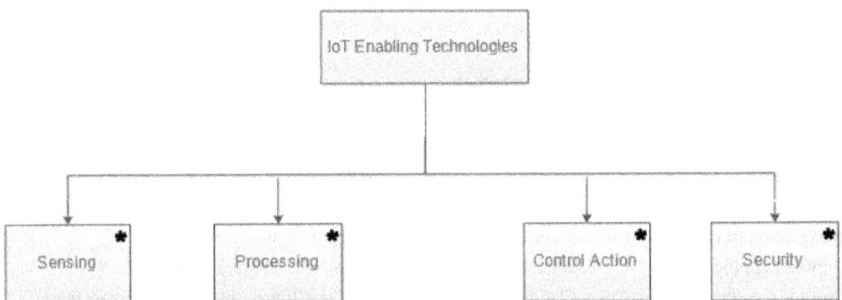

FIGURE 20.2 IoT enabling technologies.

coordinator is to collect data from all nodes, and it is the responsibility of the coordinator to connect the WSN to the internet [5]. Take the example of a weather monitoring system, which uses the WSN. In this system, nodes collect the temperature, humidity and other data, which get aggregated and appropriate action is then then performed on the gathered data after proper analysis.

c. **Communication protocols:** Communication technologies have improved and allowed sensors embedded in smart objects to send and receive data over the cloud for collecting, storing and analyzing data. To relay data, protocols to allow IoT sensors include wireless technologies such as RFID, WiFi, Bluetooth and ZigBee, as well as satellite connections and mobile networks using GSM, 3G, LTE or WiMAX [6].

d. **Embedded systems:** Embedded systems are the part of IoT. Providing internet access to an embedded device converts it into an IoT device. The basic definition of IoT is smart devices or embedded devices that are connected or controlled through the internet. There are many features of an embedded system that are important for IoT, such as the following:

1. real-time computing
2. low power consumption
3. low maintenance
4. high availability

The embedded system hardware can be of any type, such as an integrated circuit (IC). The IoT uses many embedded boards, such as Raspberry Pi, Arduino, Node MCU, Intel Galileo, etc. All these boards are small and smart. The most-used boards among these are Raspberry Pi and Arduino.

20.4 CLOUD COMPUTING VS. FOG COMPUTING AND HOW THE INTEGRATION OF BLOCKCHAIN IN IoT IS MORE SECURE THAN CLOUD-BASED IoT

The cloud is used for storing the data that is collected by IoT devices. It also has the capability of analyzing the data and extracting some useful information from it [7]. Fog computing is much the same as cloud computing. The major difference is the response time. If the data from the device goes to the cloud, the cloud first stores the data, and after that it calculates information from that data and sends back the response. However, if the data is sent to the fog, the fog immediately fetches the information and gives back a quick response [8].

Blockchain is the concept of decentralization in which data is not stored in a centralized place. Cloud computing is based on a central server storing the data. Blockchain is integrated with IoT, which is a more secure network than the cloud with IoT. Table 20.2 discusses parameters such as security, privacy, flexibility, cost, etc., and explains how blockchain is more beneficial than the cloud.

TABLE 20.2
Comparison between Cloud and Blockchain

Parameters	Cloud with IoT	Blockchain with IoT
Privacy	Privacy of data is very important, and cloud computing has many solutions for this, but after having these solutions it can't fulfil the privacy criteria as data can leak.	The concept of immutability helps in the privacy of data as no one can alter it.
Security	The cloud seems to be secure, but many insecure issues are found in IoT using the cloud.	Blockchain is very secure compared to the cloud because the attack happens only when more than 51% of the users allow the change.
Latency	Already discussed above that the cloud requires more response time.	Latency is also a challenge in blockchain as mining is a time-intensive process.
Cost	There are many factors that increase the cost: bandwidth, updating hardware/software, etc.	Blockchain's execution cost is high, but the overall cost in terms of bandwidth and updating blockchain is more feasible.
Flexibility	The centralized system is easier.	It is difficult to deal with decentralization.
Losses and Risks	There are more attacks in this process.	There is no history of attacks in blockchain.
Payment	The method of payment is very limited and often used.	There are many payment methods in blockchain, and the popular example of digital currency is Bitcoin.

20.5 BLOCKCHAIN TECHNOLOGY

A blockchain is a combination of blocks that store data publicly and in sequential order. The entire information is stored and transferred among the other nodes using cryptography. Cryptography ensures that the user's privacy is maintained and data cannot be altered. Information in the blockchain is not controlled by a central authority. Unlike modern financial institutions, nobody controls the data within a blockchain. A typical blockchain network is a public blockchain in which, as long as the user has access to the network, they have access to the data within the blockchain [7]. Everyone in the network has a copy of the blockchain, which is used for ensuring that the data remains untampered with.

20.5.1 Building a Blockchain

A block is based on three things: data, a hash of a particular block and the previous block hash, which is shown in Figure 20.3. Data is what that block contains, and it is dependent on the type of blockchain. A Bitcoin cryptocurrency transaction is based on a technology called blockchain, and blockchain holds the information about the

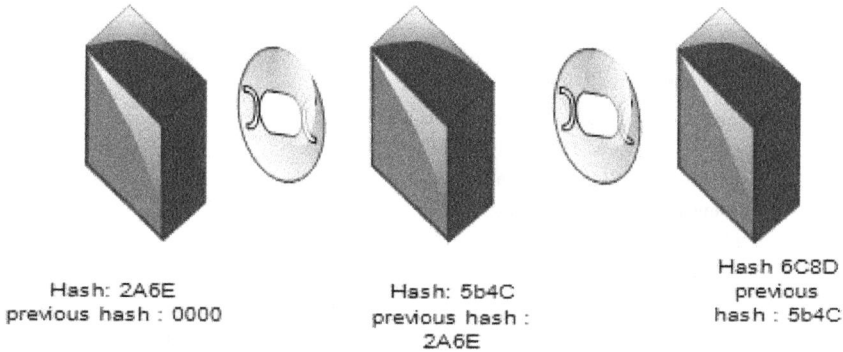

Hash: 2A6E
previous hash : 0000

Hash: 5b4C
previous hash :
2A6E

Hash 6C8D
previous
hash : 5b4C

FIGURE 20.3   Blocks connected in blockchain.

sender, receiver and the amount of currency exchanged between them. A block also contains a hash value in hexadecimal which is generated by hash cryptography SHA 256. SHA 256 means the secure hash algorithm and 256 bits mean 64 hexadecimal characters. There are five requirements of a hash algorithm: one-way, deterministic, fast computation, the avalanche effect, and starting with a collision. Once a block is created, the first thing is to calculate the hash of a block. If there is some alteration inside the block, it will cause the hash to change. Hash is the only thing that tells the user about the change in that particular block. The third element of the block is used to make the chain as it contains the unique hash value of the previous block. The first block is known as the Genesis block. The previous hash attribute of first block is zero. Blocks are cryptographically linked together.

The basic structure of a blockchain with blocks linked together is shown in Figure 20.3.

20.5.2   BASIC TERMINOLOGY OF BLOCKCHAIN

a. **Hyper ledger:** Ledger is a kind of database where transactions that have been confirmed are recorded. Blockchain platforms do not use a centralized database. Instead, each node has a copy of the ledger [9]. Cryptocurrency such as Bitcoin stores only balance information in the distributed ledger. Blockchain platforms such as Ethereum can store any kind of information, such as identity information, patient information, real estate information, etc. in the distributed ledger.

b. **Miners:** Miners are nodes that collect all the transactions which people send to each other over the network, and it's the job of the miner to transfer only valid transactions. Each miner takes a number of these collected transactions and puts them in a newly formed block. These transactions are numbered tx0, tx1, tx2, …, txn. The first transaction, tx0, is called the coinbase transaction.

How mining Works: The objective of the miner is to find some nonce that generates some random number that meets some criteria. Hash should

TABLE 20.3
Structure of a Block

Block #5	This tells the index of a block in a blockchain.
Nonce 10	Used in mining for meeting the condition of hash calculation.
Data	Contains the actual data.
Previous Hash	Hash of the previous block.
Hash	Hash of the current block.

be less than the specific target. Any hash above the target is valid. The miner is going to get information about three fields in a block: block number, data and previous hash. The nonce is going to try all possible values in order to produce a hash that meets the criteria. The structure of a block is given in Table 20.3, which contains block number, the hash of the previous block, the hash of the block, nonce value and data content in that block. The hash number is the unique identifier of a block, and every block has particular criteria for hash numbers. Changing the value of nonce of a block changes the value of hash. The structure of a block is given below, and nonce value is a random value and it is added before the hashed value of current block.

c. **Ethereum:** Ethereum came up with the idea that blockchain could be used for no purposes beyond financial transactions. It maximizes the usage of the blockchain application. Ethereum is an open-source distribution platform that allows decentralized applications to be built on it with the help of smart contracts functionality. Ether is a cryptocurrency token of the Ethereum network which is used to pay for transaction fees [9].

d. **Smart contracts:** Smart contracts are immutable pieces of code that run on the Ethereum network to perform a certain transaction or task, and when these contracts are running, or the blockchain acts like a self-operating programme, execution happens automatically when specific conditions are met. Codes in this smart contract are written in a special contract language called Solidity [10]. Solidity language is preferable to the other languages because the cost is low compared to others. The syntax of Solidity is the same as the JavaScript.

20.6   HOW BLOCKCHAIN CAN HELP IN IoT

Blockchain is the concept of decentralization; it helps IoT in many cases when blockchain is integrated with IoT. The devices which are connected in a network help the IoT to track all devices in the network. The installation and managing cost of IoT network is high when compared with the cost when these both technologies are combined. There are two main challenges in IoT while transferring data from one device to another: security and privacy. Blockchain achieves these two features – security and privacy – as it transfers the data by using cryptographic algorithms. Blockchain, when integrating with IoT, also protects the IoT network from various attacks [11]. For example, it prevents the IoT network from the man-in-middle attack because blockchain doesn't follow the technique of single-thread communication [12]. The smart contract is an agreement that can be created and executed when a particular condition will be satisfied. For example, sensors are used to fetch data, and that data is sent to the blockchain. With the help of data that is fetched, smart contracts will execute and do the desired work [13].

20.7 FACTORS AFFECTING IoT RESEARCH TECHNOLOGIES

The main effects of the key technologies are the factors described below. There are still many research issues that require attention: identification of things, architecture of IoT, communication between the devices, network technology and discovery, software in IoT and required hardware, as shown in Figure 20.4.

20.7.1 IDENTIFICATION OF THINGS

Identification technology is a globally unique identifier of the object [4]. In the identification domain, there are still many topics in which further research is needed:

- Global ID schemes
- Identity management
- Encoding and encryption of identity

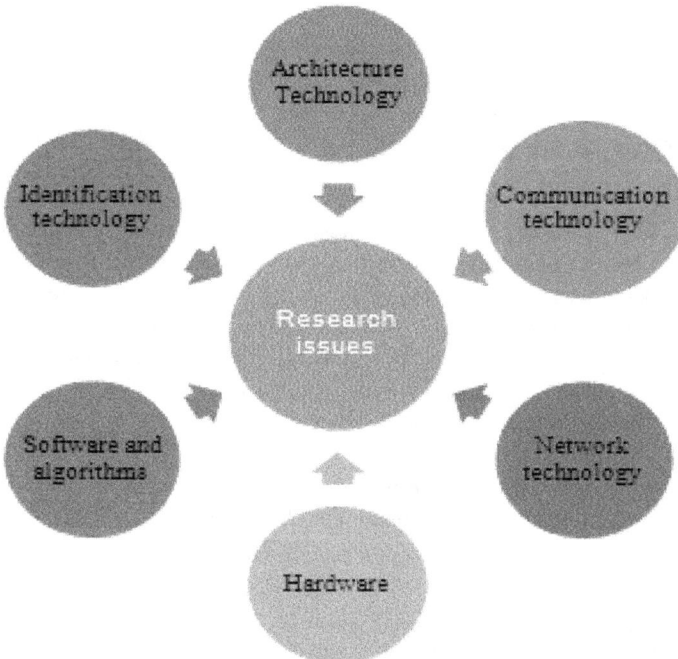

FIGURE 20.4 Factors affecting IoT research technologies.

Three categories that come under identification of things are object recognition, position and geography identification [14]. Authentication, repository management using identification authentication, and addressing schemes need to be studied, and more research should be done on the revocable person.

20.7.2 THE ARCHITECTURE OF IoT

Trillions of things are connected; it is necessary to have an adequate architecture that permits secure connectivity, control, communication and useful applications. Many objects in IoT have no proper architecture. Often devices need to connect and share data, and sometimes they need to be separated and protected [15].

In architecture design domain, some of the issues that need attention are as follows:

- The design of distributed open architecture has to cross through various barriers and finally be compatible with end-to-end devices.
- IoT is the combination of heterogeneous systems, and the interoperability between this system is important.
- The disruption in the physical level is transparent and resilient.
- Decentralized architecture is also a requirement.
- Error in one part shuts down the whole architecture [15].

IoT architecture research issues to be addressed

- Intelligent architecture is needed while moving data from a network to a user.
- Cloud computing technology, which we discussed in the previous section, also has an issue regarding the security of data.

20.7.3 COMMUNICATION BETWEEN DEVICES

Communication between devices means how two devices talk to each other. In IoT, the devices are now more compatible because of the increasing efficiency of sensors. Now sensors can communicate with other sensors, machines, and so on, but there are also many challenges in the communication scalability, interoperability and ensuring a return on investment for network operators. In this context, the communication needs will change, and to cater to the connectivity demands of emerging devices, new architectures and new bandwidths are required.

Issues to be addressed

- Energy-efficient communication.
- Software defined radios to remove need for hardware upgrades when new protocols emerge.
- Connection of the devices without a wireless network.

Topics in communication requiring consideration

- Things communicate with other things by sharing information.

- Communication is done with the help of the sensors, and the sensors collect the information and present the physical world information to the Internet world.
- There are two components in IoT: The first is sensing, which means it is sensing what is happening in the world. The second is sensing and taking action means it not only sensing but also takes action on it.
- To interact with humans, communication is needed.
- Data mining and services are providing communication.
- To provide the unique identity, IoT requires communication.

20.7.4 NETWORK TECHNOLOGY

IoT is a network of things that can communicate with each other. In a network structure, IoT architecture is built on top and the network structure integrates two technologies – either wired or wireless in a transparent manner. In the current scenario, wireless network technologies are in the limelight.

Further research is needed on the following:

In the domain of network technology, one of the challenges is the network-on-chip communication architecture. That means that we have to make chips that are capable of talking to each other.

Another research area is to build a system that is power-aware, which means that if we are installing in a remote area, the system should be aware and should be switching off and on accordingly.

Scalable communication infrastructures are another important research area. For example, if we add ten sensors today, it should still be working. This means that if there are any changes to the chip, the chip should still support and work without any error.

Issued to be addressed include the following:

- Network technologies
- Wireless network sensors
- Self-activating networking and computing
- Dynamically developed infrastructure which is capable of support small- and large-scale area dynamically
- At the network level, a unique identification mechanism
- Unspecified networking
- Both the technologies IP and post-IP

20.7.5 SOFTWARE SERVICE AND ALGORITHMS

Everything in today's world has computing, and one of the best ways to control computing is through programming. We need a software service algorithm that will manage everything related to IoT. For example, we need an algorithm to configure all sensors in a college. IoT software helps to manage applications, resources and services, and it also fulfils the requirements of IoT as imagined. To achieve all these things, a self-configuring and auto-recovery software is needed.

Encapsulation means to bind up. This feature is important as IoT is the combination of heterogeneous devices. This type of software also helps to execute in remote locations. In IoT, the software should be robust in nature. Distributed intelligence is a term used for such execution.

Issues to be addressed include the following:

- Identifying and obtaining a service
- Exchange of information, interoperability with sensors, etc
- Intelligent agents, which means two smart objects that interact with each other
- Human-machine interaction done with the help of software
- Self-management techniques to reduce the increasing complexities
- Self-adaptive software for optimization and configuration
- Software that uses less computing power and less storage space

20.7.6 HARDWARE

Any additional device added in the IoT environment is hardware. IoT devices have the capability to manage hardware configuration; this capability is known as self-management. IoT hardware also has the capability of self-adaptiveness and uses resources such as communication bandwidth, energy, medium access, etc. [16]. Research and development are needed in the area of hardware that operates at low-level power and has context-switching architecture because in this area configuration changes dynamically from time to time to introduce changes to the device. [17].The term *cognitive* means to understand the environment, and with the help of artificial intelligence hardware can dynamically understand its environment.

Some issues of hardware to be addressed are as follows:

- Nanotechnologies – miniaturization of the hardware needed to construct
- Sensor technologies – embedded sensors, actuators
- Manufacturing cost – reduction of hardware costs
- Performance of the hardware – high in regard to identification/authentication
- Communication – antennas, energy

20.8 CONCLUSION

This chapter tells how IoT is more efficient than a traditional network. Cloud computing, where data is stored, is the main feature of IoT and analyzed, but this chapter shows that fog computing is more beneficial than the cloud. The main purpose of this chapter is to show the factors affecting research technologies and the transaction system of IoT. Cloud computing is based on a centralized system, which is less secure than a decentralized system. To make the IoT data more secure, the concept of a decentralized system is achieved by blockchain. This chapter discussed concepts related to the possible deployment of blockchain for security enhancement in IoT-based networks.

REFERENCES

1. Chen, H., X. Jia, and H. Li. *A brief introduction to IoT gateway*. In *IET International Conference on Communication Technology and Application (ICCTA 2011)*. 2011. IET.
2. Khan, R., et al. *Future internet: The internet of things architecture, possible applications and key challenges*. In *2012 10th international conference on frontiers of information technology*. 2012. IEEE.
3. Ali, Z.H., Ali, H.A., and Badawy, M.M., *Internet of things (IoT): Definitions, challenges and recent research directions*. International Journal of Computer Applications, 2015. **128**(1): pp. 37–47.
4. An, J., Gui, X.-L., and He, X., *Study on the architecture and key technologies for Internet of Things*. Advances in Biomedical Engineering, 2012. **11**: p. 329.
5. Maraiya, K., Kant, K., and Gupta, N., *Wireless sensor network: A review on data aggregation*. International Journal of Scientific & Engineering Research, 2011. **2**(4): pp. 1–6.
6. Al-Fuqaha, A., et al., *Internet of things: A survey on enabling technologies, protocols, and applications*. IEEE communications surveys & tutorials, 2015. **17**(4): pp. 2347–2376.
7. Memon, R., et al., *Cloud-based vs. blockchain-based IoT: A comparative survey and way forward*, in *Frontiers of Information Technology & Electronic Engineering*. 2019, Springer.
8. Ratta, P., and Kour, A., *Based on the architecture difference between cloud and fog computing in IoT*. International Journal of Information and Computing Science, 2018. **5**(3). ISSN NO: 0972-1347.
9. Sajana, P., Sindhu, M., and Sethumadhavan, M., *On blockchain applications: hyperledger fabric and ethereum*. International Journal of Pure and Applied Mathematics, 2018. **118**(18): pp. 2965–2970.
10. Alharby, M., and Van Moorsel, A., *Blockchain-based smart contracts: A systematic mapping study*. arXiv preprint arXiv:1710.06372, 2017.
11. Khan, M.A., and Salah, K., *IoT security: Review, blockchain solutions, and open challenges*. Future Generation Computer Systems, 2018. **82**: pp. 395–411.
12. Atlam, H.F., et al., *Blockchain with internet of things: Benefits, challenges, and future directions*. International Journal of Intelligent Systems and Applications, 2018. **10**(6): pp. 40–48.
13. Reyna, A., et al., *On Blockchain and its integration with IoT. Challenges and opportunities*. Future generation computer systems, 2018. **88**: pp. 173–190.
14. Qiuping, W., Shunbing, Z., and Chunquan, D., *Study on key technologies of internet of things perceiving mine*. Procedia Engineering, 2011. **26**: pp. 2326–2333.
15. Hammoudi, S., Aliouat, Z., and Harous, S., *Challenges and research directions for internet of things*. Telecommunication Systems, 2018. **67**(2): pp. 367–385.
16. Athreya, A.P., B. DeBruhl, and P. Tague, *Designing for self-configuration and self-adaptation in the Internet of Things*. In *9th IEEE International Conference on Collaborative Computing: Networking, Applications and Worksharing*. 2013. IEEE.
17. Vermesan, O., P. Friess, and P. Friess, *Internet of things: Global technological and societal trends*. Vol. 37. 2011: River Publishers Aalborg, Denmark.

21 Identification of Background Factors Associated with Prevalence of Common Mental Disorders among Adolescent Students

Vivek Sharma
Lovely Professional University

Neelam Rup Prakash and Parveen Kalra
Punjab Engineering College

CONTENTS

21.1 Introduction 306
21.2 Subjects 306
21.3 Materials and Methods 306
21.3.1 Depression, Anxiety and Stress 306
21.3.2 Background Factors 307
21.4 Data Analysis 307
21.5 Results 311
21.5.1 Depression 311
21.5.2 Anxiety Disorder 312
21.5.3 Mental Stress 312
21.6 Conclusion and Future Work 313
21.7 Ethical Considerations 313
21.8 Funding 314
21.9 Conflict of Interest 314
References 314

21.1   INTRODUCTION

Adolescence is an age of rapid transformation from a child to a full member of society. This period also involves life-changing decisions such as occupational selection, rapid environment adaptation and increased responsibility. With the increased burden of work and expectations, the prevalence of mental disorders (i.e. depression and anxiety) is also on the rise among this group. If these mental disorders are not detected in time and are left untreated, they can manifest into severe mental disorders, substance abuse and finally suicidal behaviour [1]. Early diagnoses and treatment can reduce these alarming numbers. The suicide rate is highest among this group, i.e., 35.5 per 100,000, which can be attributed to high rate mental disorder among this group according to a World Health Organization (WHO) report. Among this group, adolescent students between ages of 16 and18 are more susceptible to the mental illness of depression [2, 3].

Economic situation, parental guidance, academic performance and social behaviour can be contributing factors to mental status [4–6]. Researchers have identified these factors as significantly associated with mental health, with prevalence rates of depression from 19.5% to upward of 71.5%, 24.4% to 66.9% for anxiety and 21.1% to 60.8% for mental stress assessed mostly by the DASS-21 psychometric self-assessment scale [7]. In most of these studies, the psychometric self-assessment test was administered in a group setting without any mental health professionals, which may account for variation in prevalence. The aim of this study was first to identify prevalence (i.e., frequency) of common mental disorders, i.e., depression, anxiety and stress, among adolescent students based on psychiatric evaluation by trained mental health professionals. Second, the study aimed to identify associations between prevalence of mental illness and various background factors such as gender, academic performance, study time, family profile. Finally, the last objective is to study the variation in prevalence of common mental illness within these background markers.

21.2   SUBJECTS

The design of this study was approved by ethical committee of Punjab Engineering College. Permission for the study was given by the director of public instructions, Chandigarh in writing and again from school principals. In total, 204 students (60% female) of class 11th and 12th from four government schools between ages of 16 and 18 years volunteered for this study. All participants were informed about the purpose and methods used in study, and their informed consent was taken. In the case of minor subjects, informed consent was received from their parents. Strict confidentially about the students' participation in study was maintained. No information about participants was given to any school teaching staff.

21.3   MATERIALS AND METHODS

21.3.1   DEPRESSION, ANXIETY AND STRESS

The Depression Anxiety Stress Scale (DASS) is a clinical psychometric self-assessment psychological scale developed by a research group at the University of New

South Wales [7]. DASS-21 consists of 21 questions designed for assessment of depression, anxiety and stress (seven each in three categories), using clinical markers such as hopelessness, lack of interest, negative thoughts, etc. for depression; anxious feelings, autonomic arousal, situational reactions, etc. for anxiety; and lack of arousal, increased mood swings and hyper reactions in the previous two weeks for mental stress [8]. Using DASS-21 cumulative scores, mental health in terms of depression, anxiety and stress can be classified into five subcategories: normal (0–3), mild (5–6), moderate (7–10), severe (11–13) and extremely severe (>14) [8]. DASS-21 can be self-administered, but results should be analyzed by a trained mental health professional. Also, DASS-21 has a high Cronbach rate, with 0.91 for depression, 0.84 for anxiety and 0.90 for stress [9]. Also, DASS-21 scores for depression and anxiety are highly correlated with two common clinical scales: Beck Depression Inventory (BDI II) and Beck Anxiety Inventory (BAI) [10]. The mental health status of each participant was assessed by trained mental health professionals based on one-on-one interviews using the DASS-21 scale, which is open access permitted by GNU licence.

21.3.2 BACKGROUND FACTORS

Common mental disorders such as depression and anxiety are more prevalent in females compared to males among youths [4–6]. Background factors such as lower income group, low income group low academic performance, lack of social interactions, lack of parental support and recent internet addiction, etc. are contributing factors for these disorders [4–6]. For identification of significant background factors associated with mental health a questionnaire-based study was also conducted. A computerized version of the questionnaire consisting of 11 such risk markers was also designed. No personal information, such as name, address or phone number, was collected from any individual.

21.4   DATA ANALYSIS

The mental status of each participant was classified into normal (i.e. healthy) or into subcategories of depression, anxiety, and stress based upon the scores of DASS-21. For identification of the prevalence of depression, anxiety and stress, descriptive statistical analysis was performed on collected mental status data evaluated by DASS-21. Further, for identification of significant association background factors and mental disorders, the chi-square test with $p<0.05$ was applied on background data and mental status (i.e. normal or level of depression, anxiety, and stress) and evaluated based upon cut-off scores according to the manual of DASS-21. All data analysis was performed using the Statistical Package for the Social Sciences (SPSS) IBM version 21 for windows 10. Prevalence of mental disorders based upon survey was tabulated and is shown in Figures 21.1 to 21.3. The results of the chi-square test estimation of significant association of background factors with mental health status at $p<0.05$ are listed in Table 21.1. Results via descriptive analysis between subcategories of background with mental disorders were also tabulated and mentioned in Table 21.2 for depression, anxiety and stress.

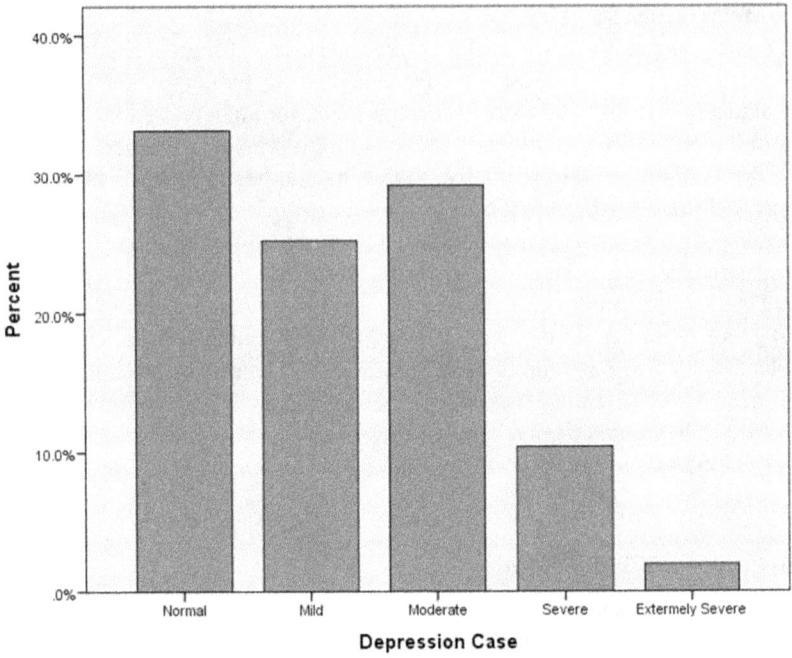

FIGURE 21.1 Prevalence of depression among students.

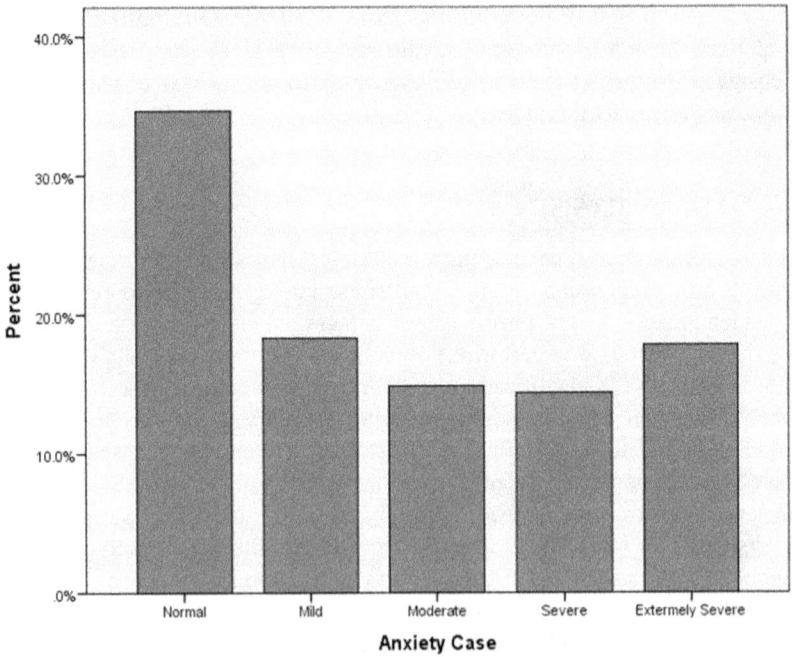

FIGURE 21.2 Prevalence of anxiety among students.

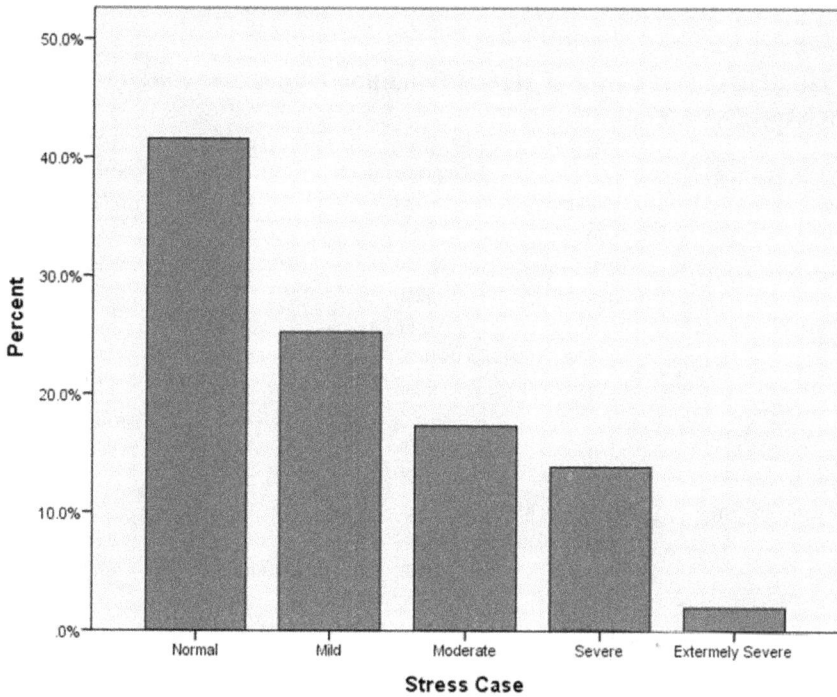

FIGURE 21.3 Prevalence of stress among students.

TABLE 21.1
Result of Chi-Square Test for Association of Background Factors with Mental Status

	df	Depression		Anxiety		Stress	
		Chi-sq	P Value	Chi-sq	P Value	Chi-sq	P Value
Sex	4	7.072	0.132	10.147	**0.038**	13.361	**0.01**
Locality	4	7.969	0.093	12.978	**0.011**	8.46	0.076
Family income	20	29.525	0.078	54.427	**0.001**	74.592	**0.002**
Father education	20	27.215	0.129	37.428	**0.01**	58.196	**0.001**
Mother education	20	29.264	0.082	53.566	**0.001**	42.658	**0.002**
Mother working	4	9.211	0.056	2.629	0.622	1.217	0.875
Social activities	16	38.243	**0.001**	55.995	**0.001**	44.425	**0.002**
Study time	16	26.730	**0.044**	50.428	**0.001**	34.489	**0.005**
Social Media time	16	26.33	**0.049**	56.179	**0.001**	36.964	**0.002**
% Scores	16	44.054	**0.0001**	41.879	**0.0001**	31.51	**0.012**
Parental time	16	28.89	**0.024**	32.793	**0.008**	48.265	**0.0001**

TABLE 21.2

Variation in Prevalence of Mental Condition Associated with Background Factors

		Depression	Anxiety	Stress
Sex	Female	**0.708**	**0.708**	**0.658**
	Male	0.61	0.573	0.476
Locality	Semi-urban	**0.734**	**0.724**	**0.667**
	Urban	0.62	0.6	0.522
Father education	10th	0.466	0.727	0.602
	12th	0.38	0.5	0.4
	Below 10th	**0.80**	**0.74**	**0.90**
	Grad	0.559	0.412	0.382
	PG	0.75	0.833	0.833
	10th–12th	0.55	0.74	1
Family income	Below 1 lakh	**0.8**	0.692	0.522
	Above 10 lakhs	0.532	0.681	0.66
	3–5 lakhs	0.552	**0.862**	**0.759**
	5–7 lakhs	0.483	0.583	0.533
	1–3 lakhs	0.79	0.4	0.4
	7–10 lakhs	0.733	0.733	0.6
Mother education	Below 10th	**0.875**	**0.845**	**0.785**
	12th	0.553	0.658	0.684
	10th	0.55	0.8	0.5
	Graduate	0.553	0.684	0.447
	PG	0.667	0.5	0.333
	10th–12th	0.475	0.825	0.775
Mother working	No	0.40	**0.654**	0.556
	Yes	**0.60**	0.646	**0.592**
Social activities	0/Nil	**0.70**	**0.69**	**0.696**
Time spent per week	10 above	0	0.5	0.2
	2–5	0.615	0.667	0.615
	5–10	0.667	0.533	0.567
	0–2	0.489	0.685	0.576
Study time	0/Nil	0.48	0.667	0.587
	0–5	**0.692**	0.615	**0.692**
	Above 30	0.1	**0.833**	0.5
	5–0	0.304	0.565	0.609
	10–30	0.69	0.641	0.487
% Scored	50–60	0.5	0.536	0.464
	60–70	0.609	0.609	0.565
	70–80	0.567	0.633	0.467
	Above 80	0.6	0.89	0.6
	Below 50	**0.652**	**0.9**	**0.62**

TABLE 21.2

Variation in Prevalence of Mental Condition Associated with Background Factors (*Continued*)

		Depression	Anxiety	Stress
Time spent on	0/Nil	0.569	0.738	0.662
internet per week	2–10	0.446	0.714	0.696
	Above 10	**0.815**	**0.744**	**0.8**
	0–2	0.791	0.442	0.419
Parents time	0/Nil	**0.818**	**0.688**	**0.655**
(time spent with	2–5	0.52	0.7	0.72
parents)	5–10	0.552	0.517	0.379
	Above 10	0.5	0	0.5
	0–2	0.579	0.667	0.526

21.5   RESULTS

21.5.1   DEPRESSION

As per descriptive analysis prevalence of depression among the sampled students plotted in Figure 21.1, a high prevalence of 12.4% of severe to extremely severe subcategories depression was observed among the sampled population of students. Another 29.2% students with borderline (moderate) depression were also identified and should be counselled by school counsellors and parents. A higher prevalence of depression was observed in females participants (70.8%) compared to males (61.0%). In students from urban areas the prevalence of depression was lower (62.7%) compared to rural students (73.4%). Students with a lower family education level of both father and mother (below class 10th) have higher prevalence (80.0% and 87.5%) of depression compared to all other groups. Higher prevalence of depression was observed in students with working mothers (60%) compared to non-working mothers (40.0%). The lowest prevalence of depression was observed in the subcategory of students with a study routine of 10 to 30 hours per week (62.5%) and highest (83.3%) among those putting in more than 30 hours of study per week. The highest prevalence of depression was observed in among the subgroup of students with poor academic performance (69.4%). Low-income students (less than 1 lakh p.a.) have the highest (65.2%) prevalence of depression. In terms of background factors, the prevalence of depression was highest (70.0%) among students who do not participate in social activities such as yoga or sports. Students with no support from family, i.e., no or less parental guidance or family holidays, have a higher prevalence of depression (81.8%). In terms of time spent on the internet (social media), a higher prevalence of depression (81.5%) was observed in the subgroup with more 10 hours per week.

As per results of chi-square test (χ^2) for significance of background factors associated with depression, as mentioned in Table 21.1, at p <0.05, only five background factors showed statistical association with depression in the chi-square test (χ^2) at p< 0.05.

These markers mostly consist of academic performance (% scores), study time, social interactions, parental time and activities on the internet (social media). No other background factor showed as statistically associated with the level of depression at p <0.05.

21.5.2 Anxiety Disorder

As per results of descriptive analysis plotted in Figure 21.2, a higher prevalence of extremely severe anxiety (17.8%) was observed among the sampled population of students using DASS-21. These students should be provided with clinical counselling and medication on urgent basis. Severe and moderate (borderline) anxiety disorders were also observed in 29.3% (cumulatively) of students. Overall, an anxiety disorder prevalence of 65.3% was observed in the sample population. A higher prevalence of anxiety was observed in females participants (70.8%) compared to males (57.3%). In students from rural areas the prevalence of anxiety was higher (72.4%) compared to urban students (60%). Students with a lower family education level of both father and mother (below class 10th) have a higher prevalence (74.0% and 82.5%) of anxiety compared to all other groups. Similar prevalence of anxiety was observed in students with working mothers (64.6%) compared to non-working mothers (65.4%). In the background factor of study time, the highest prevalence of anxiety was observed in the subcategory of students with a study routine of more than 30 hours per week (83.3%). The highest prevalence of anxiety (90.0%) was observed in the subgroup of students with poor academic performance. Low-income students (less than 1 lakh p.a.) have a lower anxiety prevalence (60.9%) compared to students with a family income of 3–5 lakhs p.a., which is the highest (86.2%). In terms of the background factor of time spent per week on social activities such as yoga or sports, the prevalence of anxiety was evenly spread among all subgroups. Students with no support from family, i.e., no or less parental guidance or family holidays, have a higher prevalence of anxiety (68.8%). In terms of time spent on the internet (social media), a higher prevalence of anxiety (74.4%) was observed in the subgroup spending more than 10 hours per week.

As per results of chi-square test (χ^2) for significance of background factor associated with anxiety, as mentioned in Table 21.1, at p <0.05, only the background marker of mother working showed no statistical association with anxiety disorders at p< 0.05. The rest of the 10 background markers showed a statistical association with anxiety disorders, at p< 0.05 as per chi-square test.

21.5.3 Mental Stress

As per results of descriptive analysis plotted in Figure 21.3, prevalence mental stress as per DASS-21 was observed in 58.4% of the sample population. Of those, 2% students with extremely severe and 13.9% with severe mental stress were also identified, and these cases require urgent clinical counselling. Also, 17.3% students were classified with borderline mental stress and should receive counselling to avoid further progression.

As per results of descriptive analysis, variation in prevalence of stress among subgroups of students associated with the background factors is shown in Table 21.2. Higher prevalence of stress was observed in female participants (65.8%) compared to males (58.6%). In students from rural localities prevalence stress was higher (66.6%) compared to urban students (52.2%). Students with a lower family education level of both father and mother (below class 10th) have higher prevalence (90.0% and 78.5%) of stress compared to all other groups. Higher prevalence of stress was observed in students with working mothers (59.2%) compared to non-working mothers (55.6%). In the background factor of study time, the highest prevalence of stress was observed in students with a study routine of 0–5 hours per week (69.2%). On the basis of academic performance, the highest prevalence of stress was observed among the subgroup students with poor academic performance (62.0%). Medium-income students (family income of 3–5 lakh p.a.) have higher stress prevalence (75.9%) compared to students with a family income of below 1 lakh p.a., which was the lowest (52.2%). In terms of the background factor of time spent per week on social activities such as yoga or sports, the prevalence of stress was highest among students who don't participate in any social activities (69.6%). Students with no support from family, i.e., no or less parental guidance or family holidays, have higher prevalence of stress (65.5%). In terms of time spent on the internet (social media), higher prevalence of anxiety (80.0%) was observed in the subgroup with more than 10 hours per week. As per results of chi-square test (χ^2) for significant association of background factors with stress, as mentioned in Table 21.1, apart from two background markers, namely, mother working status and locality, all other markers showed statistical significance at $p<0.05$ with categories of mental stress assessed by DASS-21.

21.6 CONCLUSION AND FUTURE WORK

The aim of this study was to identify prevalence of depression, anxiety and stress among adolescent students and the association with various background factors. For this purpose a cross-sectional was conducted on 204 adolescent students of class 11th and 12th. As per results, the overall prevalence rate of depression, anxiety and stress was 66.8%, 65.3%, and 49.5% among the sampled population (adolescent students) respectively. Background factors of study time, academic performance, social activities, internet time and parental time showed significant association with social anxiety, stress and depression at $p<0.05$ respectively. As per result of this study, individuals with following profile are more susceptible to mental illness, i.e. female's, rural, lower income, low family education background, less academic performance, no parental time and social activities and higher internet time respectively. The reason of variation was not in the scope of this present work and should be investigated in separate study.

21.7 ETHICAL CONSIDERATIONS

All participants were informed about the purpose and methods used in study and their informed consent was received. In the case of minor subjects, informed consent was received from their parents. Strict confidentially about the student's participation in study was maintained. No information about participants was given to any school teaching staff.

21.8   FUNDING

No external funding was received for this manuscript.

21.9   CONFLICT OF INTEREST

All authors of this manuscript declare that they have no conflict of interest.

REFERENCES

1. Albano, A.M., Chorpita, B.F., & Barlow, D.H. Childhood anxiety disorders. In E.J. Mash & R. A. Barkley (Eds.), *Child psychopathology*: New York: Guilford Press; 2003. pp. 279–329.
2. Thapar, A., Collishaw, S., Pine, D.S., & Thapar, A.K. (2012). Depression in adolescence. *Lancet*, 379, pp. 1056–1067.
3. WHO. World Health Organization Mental health action plan 2013–2020. Geneva: <http://apps.who.int/iris/bitstream//10665/89966/1/9789241506021_eng.pdf>.
4. Kumar, K.S., & Akoijam, B.S. (2017). Depression, anxiety and stress among higher secondary school students of Imphal, Manipur. *Indian Journal of Community Medicine: Official Publication of Indian Association of Preventive & Social Medicine*, pp. 94–96. doi: 10.4103/ijcm.IJCM.
5. Preeti, B., Singh, K., & Kumar, R. (2017). Study of depression, anxiety and stress among school going adolescents. *Indian Journal of Psychiatric Social Work*, 8(1), pp. 6–9.
6. Singh, M.M., Gupta, M., & Grover, S. (2018). Prevalence & factors associated with depression among school-going adolescents in Chandigarh, north India. *Indian Journal of Medical Research*, 146(2), pp. 205–215. doi: 10.4103/ijmr.IJMR.Sarkar.
7. Sarkar, S., Gupta, R., & Menon, V. (2017). A systematic review of depression, anxiety, and stress among medical students in India. *Journal of Mental Health and Human Behaviour*, pp. 88–96. doi: 10.4103/jmhhb.jmhhb.
8. Lovibond, S.H., & Lovibond, P.F. (1995a). *Manual for the Depression Anxiety Stress Scales* (2nd. Ed.) Sydney: Psychology Foundation.
9. Antony, M.M., Bieling, P.J., Cox, B.J., Enns, M.W., & Swinson, R.P. (1998). Psychometric properties of the 42-item and 21-item versions of the depression anxiety stress scales (DASS) in clinical groups and a community sample. *Psychological Assessment*, 10, 176–181.
10. Lovibond, P.F., & Lovibond, S.H. (1995b). The structure of negative emotional states: Comparison of the depression anxiety stress scales (DASS) with the beck depression and anxiety inventories. *Behaviour Research and Therapy*, 33, pp. 335–343.

22 A Cloud-Based Secured IoT Framework for Log Management

Nilima Dongre
RAIT

Mohammad Atique
SGBAU

CONTENTS

22.1 Introduction 316
    22.1.1 Necessity of Log Files for the Long Term 316
    22.1.2 Desired Properties 317
22.2 Literature Survey 318
    22.2.1 IoT Security 318
        22.2.1.1 IoT Security and Privacy 318
        22.2.1.2 Access Control 318
    22.2.2 SDN-Based Secure IoT Framework 318
    22.2.3 IoT Security and Log Management 318
        22.2.3.1 Event Logging 318
        22.2.3.2 Distlog 319
        22.2.3.3 LogSafe Data Logger 319
        22.2.3.4 BlockAudit 319
    22.2.4 Cloud Log Management 319
        22.2.4.1 Cloud Log Forensics 319
        22.2.4.2 Secure Logging-In Cloud 319
22.3 Problem Definition 319
    22.3.1 Threat Model 320
        22.3.1.1 Attacks During Transmission 320
        22.3.1.2 Authenticity 320
        22.3.1.3 Privacy on the Logging Cloud 320
    22.3.2 Adversary Model 320
        22.3.2.1 Encryption 321
        22.3.2.2 Authentication 323
22.4 Proposed System Design 323
    22.4.1 Sensor Node and Sensor Devices Layer 323
    22.4.2 Edge Layer 324

        22.4.3 Anonymous Network Layer..325
        22.4.4 Cloud Layer 325
22.5 Security Analysis..325
        22.5.1 Integrity of Log Records During Transit......................326
        22.5.2 Authenticity of Log Record Generator.........................326
        22.5.3 Privacy on Logging Cloud..326
22.6 Performance Analysis...327
22.7 Conclusion 328
References...329

22.1 INTRODUCTION

The internet of things (IoT) consists of a heterogeneous computing environment of both physical and logical entities distributed across its layers, namely, sensor devices and sensor node layer, edge or gateway layer, network layer and cloud layer. At each layer these entities perform numerous operations which are logged continuously. A log can be defined as the traces of the events occurring in a network or within a system of an organization. Logs are produced dynamically and are stored locally and remotely in plain text format.

Log Data is crucial role for following reasons:

- An audit trail – determining what happened
- Log forensics
- Finding foot prints of an intruder
- Debugging and gathering information about system health
- System recovery
- System alerts

The log services are capable of storing the log data in an organized format and at the same time making it available, i.e., fast retrieval. Heterogeneous networks generate various types of logs, time stamps, formats, etc. After the secure generation of these logs, secure storage, transit, maintenance and retrieval are equally important to achieve all the desired properties.

22.1.1 NECESSITY OF LOG FILES FOR THE LONG TERM

Log management is of paramount importance in the current era, with the growing concerns about security, system and network operations. Monitoring, troubleshooting and debugging comprise the larger part of software development today. Logging makes this a much more comfortable and smoother operation. Application performance monitoring (APM) tools are good for some of the critical performance metrics. A log-security, performance or a system-monitoring perspective seems impossible without the cloud-based environment; thus logging as an approach finds visibility; otherwise deploying a secure logging infrastructure incurs a lot of financial outflow and seems too severe.

A computing environment consists of many different types of logs which are needed to maintain security [1]. The various log sources in general for any computing environment are as follows:

Log Sources:

a. System logs
b. Application logs
c. Firewall logs
d. IDS/IPS logs
e. Server logs

The various activities of log management are as follows:

a. Log collection
b. Log aggregation at a centralized location
c. Long-term storage
d. Log analysis (static and dynamic)
e. Log search and reporting
f. Log recovery

22.1.2 DESIRED PROPERTIES

There is a dire need for the appropriate use of technology to diagnose and provide solutions for security threats; this is where log management play a very important role in knowing the facts and trends of every organization. If a hacker gains access to this critical information and compromises the system – for example, a banking website – it can have a great impact on the business's reputation. Application of computational techniques in the field of security has been an area of intense research in recent years.

The following are the required properties of secure log management system:

1. **Correctness**: The data saved in the log files should exactly match the data while it was generated. In other words, it should be correct.
2. **Tamper resistance:** Any unauthorized access, reading, updating or deleting should not be allowed. If it does occur, the system should detect it.
3. **Verifiability:** Every valid log entry should contain adequate information to confirm its authenticity to make sure that the logs are not altered.
4. **Confidentiality:** Log records contain sensitive information about the organization, and an attacker can collect this. Hence, log entries should be securely saved such that they are not available or searchable to anyone in the network.
5. **Privacy:** Log entries should not be traceable by the intruder during transmission and should be saved securely and should have secure retrieval protocol.

22.2 LITERATURE SURVEY

The IoT computing environment is voluminous and evolving with new and more complex technologies. The fundamental elements of log management remain the same. The multi-dimensional facets of the enormous data created and their security is closely related with log management. The following discussion explores log management issues and solutions from technological point of view.

22.2.1 IoT Security

22.2.1.1 IoT Security and Privacy

The research paper identifies usability and security as two major concerns of the IoT ecosystem [2, 3]. The exponential increase in the attack surfaces of IoT devices amidst the heterogeneous environment poses a huge challenge to users and stakeholders. An IoT command centre is connects all the devices and their relevant entities. Such a command centre is cloud enabled for the purpose of connectivity. This centralized connected platform serves the purpose of data sharing and storage for all the connected devices. But it also vulnerable as a single point of failure. Apart from this, well-defined rules for common security architecture maintain confidentiality, integrity and availability, using traditional measures such as encryption and access control.

22.2.1.2 Access Control

The author addresses the privacy and security concerns of the exponentially growing number IoT devices [4]. Blockchain technology is used to restrict address authentication and authorization. The smart home case study is used to monitor the temperature and the intrusion detection. Four-tier architecture is proposed to manage the data access and data storage in the cloud. The advantage is that a blockchain-based system is light weight and at the same time is highly secure in the sense that it provides secure access to IoT devices using the immutability of block chain. The drawback is that not all the data can be secured with block chain.

22.2.2 SDN-Based Secure IoT Framework

The authors explore the multi-dimensional security aspects of the IoT [5]. The proposed software-defined networking (SDN) based secure IoT framework is expensive, but it addresses security issues and their attacks models.

22.2.3 IoT Security and Log Management

22.2.3.1 Event Logging

Smart home automation needs a secure and reliable event logging system [6]. The authors propose a host-based conceptual framework for storing and processing the data. Unlike cloud-based security for privacy, a forward secure event logging system satisfying the needs of such a real-time environment is implemented. This implementation is further helpful for fault detection, forensics and accounting data.

22.2.3.2 Distlog

The author propose a distributed logging scheme for IoT forensics [7]. Attackers attempt to tamper with the log files to erase their activity footprints which can be revealed during forensic analysis. The proposed modified information dispersal algorithm (MIDA) ensures log availability. The logs generated by the IoT devices are aggregated, compressed and encrypted. Further, the encrypted logs are fragmented and stored over a distributed storage nodes after proper authentication. The drawback is that the scheme is computer intensive.

22.2.3.3 LogSafe Data Logger

LogSafe [8] is a secure and scalable distributed cloud-based data logger for IoT devices. It focuses on storing the logs securely to preserve them for future forensic analysis. The Intel Software Guard Extension (SGX) is used to store the IoT devices logs, which satisfies the security goals – confidentiality, integrity and availability. The advantage of the system is that it is capable of handling a huge amount of log data with a very high data transmission rate. The implementation depends on the capability of the SGX model.

22.2.3.4 BlockAudit

The author focuses on blockchain-driven security of audit logs generated during auditing, storing and tracking changes made to the data of any enterprise [9]. BlockAudit is a scalable and tamper-proof system leveraging the design properties of audit logs and ensuring security by using blockchain. The implementation shows that conventional audit logs can be seamlessly transitioned into BlockAudit with higher security, integrity and fault tolerance.

22.2.4 CLOUD LOG MANAGEMENT

22.2.4.1 Cloud Log Forensics

The authors explore the challenges faced during the cloud log forensics [10]. They identify log security and integrity as challenges in cloud log forensics. In the taxonomy the accessibility and integrity of log data to be collected is highlighted as a major concern for cloud log forensics.

22.2.4.2 Secure Logging-In Cloud

The authors identify the need of storing the log files over a much longer period, maintaining their security in terms of confidentiality, integrity and privacy [11]. With the use of the cloud, the capital incurred in setting up of an infrastructure is considerably reduced. The authors propose that the architecture should use the cryptographic protocols to deal with storing, maintaining and retrieving the log records.

22.3 PROBLEM DEFINITION

All activities happening in the IoT layers and their computing entities are captured as records in a log file, including all user actions. An attacker can interrupt the logging service, damage log files, intercept the log information and impersonate other entities. The challenges and security concerns to address are confidentiality breaches, active/passive sniffing, privacy breaches, unauthorized access and altered data. Log security is achieved by providing privacy during transmission, security and tamper resistance when logs are at rest and in transit over a network, and verifiability of log validity and correctness.

IoT has a bigger attack surface and hence is more vulnerable to attackers. A malicious attacker seeks to break into a system without leaving any traces of their activities behind. The attacker's modus operandi is to disrupt the operability of the system by tampering with the log file, and in turn the logging service, so as to expose the system to vulnerabilities, thus attempting to reveal the sensitive and confidential information stored in the log files. The log files contain all the recorded data, which reveal the identity of its user, the events occurring, and user and system passwords. Most importantly the log records hold information about the system health processes and their events.

22.3.1 THREAT MODEL

Intrusion detection and recalls are network security tools that organizations use to monitor the security. We do not say that an organization's security cannot be breached. However, not all the logging clients and log generators can be compromised at the same time; only a predetermined fraction of them can be compromised simultaneously. Also, if compromise takes place it can be identified easily.

We assume a Dolev-Yao [1] attacker model, which typically makes the attacker capable of intercepting the message during the transit and then synthesizing, replicating and replaying the message. Lastly, the attacker can masquerade as a legitimate participant of the network or pretend to be a legitimate host.

22.3.1.1 Attacks During Transmission

There are two types of attacks: (a) active sniffing, (b) passive sniffing.

The intruder may get illegal access to a communication channel. The attacker can not only retrieve the information but also replay or alter the data which is in transit from the log generator to the logging cloud or while it is responding from logging cloud to the log monitor.

22.3.1.2 Authenticity

The attacker may be a legitimate participant of the network or may try to impersonate a legitimate host and begin to send log records from someone else's identity.

22.3.1.3 Privacy on the Logging Cloud

The attacker can attempt to read, delete and modify data stored in the logging cloud.

22.3.2 ADVERSARY MODEL

The following is the adversary model of the above threat model.

22.3.2.1 Encryption

(a) Even if the attacker compromises the system, without the key, the attacker cannot understand the cipher text. (b) With a log monitor, if the log chunk values do not match, it gives an error message as to which chunk is compromised; thus any read, write or delete operation cannot be performed.

The Data Encryption Standard (DES) algorithm is a Feistel-type substitution-permutation network (SPN) cipher.

a. ***Symbols Used***
1. $E_p[M]$: Message M encrypted with secret key p.
2. M_1 —— M_2:M_1 and M_2 messages are concatenated.
3. $H_i k[M]$: Computing i number of times one-way hash of message M with key k. Because the hash function is a one-way function, it is impossible to invert hash function H.
4. $H_n[M]$: Computing n times hash of Message M.
5. TS: Is a global timestamp

b. ***Description***

The initialization of the protocol is as follows:

a. Generation of key: Three master keys are randomly generated by A_0 and X_0 for ensuring log integrity and K_0 for ensuring log confidentiality. These keys ensure the requirements of the secret-sharing scheme.
b. Preparation of the log records: The log generator generates a series of messages $L_1, L_2, ..., L_n$, where each L_i message consists of a group of log records. It is assumed that these log messages are transmitted over an authenticated network to the logging client.
c. Uploading of logs: The log messages are uploaded by the logging client in batches of size n. The batch of n log records is prepared. At the beginning itself the n value is determined randomly.

The working of the protocol is as follows:

1. A special reset (RST) log entry is created by the logging client before any log data is sent to it

$$L_0 = TS, \text{ log initialization,n.}$$

This initial log entry is encrypted with key K_0 and a message authentication code of the encrypted entry is computed with the key A_0.

$$MAC_0 = HA_0[EK_0[L_0]]$$

The client adds the RST log entry for the corresponding batch MAC_0 to the log entry

$$EK_0[L_0]$$

2. The next iteration is the same as above. Creation of a new set of keys by the logging client and securely erasing the previous set of keys is done while waiting for the arrival of the next log message.

$$A_1 = H[A_0], X_1 = [X_0] \, AND \, K_1 = H[K_0]$$

3. On arrival of L_1rst log message, a record is created by the logging client.

$$M_1 = L_1 \, \| \, HA_0[E_{K0}[L_0]]$$

Message M_1 is encrypted with key K_1 and a message authentication code is computed for the resulting data as

$$MAC_1 = HA_1[EK_1[M_1]]$$

Further, an aggregated message authentication code is also computed.

$$MAC_1 = Hnx_1[MAC_0 \| MAC_1 \| n]$$

Thus, the log batch entry is

$$EK_1[M_1], MAC_1$$

The next set of keys is created.

$$A_2 = H[A_1], X_2 = H[X_1] \, AND \, K_2 = H[A_1]$$

4. For every ith new log data entry Li receiving at the logging client, subsequent log entries are created.

$$EK_i[M_i], \, MAC_i, \text{ where}$$

$$M_i = L_i \, \| \, MAC_{i-1} \, and \, MAC_i = HA_i[EK_i[M_i]]$$

An aggregated message authentication code is also created.

$$MAC_{i=} = H_{ni} + 1 X_i[MAC_{i1} \ldots \ldots \ldots MAC_i] \ldots \ldots \ldots {}_{n \, l+1}.$$

On generation of MAC_{i+1}, MAC_i is securely deleted, including secure erasure of the keys A_i, X_i and K_i, followed by generation of new keys nally.

$$Ai + 1 = H[A_i], X_{i+1} = H[X_i] \, and \, K_{i+1} = HK_i$$

On creation of the last log entry Mn for the current batch from the last log data Ln, it creates a special log-close entry and an aggregated message authentication code MACn1 followed by secure deletion of MAC and secure erasure of the keys.

$$LC = EKn + 1[TS, Log\text{-}close \text{ and } MACn], HAn + 1[EKn + 1]TS, log\text{-}close \text{ and } MACn]]$$

$$LC = EKn + 1[TS, Log\text{-}close \text{ and } MACn], HAn + 1[EKn + 1]TS, log\text{-}close \text{ and } MACn]]$$

22.3.2.2   Authentication

It is advisable to maintain the cloud and the organization infrastructure secure by the use of passwords, biometrics, etc. to ensure that privacy is achieved and tampering is prevented, thus ensuring the tamper-resistance property of the log file. Thus, correctness, tamper resistance, verifiability, confidentiality and privacy are attained.

22.4   PROPOSED SYSTEM DESIGN

Traditional environments do not handle data through a complex and dynamic process. Common web structures use web services for request and responses. User datagram packet are used by the Syslog protocol to transit processed data to the log server. Hence, the reliability of log records being delivered securely during the transit or at the end becomes a major concern.

In an IoT perspective, the following are the limitations for secure log management:

1. Sensor data is not secure and is also without authentication.
2. Sensor nodes are not suitable for storing large amounts of data produced by the sensors on a real-time basis.
3. Maintaining long-term storage of such data is expensive.

The existing cloud-based secure logging architecture is not suitable for the IoT platform [12]. The proposed system design framework is divided based upon the layers, namely, sensor devices and sensor node layer, edge or IoT gateway layer, anonymous network layer, and cloud or storage layer as shown in the Figure 22.1.

Log generators produce log records which are transmitted to the edge layer over the network. To provide more security log monitors merge with the edge layer. There are four divergent types of underlying components in the secured cloud-based framework shown in the block diagram:

a. Log generators
b. Logging client
c. Log monitor
d. Logging cloud

22.4.1   Sensor Node and Sensor Devices Layer

This is an end layer comprising the sensors placed on an end object. The sensors send the data to the sensor nodes. The log generator is implemented on these sensor nodes.

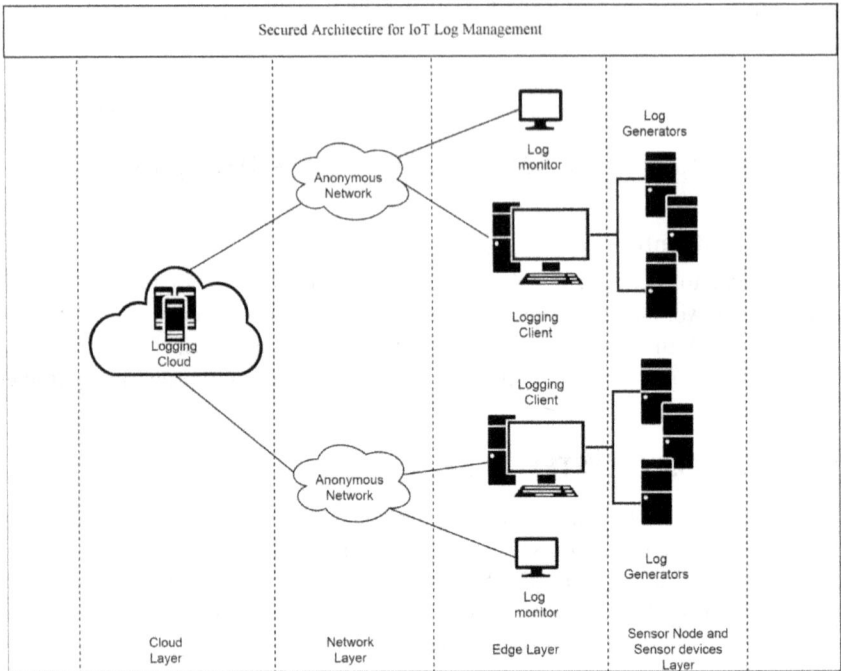

FIGURE 22.1   Secured framework for log management.

Log Generator (Figure 22.2): The sensor nodes are computing devices, called as log generators, which generate the log entries. These generators produce the log entries but do not store the logs; instead these log entries reside in these devices till they are sent securely to the logging client.

22.4.2 EDGE LAYER

The edge layer comprises the logging client and the logging monitor. An edge establishes communication between the sensor node layer and the cloud layer.

Logging Client (Figure 22.3): This collects the log entries generated by the sensor nodes or log generators. Log entries are pushed to the logging relay either on the basis of amount of data to be transmitted or on a particular schedule or in batches. The logging client can be implemented as a group of interconnected hosts.

Logging Monitor (Figure 22.4): The logging monitor monitors and reviews the log entries. Deleting logs and rotating logs are the prime responsibilities which can be

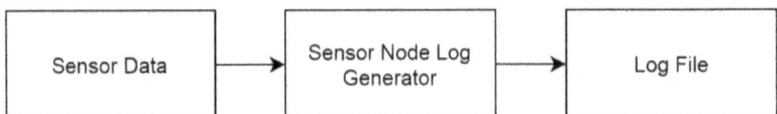

FIGURE 22.2   Sensor node as log generator [13].

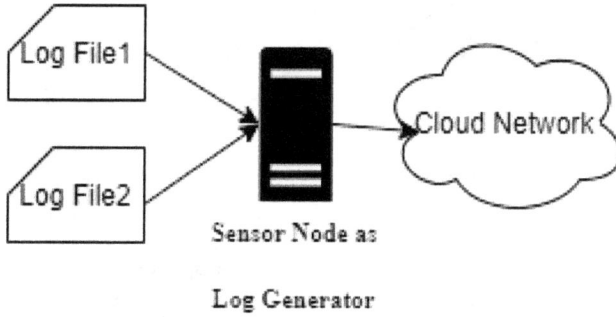

FIGURE 22.3 Logging client [13].

performed by the logging monitor to be done by the logging cloud. Requests to retrieve logs are sent to the logging cloud. The log monitor performs analysis of the log entries when required.

22.4.3 ANONYMOUS NETWORK LAYER

The anonymous network layer is not secure; hence the security of the log files from the edge layer to the cloud layer depends solely on the protocols used in this network. Thus our basic assumption is that this network is secure and authenticated. Nevertheless, the secure transmission of the log files is achieved in the proposed framework, and hence there is less botheration of the anonymous network layer.

22.4.4 CLOUD LAYER

Logging Cloud (Figure 22.5): Only legitimate users can upload the data to the cloud. The logging cloud provides not only the secure storage and retrieval but also easy maintenance which belongs to dissimilar logging hosts of different organizations and environments.

22.5 SECURITY ANALYSIS

An attacker can breach the security in three ways:

- An attacker can sniff and alter the messages over communication channel.
- An attacker can not only replicate but also replay information.
- An attacker can act like a legitimate entity of the network.

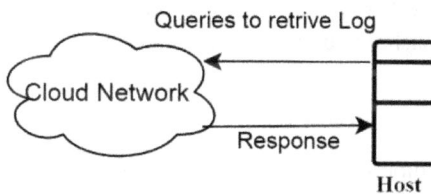

FIGURE 22.4 Logging monitor [13].

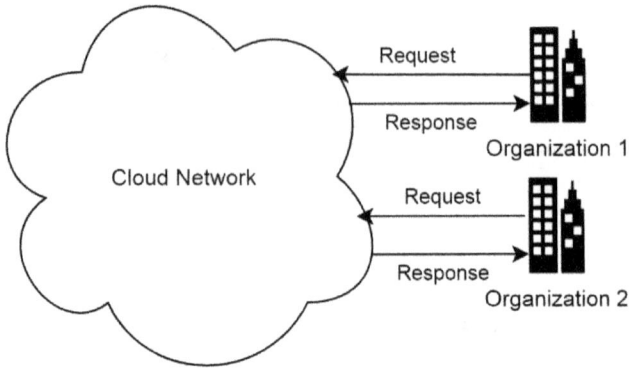

FIGURE 22.5   Logging cloud [13].

The encryption technique in Section 22.3.2 ensures the desirable properties. We shall now see the various cases in which the system is disturbed externally. The encrypted file is created by the sensor node, aka the logging client, and is sent to the logging cloud over the network. Further, if the log file is tampered with, then the validation of the stored log file on the cloud fails.

22.5.1   INTEGRITY OF LOG RECORDS DURING TRANSIT

Every message sent by the logging client to the logging cloud for storage is supposed to be encrypted and a hash generated. For the nth hash to be generated, the logging client requires the correct (n-1) hash, which in return is nothing but the concatenation of the previous hash and the message. Thus, even if an attacker intercepts the message over the network, they will not be able to understand the cipher text. Also, in addition, if the attacker wishes to convert this cipher text to plain text, the attacker does not have the required keys or the previous hash to calculate. Even if the attacker synthesizes and tries to replicate or replay messages over the network; it will not affect the stored log files, and the messages will have validation failed.

22.5.2   AUTHENTICITY OF LOG RECORD GENERATOR

The attacker may be a legitimate participant of the network or may try to impersonate a legitimate host; in this case, if they are a passive sniffer they cannot understand the cipher text; on the other hand, if they wish to decode the message, they should have all the previous hashes and the private keys as well.

22.5.3   PRIVACY ON LOGGING CLOUD

The attacker may attempt to read, delete or modify data travelling on the network. In the case of modification, the hash key changes, which does not match the saved one and hence the chunk validation fails; thus the system is safe from the attack. The attacker may attempt to read, delete or modify data stored in the cloud infrastructure as well; so the next time the user validates the log file, it shows that the validation failed, showing that the system was compromised by the attacker. The user must take all the necessary steps for replacing the compromised file and safely saving the other transactions and replacing the old secure file, i.e., execute mitigation plans.

22.6 PERFORMANCE ANALYSIS

Figure 22.6 shows the performance analysis of log insertion, including generating and updating the log, i.e., the time required to generate the log file and reach the logging client. It is seen that time increases linearly with the increase of log size. As the number of users increases, the load on the system also increases, thus giving a rise to increase in scalability; the time required intermittently also increases at the same time.

Figure 22.7 illustrates the performance analysis of generating the encrypted file. It is a nearly constant amount of time. There may be an increase in time for different sizes of logs. As the number of users increases, simultaneously the scalability of system is affected as the new encrypted text is formed from the old encrypted text only. It represents the time for verifying the validity of each log. It shows a nearly constant amount of time with the increase in log size. It also depends on the bandwidth of the

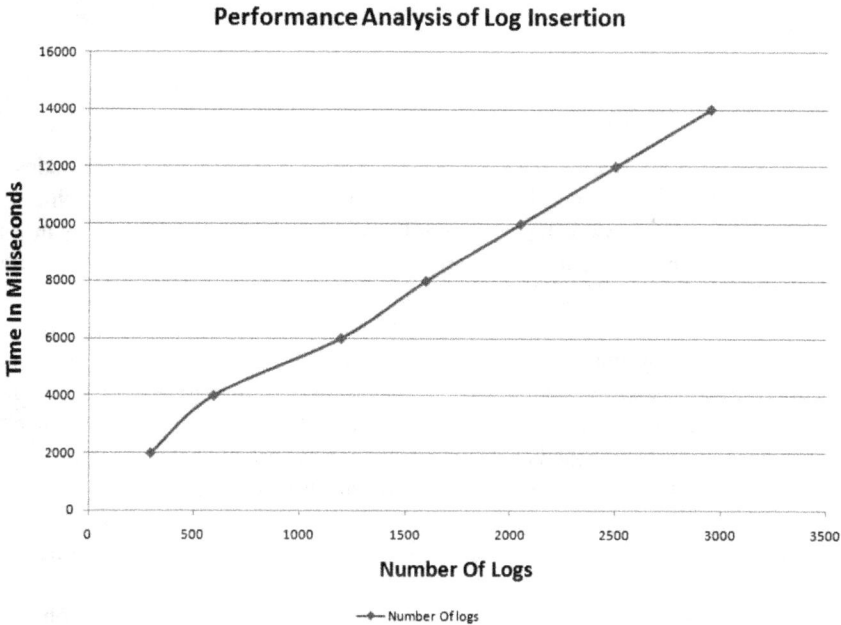

FIGURE 22.6 Log insertion.

Performance Analysis of creating a Secure file

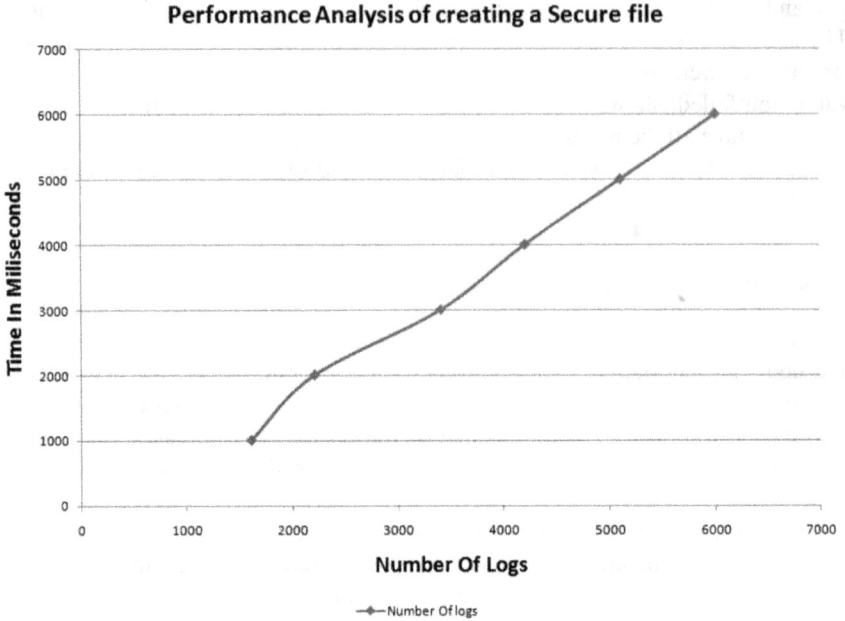

FIGURE 22.7 Creation of a secure file.

network, the number of users and the number of messages sent by the log generators every millisecond.

22.7 CONCLUSION

IoT sensors produce enormous data which cannot be stored on the sensor nodes or edge nodes. The security and integrity of log files is an expensive task. Achieving confidentiality and privacy for log record generation, transit and storage in any heterogeneous environment is of utmost importance. The cloud-based secured framework for maintaining log records in IoT takes a layered approach as log records contain crucial and sensitive data to be protected from third-party hackers. Outsourcing log management to the cloud environment is a huge saving on the capital expenses. Outsourcing log management is the default storage option for IoT because not only the log generation phase but also the storage and maintenance of log records in the cloud-based environment are taken care of. Cryptographic protocols for confidentiality are used. We propose and implement a secure cloud-based IoT framework for log management using the four IoT layers. Our results show that there is linear development for securing the log files and storing them on cloud with respect to time.

Logs serve as the footprint for all the activities done in a computing environment, serving as IoT forensics. We surveyed various log management schemes, such as distributed logging, LogSafe and event logging. The use of blockchain would definitely tighten the security of devices and, in turn, the logs.

REFERENCES

1. C. Cohen and A. Mahboubi, "Formal proofs in real algebraic geometry: From ordered fields to quantifier elimination," *Log Methods Computer. Sci.*, vol. 8, no. 2 SPEC. ISS., pp. 1–26, 2012.
2. U. Gupta, "Secure management of logs in Internet of Things," *Int. J. Adv. Network. Appl.*, vol. 7, pp. 2636–2639, 2015.
3. T. W. Paper, "IoT security & privacy," Technical White Paper, June, pp. 1–9, 2015. www.sentrybay.com.
4. P. Ghadekar, N. Doke, S. Kaneri, and V. Jha, "Secure access control to IoT devices using blockchain," *Int. J. Recent Technol. Eng.*, vol. 8, no. 2, pp. 3064–3070, 2019.
5. K. Jaswal, T. Choudhury, R. L. Chhokar, and S. R. Singh, "Securing the Internet of Things: A proposed framework," *Proceeding - IEEE International Conference on Computing, Communication and Automation, ICCCA 2017*, vol. 2017January, pp. 1277–1281, 2017.
6. S. Avizheh, T. T. Doan, X. Liu, and R. Safavi-Naini, "A secure event logging system for smart homes," *IoT S P 2017 – Proc. 2017 Work. Internet Things Secur. Privacy, co-located with CCS 2017*, pp. 37–42, 2017.
7. H. N. Noura, O. Salman, A. Chehab, and R. Couturier, "DistLog: A distributed logging scheme for IoT forensics," *Ad Hoc Networks*, vol. 98, p. 102061, 2020.
8. H. Nguyen, R. Ivanov, L. T. X. Phan, O. Sokolsky, J. Weimer, and I. Lee, "LogSafe: Secure and scalable data logger for IoT devices," *Proc. – ACM/IEEE Int. Conf. Internet Things Des. Implementation, IoTDI 2018*, pp. 141–152, 2018.
9. A. Ahmad, M. Saad, M. Bassiouni, and A. Mohaisen, "Towards blockchain-driven, secure and transparent audit logs," *ACM Int. Conf. Proceeding Ser.*, pp. 443–448, 2018.
10. H. Alobaidli, Q. Nasir, A. Iqbal, and M. Guimaraes, "Challenges of cloud log forensics," *Proc. SouthEast Conf. ACMSE 2017*, pp. 227–230, 2017.
11. V. Prathyusha, Dr. N. C. Sekhar, P. I. C. Kumari, "A new approach to secure logging-In cloud with the use of cryptography techniques," *Int. J. Dev. Comput. Sci. Technol.*, @ Feb-March, no. V-2, I-2, SW-15, ISSN-2320-7884 (Online) ISSN-2321-0257 (Print), 2014.
12. I. Ray and et al., "Secure loging as a service- delegating log management to the cloud" *IEEE Syst. J.*, Vol 7, No 2, June 2013.
13. Priyadarshini. T. R, Et.al, "Delegating log management to the cloud by seucre logging as a service" *IJARCET*, Volume 3 Issue 3 March 2014. Pp- 744–748, 2014.

23 Empirical Sentiment Analysis of Social Pages

Aarti
Lovely Professional University

Raju Pal
Jaypee Institute of Information Technology

CONTENTS

23.1 Introduction 331
    23.1.1 Document-Level Analysis 332
    23.1.2 Sentence-Level Analysis... 332
    23.1.3 Entity/Feature-Level Analysis.. 332
23.2 Literature Review 332
    23.2.1 Sentiment Analysis Techniques and Packages 332
    23.2.2 Data Extraction for Sentiment Analysis 334
    23.2.3 Sentiment Analysis Survey, Classifiers and N-Grams 334
    23.2.4 Sentiment Analysis Using Big Data Tools...................................... 335
    23.2.5 Sentiment Analysis Using Websites 335
    23.2.6 Sentiment Analysis in Social Sites 336
    23.2.7 Comparison Table... 336
23.3 Comparison Charts... 336
23.4 Conclusion and Future Scope 339
References... 340

23.1 INTRODUCTION

With the growth of internet technology, we have an abundance of data online. We can use various approaches to use this data effectively – we can use the data for knowledge discovery and use that knowledge for our business needs. Nowadays, we use online statistics to judge public opinion, but with growing data we have a challenge to manage data efficiently to make most of it. Sentiment analysis is the process by which we check the opinion of a person about some specific topic or product, generally aiming at finding out whether their opinion is positive, negative or neutral. This type of analysis is generally done on text data. We can use the reviews and comments on different commercial websites and social networking websites for sentiment analysis.

There are basically three levels of analysis of text data:

23.1.1 Document-Level Analysis

In this level of analysis, a whole document is considered as text for analysis. The limitation of this is that the whole unit considered for analysis must have a single objective so that we may find out whether the sentiment of the considered document is positive, negative or neutral. This system does not perform well for multi-topic textual units.

23.1.2 Sentence-Level Analysis

In this level of analysis, the sentiments expressed in each sentence are checked instead of considering a whole document. This level is considered good for analysis of the sentiments of reviews or comments. Here the textual data may be about multiple topics. We can look at each sentence individually (which may be about the same or different topics) and check for positive, neutral or negative opinion. This level of analysis is closely related to subjectivity classification, which distinguishes sentences that express factual information from sentences that express subjective views and opinions.

23.1.3 Entity/Feature-Level Analysis

In this level, we move one step deeper than the sentence-level analysis. It is a more fine-grained method for sentiment analysis. Here, instead of looking at documents, paragraphs, sentences etc., we are concerned with understanding the opinion itself. We extract different features from the text; from these we try to understand the opinion. Therefore, it is popularly known as feature-based sentiment analysis.

Sentiment analysis is a very popular and open area for research. This topic has become far more important with the advent of social media websites, blog pages, etc. As there is huge amount of textual data available from these sources, there is a challenge to find new techniques to extract knowledge from that data.

23.2 LITERATURE REVIEW

The literature of this paper is divided into six categories: Part (a) contains the papers for the basic sentiment analysis and considers a few techniques for sentiment analysis. Part (b) contains the papers for data extraction from social media which can later be used for sentiment analysis. Part (c) contains the survey paper on sentiment analysis and the classifiers and other important elements that can be useful in increasing the accuracy of sentiment analysis. Part (d) contains the papers of sentiment analysis using big data tools. Part (e) contains the literature regarding performing the sentiment analysis by extracting data from websites. Part (f) contains the papers of sentiment analysis of the social data. Parts (g) and (h) contain a small comparison of all the papers.

23.2.1 Sentiment Analysis Techniques and Packages

Jeonghee Yi (2003) [1] stated that sentiment analysis about a particular topic can be performed using the sentiment analyzer. Here they did not consider the whole document to be one topic; instead they concentrated more on feature-extraction techniques. They used the techniques of natural language processing to analyze the sentiment in a given document. They mainly considered the datasets of digital camera and music review articles. For analysis they used the sentiment lexicon and the sentiment pattern database. Performance was evaluated for different web reviews using different sentiments for different subject. The results on general web pages are better than those of state-of-the-art algorithms.

Lipika Dey (2009) [2] took noisy data, pre-processed the data using different techniques and performed sentiment analysis. All previous techniques consider the data to be linguistically correct, but in this technique a new WordNet package was created for noisy data. This technique could be considered useful because the most of data on social microblogging sites is noisy data. So this technique could provide more efficient results.

Jingbo Zhu (2011) [3] worked on reviews for a restaurant and used single-aspect and multi-aspect techniques for sentiment analysis. This method is designed for Chinese restaurants but the same techniques could be followed for English.

Chenghua Lin (2012) [4] created a new algorithm, JST and reverse JST, for classification purposes. It is capable of identifying the topic or context of a given text. All previous approaches in this field of sentiment classification are supervised, but this approach is weakly supervised, which makes it more capable with less training data.

Cane Wing-ki Leung (2011) [5] proposed a novel Probabilistic Rating infErence Framework (PREF), for mining user preferences from reviews and then mapping such preferences onto numerical rating scales. They compared the sentiment analysis done on snippets of data using previous techniques and also studied the effect of size of dataset on analysis.

Alena Neviarouskaya (2011) [6] developed a new lexicon in this paper. Instead of considering the words contained in famous sentiment analysis package WordNet, they have widened the scope by adding a techniques to give scores to the synonyms and antonyms as well. They also expanded it through deriving the words or compounding the known lexical units. In the paper, the importance of considering modifiers, contextual valence shifters, and modal operators (which are integral parts of the SentiFul lexicon for robust sentiment analysis) are also discussed.

Danushka Bollegala (2016) [7] tried to develop an automated system for sentiment analysis. They created a thesaurus which made it possible to evaluate the sentiments in multiple domains. One major problem in cross-domain implementation is that the words in the training domain may have different influence those in the test domain. They used techniques to build feature-expansion methods by which the same thesaurus can be used for multiple domains. There is also a comparison between WordNet and the thesaurus used in this project. It outperforms in some methods but not in all the cases. This is an open topic to generalize more topics under this technique.

Uros Krcadinac (2013) [8] created a java-based package for sentiment analysis based on WordNet as no previous packages based on WordNet were able to vary the efficiency of the native package. In this the data is considered to be in two main parts: word lexicon and emotion lexicon. The word lexicon is based on WordNet. The emotion lexicon manages certain emotions used in the social network these days. The output of Synesketch is classified according to the Ekman emotion clas-sification. It also contains visual libraries to represent the results.

Lorenzo Gatti (2016) [9] created SentiWords to have high precision and larger coverage in sentiment analysis. In this he created a prior polarity lexica with which they could have high precision and larger coverage. They also compared it to previ-ous techniques and its results outperformed many of the previous techniques.

Abinash Tripathy (2017) [10] used a hybrid approach considering neural net-works and classification techniques for sentiment analysis. They used a hidden layer of neural networks in which they defined number of neurons. The authors considered precision, recall, f-measure and accuracy to evaluate the performance, and it is found that this technique performs better.

23.2.2 DATA EXTRACTION FOR SENTIMENT ANALYSIS

Chris Howden (2014) [11] tried to build up the gap in the online social network and forensics. They built a system in which they used Twitter data (i.e. one of the most popular online social networks) for forensic purposes. The main explanation in this paper is about how the OAuth Twitter application can be used to extract personal information.

Rehab M. Duwairi (2015) [12] in his paper programmed a software that helped in faster and easier extraction of Facebook data. Real use monitoring (RUM) is com-posed of three basic elements: a server running the crawling agents, a PHP applica-tion which implements logic and an Apache interface that manages the data flow through the web. They made a user-friendly interface which makes the data scrap-ping very easy.

23.2.3 SENTIMENT ANALYSIS SURVEY, CLASSIFIERS AND N-GRAMS

Bo Pang (2008) [13] discussed in detail all the concepts of opinion and sentiment analysis. They also considered the facts that make opinion mining difficult and the general challenges in sentiment analysis. They included diverse topics and meth-ods for sentiment analysis while mainly focusing on neuro-linguistic programming (NLP) technique.

Hsinchun Chen (2010) [14] discussed past implementations in the field of sen-timent analysis and problems that may occur in different perspectives. They con-cluded that sentiment analysis techniques have not reached high perfection levels because much work remains to be done on this topic.

Anuj Sharma (2012) [15] compared different feature-extraction and machine-learning techniques for sentiment analysis working on movie datasets. They found that support vector machines (SVMs) perform best in machine-learning techniques and Naïve Bayesian is the best classifier for sentiment analysis.

Dhiraj Gurkhe (2014) [16] tried to make a system that could automatically detect the sentiments from textual data taken from a social media platform. A human can understand the sentiments using the intelligence, but with automated systems we use natural language processing systems. For classification purposes, the test data is pre-processed and a feature vector of test data is formed. This test data is then fed into the Naïve Bayes algorithm along with the training data to calculate the probability.

Walla Medhat (2014) [17] made a survey starting from the basics of the sentiment analysis. They explained various approaches that can be used for sentiment analysis using computers. There is also explanation of which type of system performs better in particular conditions. They concluded that Naïve Bayesian and SVM are the most frequently used methods for sentiment analysis.

Fotis Asiopos (2016) [18] tried to sort a big challenge in social media – to create a system that is language independent. It has an advantage over previously used techniques in that we are not constrained by a particular vocabulary. It can be applied at any level of sentiment analysis.

23.2.4 SENTIMENT ANALYSIS USING BIG DATA TOOLS

Sudipto Shankar Dasgupta (2015) [19] took a step forward in sentiment analysis technique. They tried to make use of Hadoop open source technologies to help in faster processing of data and extracting sentiment from the text. The data he used is taken from Facebook. The extracted text is first cleaned and then passed through various procedures before the final step of sentiment classification. This procedure provided faster results and achieved accuracy of about 67.6%.

Devendra K. Tayal (2016) [20] used a Bloom filter in Hadoop to implement sentiment analysis and operate it at a high speed. These days social media platforms are vast and comments appear with a great frequency as soon as something is posted on some business page. So we need to implement techniques for fast and high-quality knowledge extraction from data.

23.2.5 SENTIMENT ANALYSIS USING WEBSITES

L. Zhang (2010) [21] explained how they created a system for business intelligence in which they categorized the information crawled from website, first at document level and then at entity level. Finally they were able to create an entity-level smart system for business intelligence.

Shoiab Ahmed (2015) [22] used rule-based classifiers for sentiment analysis. He used the techniques of word count, part-of-speech (POS) tagging and certain steps of data extraction and cleaning before actual evaluation. The proposed approach was tested on online books and political reviews and demonstrates the efficacy through Kappa measures, which have 97.4% accuracy and a lower error rate. Comparative experiments on various rule-based machine-learning algorithms have been performed through a ten-fold cross validation training model for sentiment classification.

Shoiab Ahmed (2015) [23] built a technique for sentiment analysis in which they used clean data from websites. In this technique they first emphasised cleaning the raw data extracted from website. They used POS tagging before making a final analysis. They built a system in which you can extract the data from websites and use that to carry forward the sentiment analysis. They used seven classes to categorize the text.

23.2.6 SENTIMENT ANALYSIS IN SOCIAL SITES

Rahim Dehkharghani (2014) [24] tried to use causal rules for the analysis of a topic. They extracted the data from a Twitter page on a certain topic and built an automated system based on causal rules to analyze the opinions. It provided acceptable results when compared to human intelligence systems, but the limitation is that we have to create separate rules each time we work on different topic pages. In this paper they used a Kurdish issue in Turkey for the analysis.

Alexandra Balahur (2015) [25] performed the sentiment analysis using SVM on English and Spanish tweets. They compared the system for a single language and for multiple languages. They concluded that stopwords removal and stemming decrease the performance rate in multilingual systems, whereas n-grams help us in understanding the language better.

Matthijs Meire (2016) [26] used auxiliary data for the analysis of sentiment. They used user personal information from with Facebook and user history (i.e., how he has reacted to the previous posts on different aspects) as the pre-information. Also, they used the number of likes or other emoticons a post had along with the sentiment analysis of the text comments as the post information. They combined the results of this both, pre and post data, to carry out the final results for the analysis. When the user post a comment, the analyst can detect the mood from the user's comment history. This technique was successful in adding more valuable and effective parameters in sentiment analysis.

Cambria (2016) [27] explained various procedures and techniques that can be followed in sentiment analysis. He also tried to explain that, in this era of internet, we need to understand the emotions of the people for better commercial results instead of depending only on the IQ statistical methods. Today we can grab a lot of data from social networking sites and discern valuable knowledge for commercial purposes.

Avinash Chandra Pandey (2017) [28] used a hybrid approach for sentiment analysis. In this they used two algorithms: k-means and cuckoo search. They considered emoticon symbols in the analysis. The method performs well in some of the cases, but there is still opportunity for improvement.

23.2.7 COMPARISON TABLE

In Table 23.1, we compare in detail the techniques used by various authors to proceed with their work. We check out whether the technique used is applicable to all types of data or is domain oriented.

23.3 COMPARISON CHARTS

In Figure 23.1, we described the literature survey we carried out from different publications. During our literature survey we took most of the papers from IEEE. We took literature from Springer and Elsevier also. We were also able to find some good literature from some other journals.

TABLE 23.1
Literature Survey

Reference	Task	Data	Domain Oriented	Algorithm/Technique Used
			Sentiment Analysis Packages and Techniques	
[1]	Worked on Sentimental Analysis using the Wordnet dictionary	reviews(camera reviews)	Y	For each feature considered likelihood score .
[2]	Worked on Sentimental Analysis using noisy data.	crawled data from websites	Y	for data extracted used two level preprocessing
[3]	Used multi aspect and single aspect technique for restraunt reviws.	used restraunt reviews from DianPing.com	Y	Used multi aspect technique for processing reviews.
[4]	Used the joint sentiment technique for classification of text data	used publicay available movie reviews from MR	Partialy	Used JST and reverse-JST for detecting a topic from text.
[5]	Worked on a technique to rate the reviews on a numerical scale	scale dataset of Pang and Lee(domain-movies)	Y	PREF - Probabilistic Rating infErence Framework
[6]	Studied Wornet package and tried to improve it using the synonomy and antonomy relations.	NA	Y	created rules for synonomy and antonomy relations
[7]	Used feature expansion techniqiues to make the algorithm work on multiple domains.	used the data prepared by Blitzer	N	Used the technique of feature expansion to make the technique work on mor than one domain
[8]	Created a java application for sentimental analysis that is based on the wordnet package. Moreover it contains good visualizers for representing the tasks.	used some preprepared corpus	Y	Evaluted the results for text data and emotion data seperately and combined them for final result.
[9]	Sentimental analysis based on the previous knowledge instead of selecting new algorithm .	Anew, The Harvard General Inquirer dictionary	Y	Used previous knowledge instead of applying some random algorithm.
[10]	Worked on a hybrid approach for sentimental analysis	acl IMDb dataset, polarity dataset	Y	Used SVM and ANN as hybrid approach for sentimental analysis
			Data Extraction	
[11]	Worked on extraction of the personal data of a user from social sites and using that data for forensic purposes.	random data from twitter	N/A	Used OAUTH apps for extraction of data from social sites.
[12]	Extracted the data from facebook using an automated code of Php	random data from Facebook	N/A	Facebook Graph API used to extract data
			Sentiment Analysis Survey, Classifiers and N Grams	
[13]	Explained in details challenges and difficulties in sentiment analysis	N/A	N/A	discussed all previous techniques.
[14]	Analyzed the previously techniques and worked on walmart data for opinion mining.	US corporation, Wal-Mart	Y	discussed the previous techniques and created a simple new one for opinion mining.

(continued)

TABLE 23.1

Literature Survey (*Continued*)

[15]	compared different feature extraction and machine learning techniques for sentiment analysis	mostly movie review daatasets from IMDB	N/A	Compared previous machine learning and feature selection techniques.
[16]	Performed sentimental analysis using Naïve Bayesian	Data Extracted using Twitter API	Y	Naïve Bayesian
[17]	Performed survey on the sentimental literature	N/A	Y	Compared multiple techniques.
[18]	Explained how we can use N-Grams to increase the efficiency of sentimental analysis.	Different Corpora for Different Languages	N	N gram for mutilingual and multitopic data.
Sentiment Analysis using Big data Tools				
[19]	Used open source tools of Big Data for fast processing of data in sentimental analysis.	Data extracted using Facebook API	Y	Used R and Spark for fast processing.
[20]	Used Bloom Filter in Hadoop for fast processing	Used data of Wordnet	Y	Bloom filter and sentimental tree structure.
Sentiment Analysis in Websites				
[21]	created a system for Business Intelligence.	crawled data from website	Y	The information crawled from website and then firstly analysed at document level and then at entitiy level.
[22]	Used reviews from the commercial websites and evaluated sentiments.	Data collected using JSoup Web Crawler.	Y	Performed classification using SentiWordNet
[23]	Used reviews from websites and classified them in seven different classes	Extracted data using web crawler	Y	Performed classification using SentiWordNet
Sentiment Analysis in Social Media				
[24]	Used causal rules for sentiment analysis	Extracted data from Twitter	Y	Created cause effect rules for sentiment Evaluation.
[25]	Performed the sentimental analysis using SVM on multilingual tweets.	corpus training set of TASS 2013, data set of tweets used SemEval 2013.	Y	Created two different techniques on for english and second for spanish tweets.
[26]	Explained that how Auxillary data can be used for increading the efficiency of sentiment analysis	Extarcted data using OAUTH applications in social media	Y	Evaluated the results based on that how the user has previously acted on some other reviews.
[27]	A Chapter from book explaining about how much social data has penetrated to business solutions	N/A	N/A	Explained different level penetration of social data.
[28]	Used cukkoo search and kmeans for sentimental analysis	Twitter data API	Y	Combined cukko search and k-means for sentimental analysis

In Figure 23.2, we compared the literature annually, describing the number of papers published each year. The topic of sentiment analysis came into the limelight after we started seeing a large amount of data being collected on social networking sites. So mostly the literature of sentiment analysis dates from after 2003.

Publications

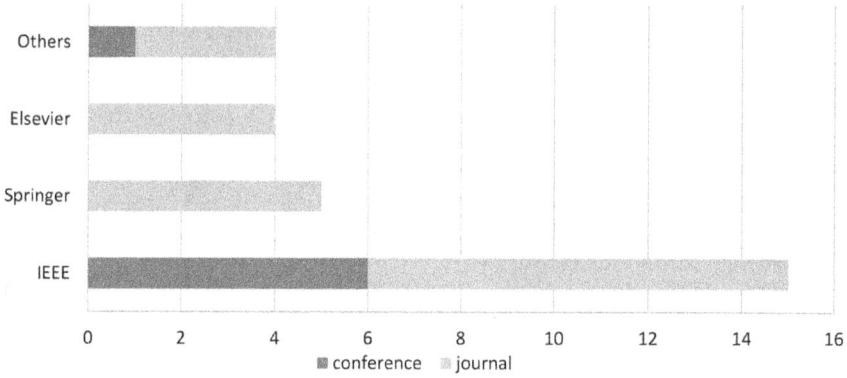

FIGURE 23.1 Comparison related to publication.

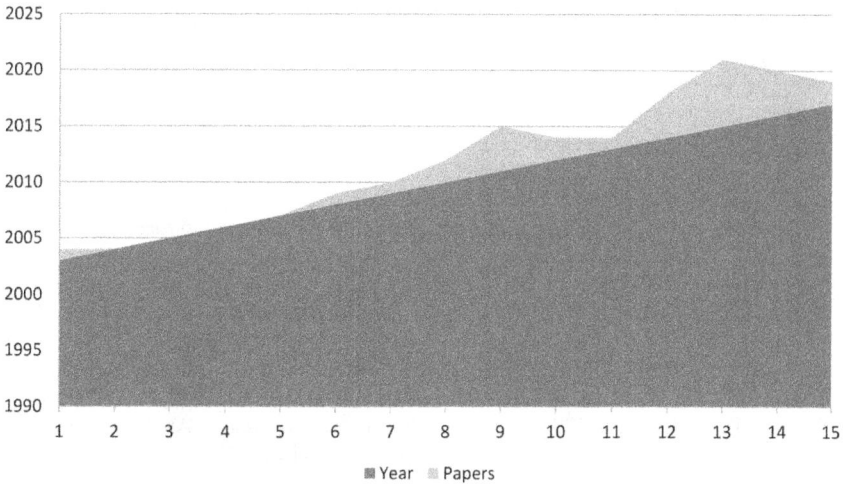

FIGURE 23.2 Comparison related to years.

23.4 CONCLUSION AND FUTURE SCOPE

Sentiment analysis is a prominent field these days. Many organizations as well as individuals want to have analysis of the data on social media. Various techniques have been used for sentiment analysis, but mostly the techniques are not domain independent. So far the techniques are for specific domains. We can work on building a domain-independent technique.

REFERENCES

1. Yi, J., Nasukawa, T., Bunescu, R. & Niblack, W. (2003). Sentiment Analyzer: Extracting Sentiments About a Given Topic Using Natural Language Processing Techniques. *Proceedings – IEEE International Conference on Data Mining, ICDM*, 427–434. 10.1109/ICDM.2003.1250949.

2. Dey L. & Haque S.K. (2008). Opinion Mining from Noisy Text Data. *International Journal on Document Analysis and Recognition (IJDAR)*, 12(3), 205–226. 10.1007/s10032-009-0090-z.

3. Zhu J., Wang H., Zhu M., Tsou B. & Ma M. (2011). Aspect-Based Opinion Polling from Customer Reviews, *IEEE Transactions on Affective Computing*, 2(1), 37–49. 10.1109/T-AFFC.2011.2.

4. Lin C., He Y., Everson R. & Ruger S. (2012) Weakly Supervised Joint Sentiment-Topic Detection from Text, *IEEE Transactions on Knowledge and Data Engineering*, 24(6), 1134–1145. 10.1109/TKDE.2011.48.

5. Leung, C.W., Chan, S.C. & Chung, F. (2011) A Probabilistic Rating Inference Framework for Mining User Preferences from Reviews, *World Wide Web*, 14(2), 187–215 https://doi.org/10.1007/s11280-011-0117-5.

6. Neviarouskaya A., Prendinger H. & Ishizuka M. (2011) SentiFul: A Lexicon for Sentiment Analysis, *IEEE Transactions on Affective Computing*, 2(1), 22–36, doi: 10.1109/T-AFFC.2011.1.

7. Bollegala D., Mu T. & Goulermas J.Y. (2016) Cross-Domain Sentiment Classification Using Sentiment Sensitive Embeddings, *IEEE Transactions on Knowledge and data Engineering*, 28(2).

8. Krcadinac, U., Pasquier P., Jovanovic J. & Devedzic V.(2013). Synesketch: An Open Source Library for Sentence-Based Emotion Recognition, *IEEE Transactions on Affective Computing*, 4, 312–325. 10.1109/T-AFFC.2013.18.

9. Gatti L., Guerini M. & Turchi, M. (2016) SentiWords: Deriving a High Precision and High Coverage Lexicon for Sentiment Analysis, *IEEE Transactions on Affective Computing*, 7(4), 409–421. 10.1109/TAFFC.2015.2476456.

10. Tripathy A., Anand A. & Rath S. (2017). Document-Level Sentiment Classification Using Hybrid Machine Learning Approach, *Knowledge and Information Systems*. 10.1007/s10115-017-1055-z.

11. Howden C., Liu L., Li Z., Li J. & Antonopoulos N. (2014). Virtual Vignettes: The Acquisition, Analysis, and Presentation Of Social Network Data, *Science China Information Sciences*, 57, 1–20. 10.1007/s11432-014-5069-9.

12. Faqeeh M. (2014). RUM Extractor: A Facebook Extractor for Data Analysis. 10.13140/RG.2.1.3267.3367.

13. Pang B. & Lee L. (2008) Opinion Mining and Sentiment Analysis, *Journal of Foundations and Trends in Information Retrieval*, 2(1), 1–135.

14. Chen H.-C. & Zimbra D. (2010). AI and Opinion Mining, *IEEE Intelligent Systems*, 25, 74–80. 10.1109/MIS.2010.75.

15. Sharma, A. & Dey, S. (2012). A comparative study of selection and machine learning techniques for sentiment analysis. Proceeding of the 2012 ACM Research in Applied Computation Symposium, RACS 2012. 1–7. 10.1145/2401603.2401605.

16. Gurkhe D., Pal N. & Bhatia R. (2014). Effective Sentiment Analysis of Social Media Datasets using Naïve Bayesian Classification, *International Journal of Computer Applications*, 99, 1–4. 10.5120/17430-8274.

17. Hassan Y, Ahmed, M. W. & Mohamed H. (2014). Sentiment Analysis Algorithms and Applications: A Survey, *Ain Shams Engineering Journal*, 5. 10.1016/j.asej.2014.04.011.

18. Aisopos F., Tzannetos D., Violos J. & Varvarigou T. (2016). Using N-Gram Graphs for Sentiment Analysis: An Extended Study on Twitter. *IEEE Second Intenatinal Conference on Big Data Computing Services and Applications*, 44–51. 10.1109/BigDataService.2016.13.
19. Dasgupta S. S., Natarajan S., Kaipa K. K., Bhattacherjee S. K. & Viswanathan A. (2015) Sentiment Analysis of Facebook Data Using Hadoop Based Open Source Technologies. *IEEE International Conference on Data Science and Advanced Analytics (DSAA)*, Paris, 1–3, 10.1109/DSAA.2015.7344883.
20. Tayal D.K. & Yadav S.K. (2016) Fast Retrieval Approach of Sentimental Analysis with Implementation of Bloom Filter on Hadoop. In *International Conference on Computational Techniques in Information and Communication Technologies*.
21. Zhang L., Bao S., Guo H., Zhu H., Zhang X., Cai K., Fei B., Wu X., Guo Z. & Su Z. (2010). EagleEye: Entity-Centric Business Intelligence for Smarter Decisions. *IBM Journal of Research and Development*, 54, 1. 10.1147/JRD.2010.2069710.
22. Ahmed S. & Danti A. (2016) Effective Sentimental Analysis and Opinion Mining of Web Reviews Using Rule Based Classifiers. In Behera H. & Mohapatra D. (eds) *Computational Intelligence in Data Mining—Volume 1. Advances in Intelligent Systems and Computing*, 410. Springer, New Delhi.
23. Shoiab A. & Ahmed D. (2015) A Novel Approach for Sentimental Analysis and Opinion Mining based on SentiWordNet using Web Data. In *International Conference On Trends In Automation Communications and Computing Technology*.
24. Dehkharghani R., Mercan H., Javeed A. & Saygin Y. (2014). Sentimental Causal Rule Discovery from Twitter, *Expert Systems with Applications*, 41, 4950–4958. 10.1016/j.eswa.2014.02.024.
25. Balahur A. & Perea-Ortega J. (2015). Sentiment Analysis System Adaptation For Multilingual Processing: The Case of Tweets, *Information Processing & Management*, 51, 547–556. 10.1016/j.ipm.2014.10.004.
26. Meire M., Ballings M. & Van den Poel D. (2016). The Added Value of Auxiliary Data in Sentiment Analysis of Facebook Posts, *Decision Support Systems*, 89. 10.1016/j.dss.2016.06.013.
27. Cambria E. (2016). Affective Computing and Sentiment Analysis. *IEEE Intelligent Systems*, 31, 102–107. 10.1109/MIS.2016.31.
28. Pandey A.C., Singh D., Rajpoot D.S. & Saraswat M. (2017) Twitter Sentiment Analysis Using Hybrid Cuckoo Search Method, *International Journal On Information Processing And Management*, 53(4), 764–779.

Index

A

analog-to-digital converters (ADCs), 99
actuator, 203
agriculture-based IoT technology, 240
analysis prevalence of depression, 311
application performance monitoring (APM), 316

B

big data and PM, 151
Bitcoin, 279
blockchains, 279

C

cloud computing, 295
common mental disorders, 307
comparator design, 100
crop water stress index (CWSI), 244
CT scan image, 170
cryptographic algorithms, 7

D

data driven medicine, 154
data memory (DM), 96
dielectric resonator antennas (DRAs), 114
direct analog synthesis (DAS), 136
direct digital synthesis (DDS), 134
DRAs design algorithm, 121
dynamic comparators, 104

E

edge layer, 324
electromagnetic interference (EMI), 30
embedded systems, 295
energy-efficient coverage protocol (EECP), 12
ensemble classifier, 175

F

face detection, 161
fractional-order (FO) system, 223, 225
fuzzy operator, 69

G

gain bandwidth (GBW) product, 143
GLCM algorithm, 173
graphics processing unit (GPU), 157

H

Harris Hawks Optimization (HHO), 235
HFSS, 123
high-boost filtering, 70

I

image enhancement, 67
industry 4.0, 281
instruction memory, 96
intelligent robotic system, 190
intrusion detection, 4
intrusion detection system (IDS), 48
IoT architecture, 292
IoT security and privacy, 318

K

k-nearest neighbour algorithm, 57

M

machine learning algorithm, 2
MATLAB, 124
memory design, 93
model order reduction (MOR), 221
morphological filter, 71
multiple-input multiple-output (MIMO), 114, 127

N

natural language processing (NLP), 1
Network Anomaly Detection and Intrusion
        Reporter (NADIR), 56
network model, 14, 268

O

operational transconductance amplifier (OTA),
        140
optical coherence tomography (OCT), 154
optical sensors, 247

P

particle swarm optimization approach, 15
pass transistor logic (PTL), 87
passive infrared sensor (PIR), 209
phase-locked loop, 132
PiCamera, 203

Pipelining, 92
plant disease, 244
Power Consumption, 87
precision medicine (PM), 149
Precision Medicine Initiative (PMI), 149
pupil segmentation, 159

R

radio-frequency identification (RFID), 32
Raspberry Pi, 202
rectangular dielectric resonator antennas
        (RDRAs), 125
regression, 258
reverberation chamber (RC), 31
ROI selection, 160

S

SAR ADC, 101
signature-based NIDS, 48
soil-mapping, 242
SRAM, 94
stirring techniques, 33
supervised learning (SL), 258

supply chain process, 284
support vector machine, 58
swam intelligence, 192

T

Time-Based Inductive Machine (TIM), 56
traffic control, 2
transistor sizing, 91

U

ultrasonic ranging sensors, 247

V

vehicular ad-hoc network (VANET), 3
voltage scaling, 90
voltage-controlled oscillator (VCO), 133

W

wide dual-band rectangular souvenir, 125
wireless sensor networks (WSN), 11

For Product Safety Concerns and Information please contact our EU
representative GPSR@taylorandfrancis.com
Taylor & Francis Verlag GmbH, Kaufingerstraße 24, 80331 München, Germany

www.ingramcontent.com/pod-product-compliance
Lightning Source LLC
Chambersburg PA
CBHW060801220326
41598CB00022B/2511